Conference Proceedings of the Society for Experimental Mechanics Series



For other titles published in this series, go to
www.springer.com/series/8922

Tom Proulx
Editor

# Dynamics of Bridges, Volume 5

Proceedings of the 28th IMAC, A Conference on Structural Dynamics, 2010

Springer

*Editor*
Tom Proulx
Society for Experimental Mechanics, Inc.
7 School Street
Bethel, CT 06801-1405
USA
tom@sem1.com

ISSN 2191-5644 e-ISSN 2191-5652
ISBN 978-1-4614-2881-7 ISBN 978-1-4419-9825-5 (eBook)
DOI 10.1007/978-1-4419-9825-5
Springer New York Dordrecht Heidelberg London


Softcover reprint of the hardcover 1st edition 2011


Printed on acid-free paper

Springer is part of Springer Science+Business Media (www.springer.com)

# Preface

*Dynamics of Bridges* represents one of five clusters of technical papers presented at the 28th IMAC, A Conference and Exposition on Structural Dynamics, 2010 organized by the Society for Experimental Mechanics, and held at Jacksonville, Florida, February 1-4, 2010. The full proceedings also include volumes on Structural Dynamics and Renewable Energy, Nonlinear Modeling and Applications, Dynamics of Civil Structures, and Structural Dynamics.

Each collection presents early findings from experimental and computational investigations on an important area within Structural Dynamics. The current volume on *Dynamics of Bridges* includes studies on Dynamics of Bridges, Health Monitoring of Bridges, and Bridges – System Identification

This focus area seeks to provide a dedicated forum for researchers, designers and bridge owners to exchange ideas and concepts relating to the development, implementation and application of efficient experimental and numerical tools for the assessment, monitoring and control of the dynamic behavior of roadway, railway or pedestrian bridges, taking in mind problems of human comfort, structural performance and durability, damage detection and location, or structural rehabilitation. Particular attention is devoted to innovative applications in full scale structures.

The organizers would like to thank the authors, presenters, session organizers and session chairs for their participation in this track. The Society would like to thank the organizers of this Focus Topic, Á. Cunha, *Universidade do Porto*, R. Cantieni, *RCI Dynamics* for their efforts.

Bethel, Connecticut

Dr. Thomas Proulx
Society for Experimental Mechanics, Inc

# Contents

Proceedings of the IMAC-XXVIII
February 1–4, 2010, Jacksonville, Florida USA


# Traffic Excited Vibrations Acting on Pedestrians Using a Highway Bridge

Reto Cantieni, rci dynamics, Structural Dynamics Consultants
Raubbuehlstrasse 21B, CH-8600 Duebendorf, Switzerland

## ABSTRACT

The new Langensand Bridge Lucerne is a single span bridge with a span length of 80 m and a width of 27 m. Whereas the middle part of the bridge is used by five road traffic lanes, pedestrian and cyclist lanes are located on both cantilevers. These have a free span length of about 6 m each. The questions to be answered through the experimental investigation discussed here were: Are the pedestrians walking on the bridge cantilevers annoyed by traffic induced vibrations? If yes, what would be the optimum layout for TMD's (tuned mass dampers)? In a first step, an ambient vibration test was performed. The structure's natural vibrations were extracted from vibrations predominantly excited by the many trains passing on the 11 railway tracks underneath the bridge. The measurement point grid consisted of 52 points. In the 1...34 Hz range, 21 natural modes of the bridge could be identified. In a second step, 77 dynamic tests were performed with three heavy vehicles (equipped with steel and air suspensions respectively) being driven over the bridge in different configurations. Maximum acceleration amplitudes measured at the free cantilever end were in the range of a = 0.5...0.8 m/s2 for passages on the normal pavement and reached about a = 2.6 m/s2 for passages over a 25 mm and a 50 mm plank respectively, located at mid-span. These values were rated as acceptable. No attempts had to be taken to design a TMD system.

## THE BRIDGE

The Langensand Bridge is a single span steel/concrete composite bridge with a span length of 79.7 m and a width of 27.3 m. The skewness of the abutment axes is 17° (Fig. 2). Two steel box girders are connected to a concrete bridge deck (Fig. 1). There were two challenges with this bridge. Firstly, the bridge had to be built in two parts. The bridge link being vital to Lucerne, it was not possible to completely stop traffic for a longer period of time. Therefore, replacement of the existing bridge was performed in four steps: a) demolition of half of the old bridge, b) construction of half of the new bridge, c) demolition of the remaining half of the old bridge, and d), construction of the second half of the new bridge. Secondly, the structure had to bridge the 11 railway tracks underneath in one span instead of the two spans of the existing bridge. Using the same headway, of course.

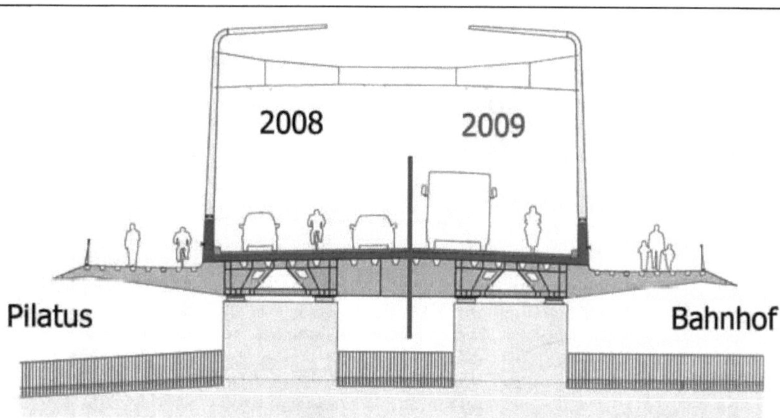


Fig. 1: Langensand Bridge Lucerne. The paper covers tests performed on the Pilatus side half of the bridge.

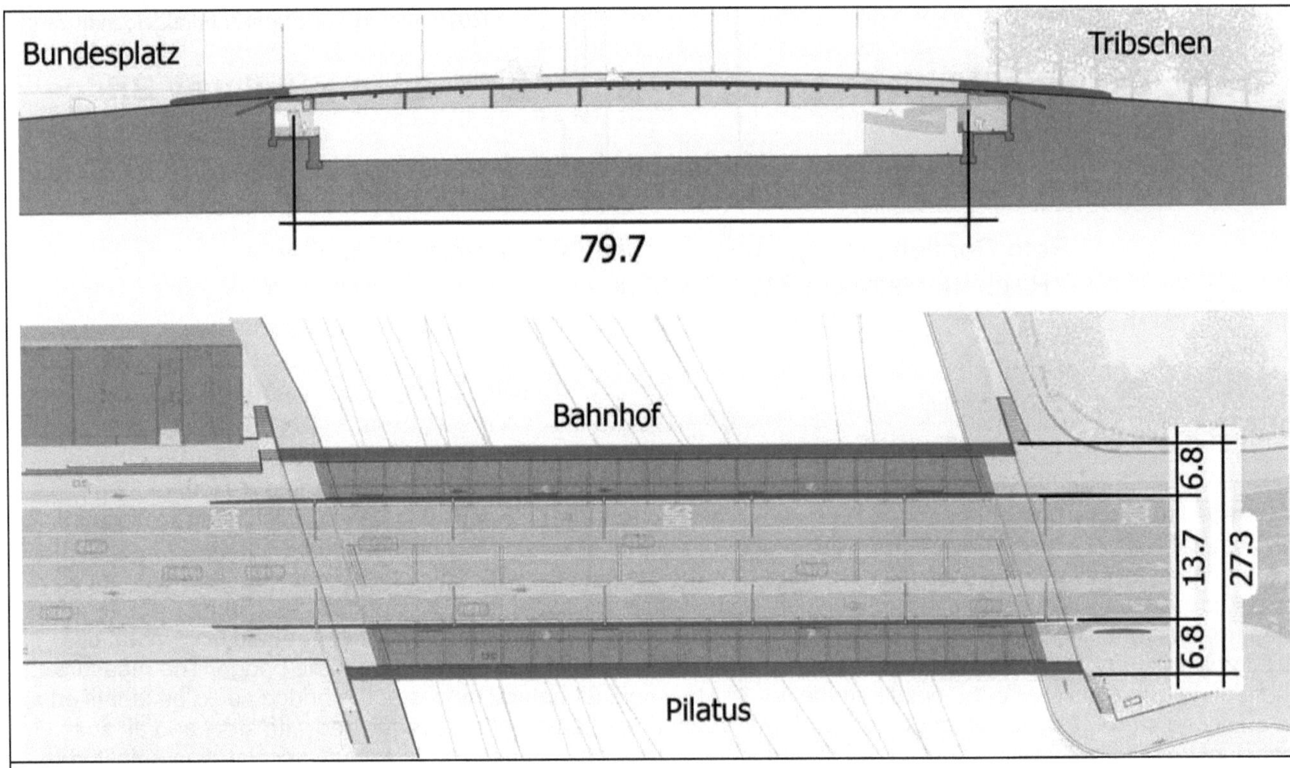


Fig. 2: Langensand Bridge, plan view and longitudinal section. Dimensions: meter. (Underlying graphics: TU Bruno Zwahlen Mayr, courtesy tec21, the Journal of the Swiss Association for Engineers and Architects.)

**AMBIENT VIBRATION TESTS; INSTRUMENTATION**

As no reliable data from an FE-analysis were available a pilot test was performed as a first step. As a result, the fundamental natural mode's frequency was determined to f = 1.28 Hz. This knowledge allowed optimum choice of the sensors to be used. Furthermore, the initially planned measurement point layout had to be revised. The pilot test, using two 3D measurement points at 0.4 of the span had revealed that the transverse as well as the longitudinal bridge behavior was much more intense than assumed. Therefore, of the 52 measurement points, half were equipped with 1D, the other half with 3D sensors.

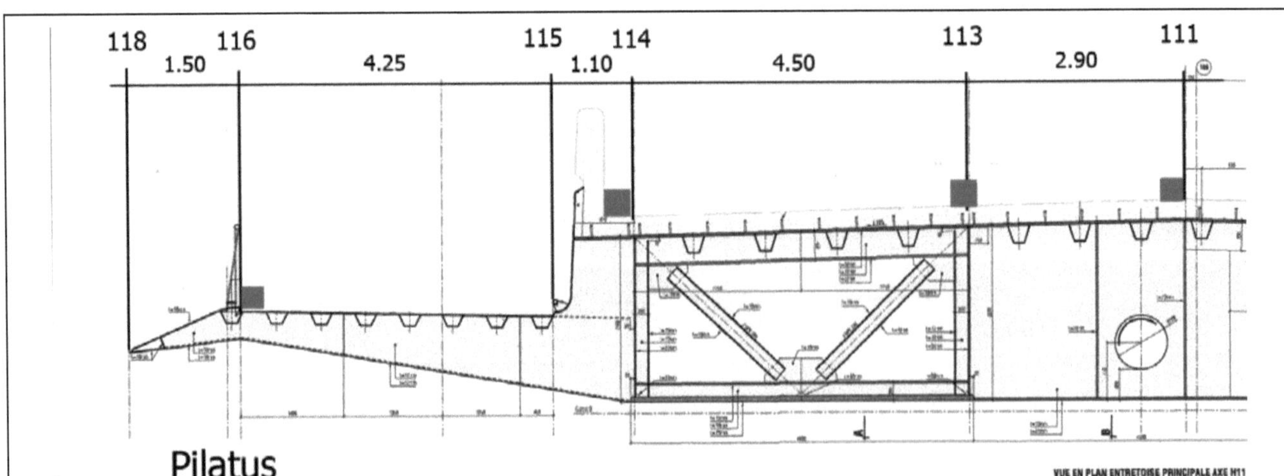


Fig. 3: Langensand Bridge, Location of the four longitudinal measurement axes. Dimensions: meter.

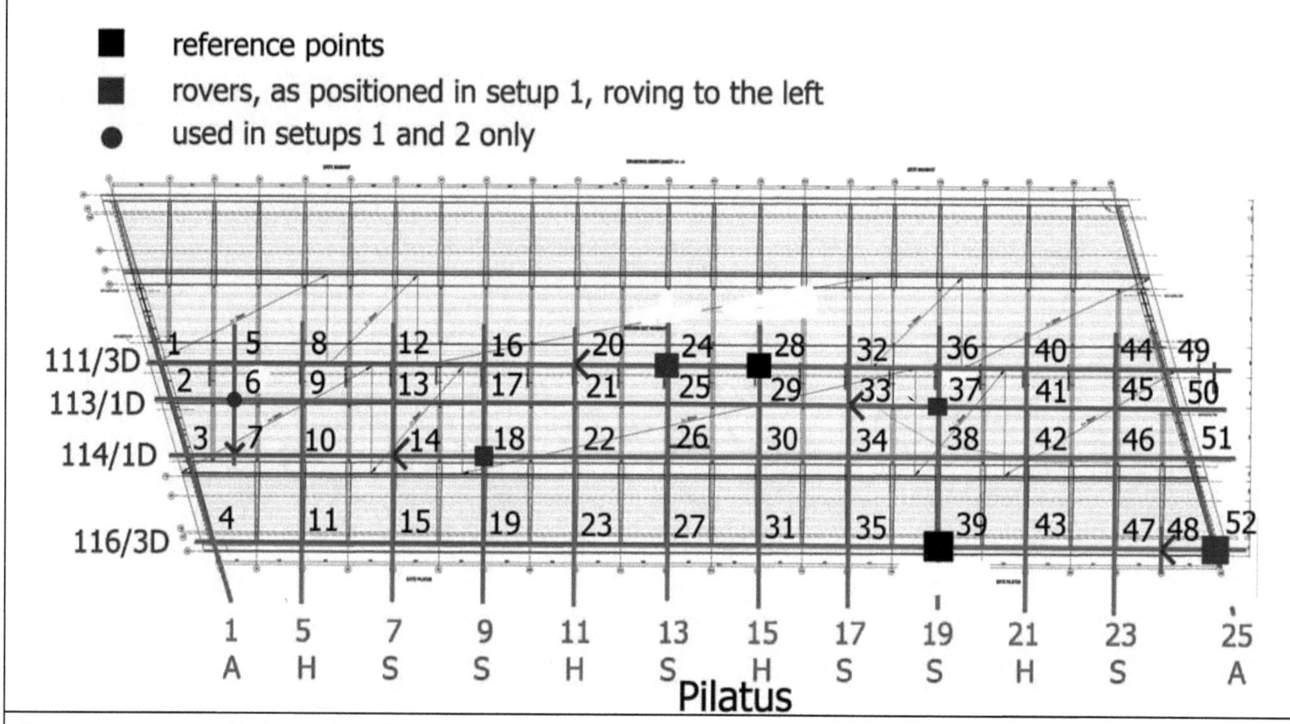


Fig. 4: Langensand Bridge. Ambient tests. Measurement point grid and indication of DOF's per grid axis.

Figures 3 and 4 show the arrangement of the measurement points and the respective DOF's. The two references chosen at 0.33 and 0.4 of the span length were 3D points. Another eight sensors were used to equip two 1D and two 3D rovers. As two remaining points could not be covered with this schedule, two additional 1D sensors were used in setups 1 and 2. As a results, of the 12 setups performed, two were with using 15 sensors and ten with 14 sensors.

| | |
|---|---|
| Sensors used: | PCB 393B31, 10 V/g (Figs. 5 and 6) |
| Frontend used: | LMS Pimento, 16 channels, 24 bit |
| Sampling rate: | 100 Hz |
| Time window length per setup: | 30 minutes |
| Number of setups: | 12 |
| Total testing time: | 8 hours |
| Outdoor temperature: | 2 degrees C |

Fig. 5: 3D measurement point.

Fig. 6: 1D measurement point.

## AMBIENT VIBRATION TESTS; SIGNAL PROCESSING AND RESULTS

Applying the Artemis Software Suite and making use of the EFDD (Enhanced Frequency Domain Decomposition) routines allowed identification of 21 bridge modes in the frequency range f = 1.27…34.04 Hz. The SVD diagrams presented in Figures 7 and 8 show that the number of projection channels was chosen to 6. The number of frequency lines for f = 0…50 Hz was chosen to 1'024.

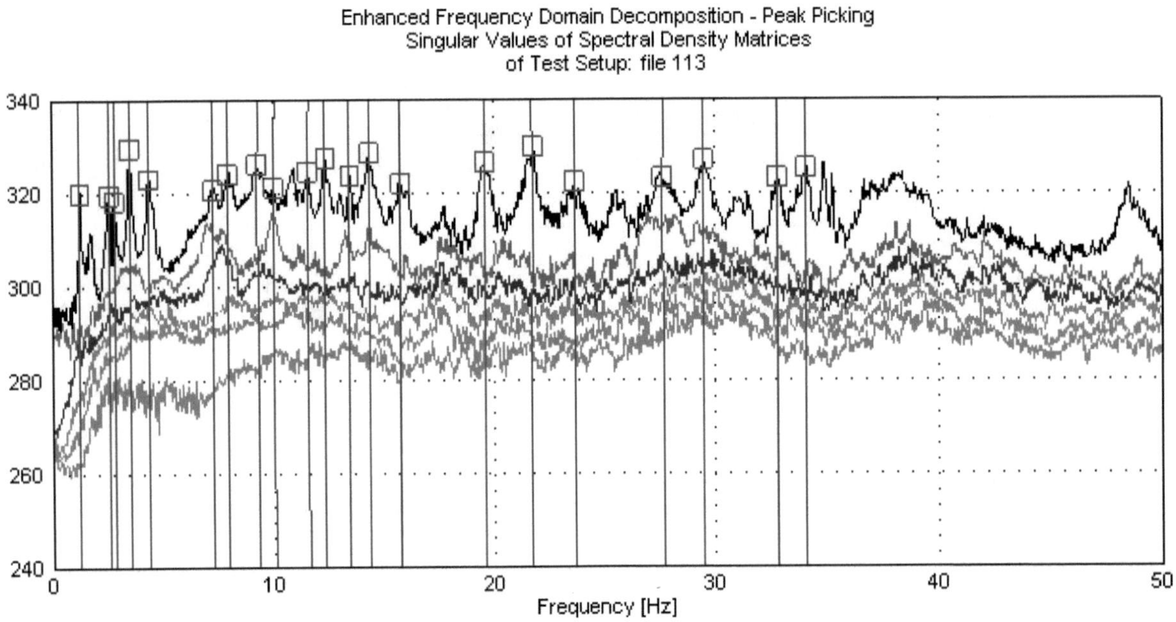


Fig. 7: 0…50 Hz EFDD SVD diagram as determined through application of Artemis Extractor routines.

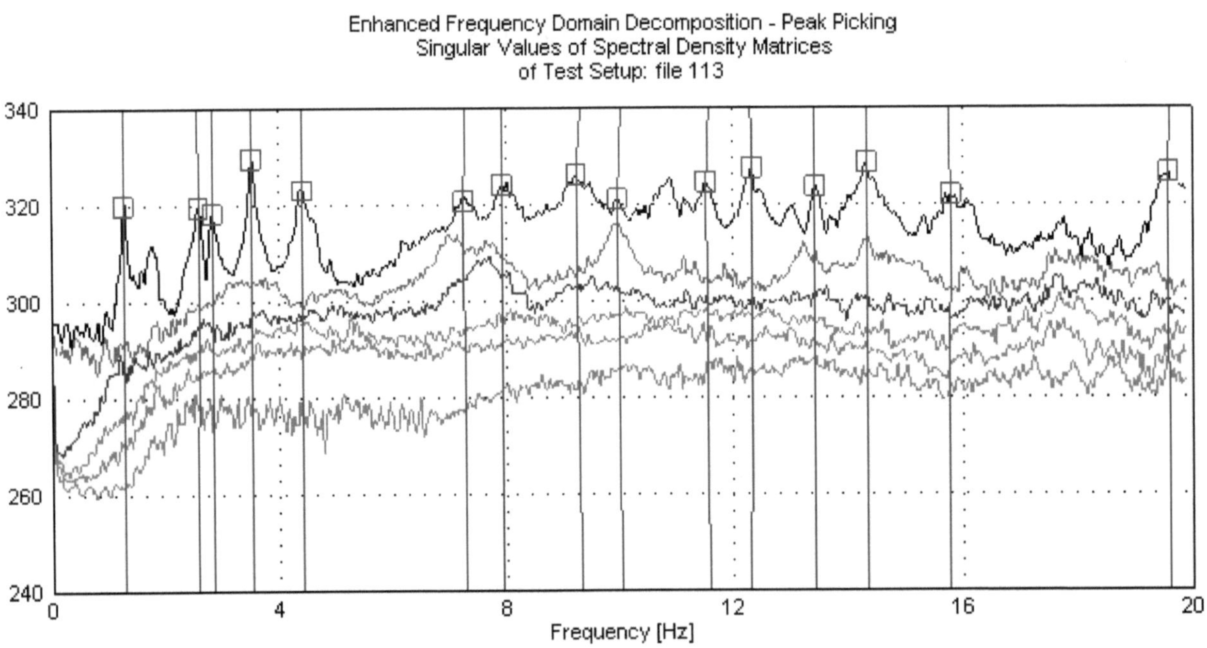


Fig. 8: 0…20 Hz EFDD SVD diagram as determined through application of Artemis Extractor routines.

Table 1: Natural modal parameters of the Langensand Bridge, Pilatus half.

| | frequency [Hz] | damping [%] | type | |
|---|---|---|---|---|
| 1 | 1.27 | 1.69 | 1. longitudinal bending | LB1 |
| 2 | 2.58 | 1.62 | 1a. transverse bending plus Torsion | TB1a |
| 3 | 2.82 | 1.38 | 1b. transverse bending plus Torsion | TB1b |
| 4 | 3.56 | 0.88 | 2. longitudinal bending | LB2 |
| 5 | 4.40 | 1.45 | 1. torsion plus 1. cantilever | TC1 |
| 6 | 7.29 | 1.80 | 3. long. bend. plus 2. transv. bend. | LB3TB2 |
| 7 | 7.95 | 1.25 | 2. cantilever | C2 |
| 8 | 9.34 | 1.74 | 2. torsion | T2 |
| 9 | 10.05 | 0.78 | 3. cantilever | C3 |
| 10 | 11.57 | 0.85 | 5a. cantilever Tribschen side | C5a |
| 11 | 12.31 | 0.89 | 5b. cantilever Bundesplatz side | C5b |
| 12 | 13.43 | 0.44 | "strange" cantilever vibration | CS |
| 13 | 14.35 | 0.63 | 7. cantilever | C7 |
| 14 | 15.75 | 0.70 | 8. cantilever | C8 |
| 15 | 19.61 | 0.69 | 10. cantilever | C10 |
| 16 | 21.71 | 0.79 | 11. cantilever | C11 |
| 17 | 23.69 | 0.87 | 12. cantilever | C12 |
| 18 | 27.71 | 0.72 | | |
| 19 | 29.48 | 0.68 | | |
| 20 | 32.78 | 0.46 | | |
| 21 | 34.04 | 0.46 | | |

**VEHICLE EXCITATION**

As no reliable information on the bridge natural dynamic behavior was available before the tests, three vehicles with different suspension system were used for the driving tests. One air-suspended 4-axle 35-kN-Truck, one steel-suspended 4-axle 35-kN-Truck and, because the bridge is used by a Bus Line, a three-axle air suspended 25-kN Bus.

Fig. 9: Steel suspended rear twin-axle.

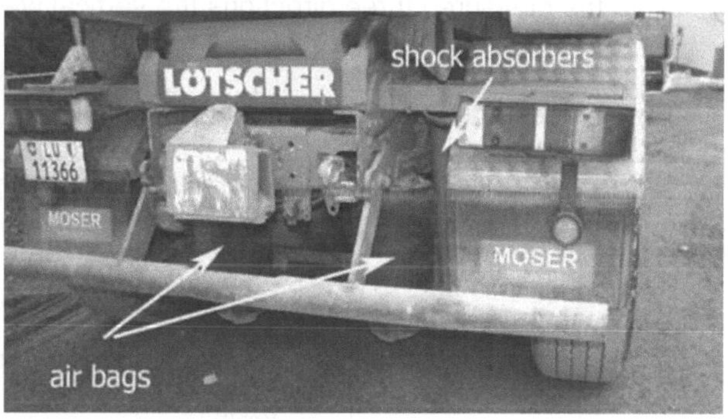


Fig. 10: Air suspended rear axle.

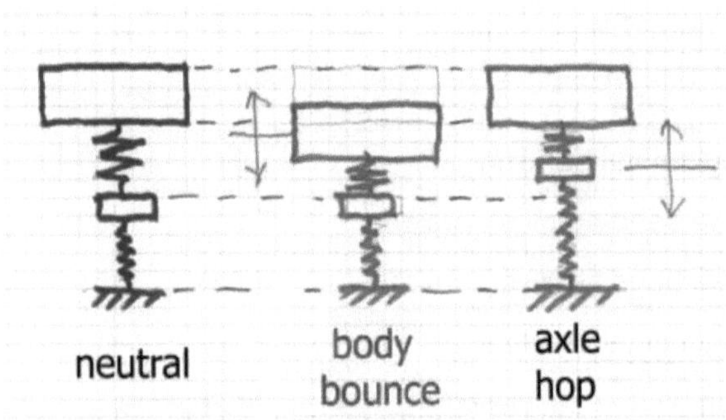


Fig. 11: 2DOF system representing a heavy vehicle's two natural modes.

Fig. 12: The three test vehicles crossing the bridge in line in the right lane direction Tribschen.

Figure 11 shows an undamped quarter-car model to explain a vehicle's basic vibrational modes. The model has four elements: a) the body mass, b) the suspension spring, c) the unsprung mass, and, d) the tire spring. Since we have a simple two-degree-of-freedom system here, this has two natural modes: a) the body bounce mode, and b) the axle hop mode. With the body bounce mode, the unsprung mass does not move compared to the large movement of the body mass. With the axle hop mode, the body mass does not move compared to the large axle movement. The modes are practically uncoupled (see the related frequencies given above). This is due to the facts that the body mass is much larger than the unsprung mass (the axle) and that the suspension spring is much softer than the tire spring.

The predominant dynamic wheel load frequencies are f = 1.5...1.8 Hz for air and f = 2.5...3.5 Hz for steel suspended vehicles in the body bounce mode and f = 10...15 Hz in the axle hop mode for both suspension systems [1].

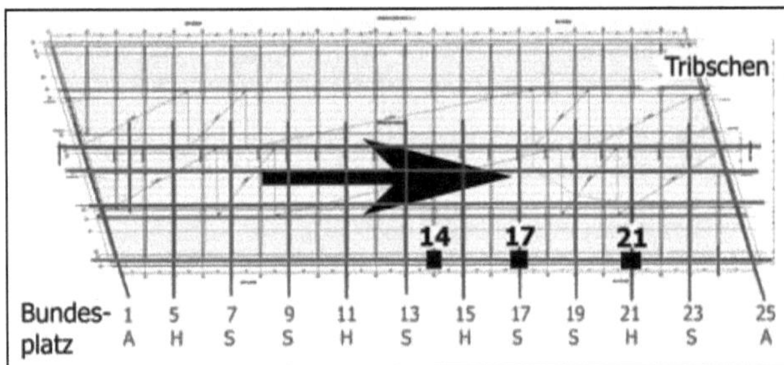


Fig. 13: Location of the three 3D measurement points MP 14, MP 17 and MP 21. The driving direction is indicated.

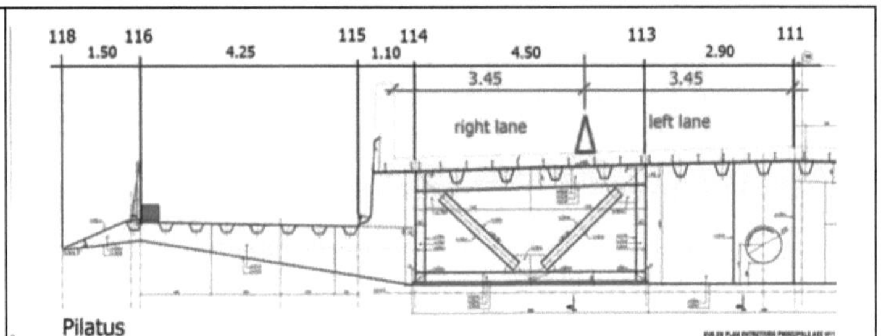


Fig. 14: Location of the measurement points in the cross section (red square); situation of the two driving lanes.

A total of 77 passages was performed with varying the vehicle speed (5 to 40 km/h), the number of vehicles (1, 2 or 3) the driving lane and the pavement roughness (without and with a 25 and 50 mm thick plank respectively at mid-span). From the measured three directions the vertical was determining and is discussed here only.

Table 2: Maximum accelerations measured at the Pilatus cantilever tip.

| | MP 14 amax [m/s2] | MP 14 who | MP 17 amax [m/s2] | MP 17 who | MP 21 amax [m/s2] | MP 21 who |
|---|---|---|---|---|---|---|
| left lane no plank | 0.60 | air 35 km/h | 0.79 | air 40 km/h | 0.75 | air 35 km/h |
| right lane no plank | 0.44 | air + bus 25 km/h | 0.50 | air + bus 25 km/h | 0.56 | bus 30 km/h |

| | MP 14 amax [m/s2] | MP 14 who | MP 17 amax [m/s2] | MP 17 who | MP 21 amax [m/s2] | MP 21 who |
|---|---|---|---|---|---|---|
| left lane 25 mm plank | 2.55 | air+steel+bus 20 km/h | 2.27 | air+steel+bus 20 km/h | 2.36 | air+steel+bus 20 km/h |
| right lane 25 mm plank | 1.31 | air + bus 20 km/h | 1.20 | air + bus 20 km/h | 0.82 | air+steel+bus 20 km/h |
| left lane 50 mm plank | 2.50 | steel 20 km/h | 2.63 | air+steel+bus 10 km/h | 2.54 | air + bus 20 km/h |
| right lane 50 mm plank | 1.69 | air + bus 20 km/h | 1.38 | air + bus 20 km/h | 1.35 | air + bus 20 km/h |

## ACCEPTANCE OF THE CANTILEVER VIBRATION LEVELS

There are many references dealing with the question of allowable vibration levels induced into pedestrians on bridges. Two of them are cited in the reference's list [2], [3]. Human beings tend to have different levels of acceptance versus vibrations. It is therefore not possible to define a given number as a firm limit. The tendency in the literature is that vibrations with amax = 0.5 m/s2 are not likely to result in complaints and that this may change for vibrations with amax larger than amax = 1.0 m/s2. For passages without a plank, the value of amax = 0.8 m/s2 has not been surpassed for the new Langensand Bridge Pilatus Half. The owner therefore decided to leave the bridge without TMD's and respond to complaints as they arrive.

## BRIDGE VEHICLE INTERACTION

The main purpose of the ambient structural identification test discussed at the beginning of this paper was to dispose of the necessary information to be able to designing TMD's without further delay in case the vibation levels induced by the heavy vehicles would not be acceptable. The measured time signals were therefore analyzed in detail in the frequency domain. The bridge natural frequencies were known in detail, of course with some flexibility due to the vehicles mass effect on these.

Inspection of the frequency spectra yields that four bridge modes are predominantly responsible for the dynamic bridge response at the cantilever tip Pilatus side:

- f4, LB2, f = 3.52 Hz, second main girder longitudinal bending mode,
- f5, TC1, f = 4.40 Hz, first main girder torsion and first cantilever bending,
- f10, C5a, f = 11.6 Hz, 5th cantilever bendig Trlbschen side,
- f12, CS, f = 13.4 Hz, "strange" cantilever vibrational shape.

According to the expectations, the first bridge longitudinal bending mode, f1 = 1.27 Hz, is not excited through the heavy vehicles. It's shape would be optimum but it's frequency is too low.

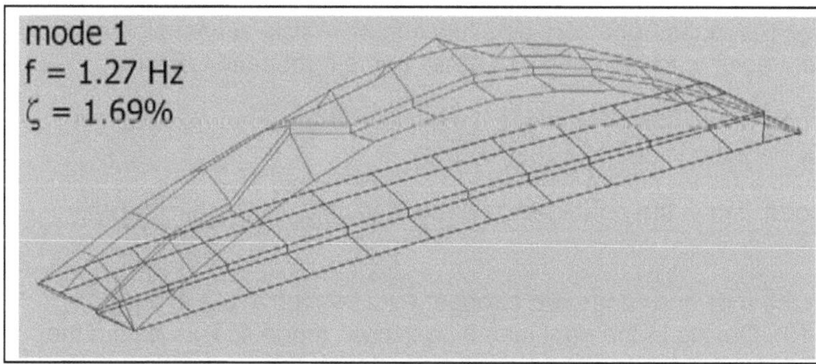


Fig. 15: Mode 1. As a result of measurement channel 14 having had some signal drop out problems the shape is not completely clean.

It is however astonishing that modes 2 and 3 (f2 = 2.58 Hz, f3 = 2.82 Hz) are not really excited. A reason for this might be that the modal amplitudes of these vibrations in horizontal transverse and longitudinal direction are comparatively large. These two shapes also show that the transverse and longitudinal modal amplitudes for the lines 113 and 114 are "faked". The software was told that they are the same as for lines 111 and 116 respectively. And they are really not the same! A full 3D measurement point grid would have been necessary to cope with this!

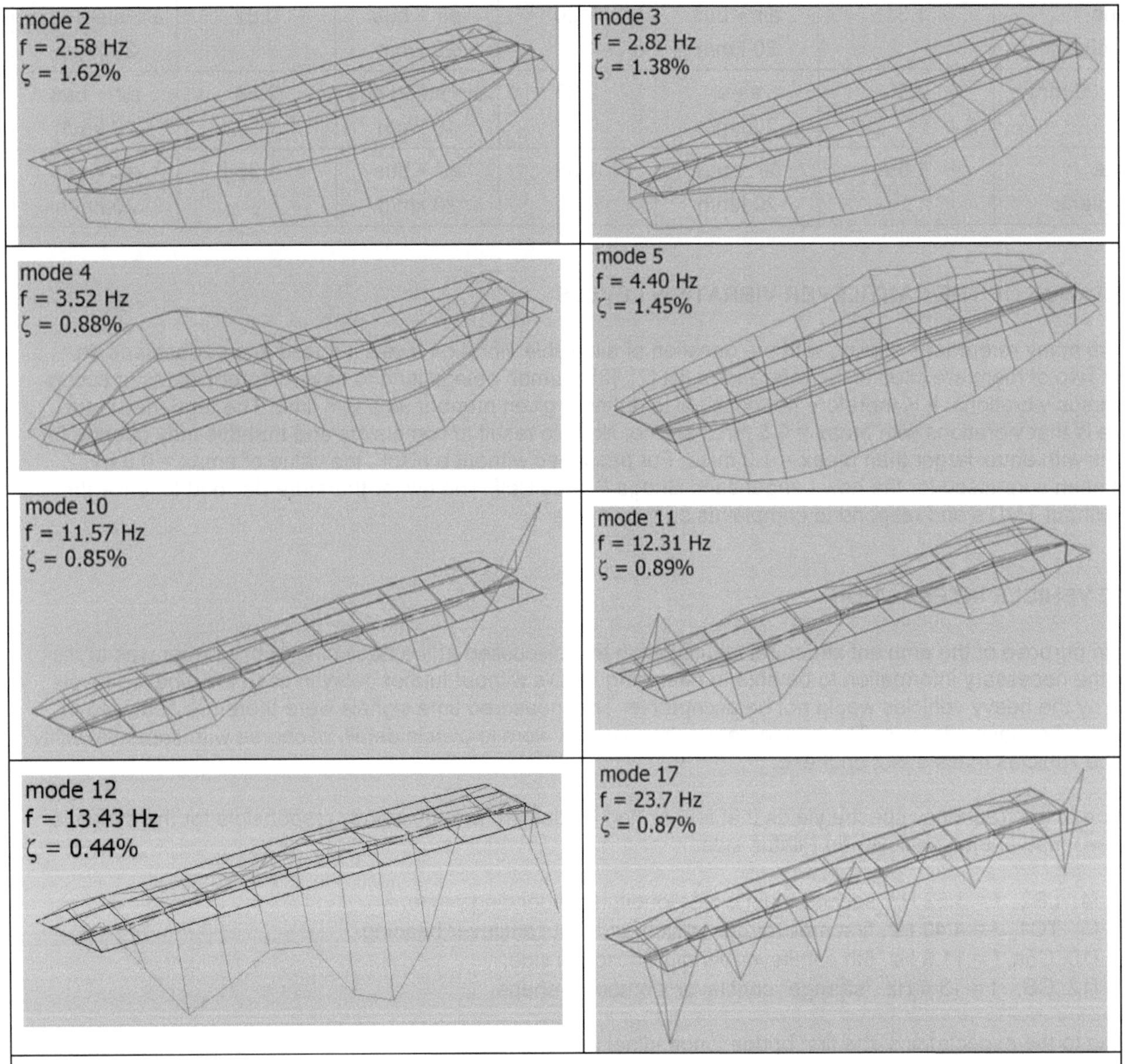


Fig. 16: Seven decisive modes and a nice one (mode 17).

The discussion below covers some specific bridge response behaviors as a function of vehicle suspension and driving axis. The diagrams show the measurement point in the left-and and amax in the right hand top line.

Modes 4 and 5 are excited through the steel sprung vehicle only. As mode 4 exhibits a transverse node at mid-span, this is not excited in point 14 but can nicely be seen in point 21 (Fig. 17).

Since mode 5 (4.40 Hz) exhibits a longitudinal nodal line in the right lane, this mode is only weakly excited with the vehicles using this lane (Fig. 18).

With the air suspended vehicle, bridge modes in the axle hop range are stronger excited than those in the body bounce range when driving in the left lane (Fig. 19). Driving in the right lane also excites mode 4. This is also the only case where some life signs of the bridge fundamental mode can be reckognized (Fig. 20).

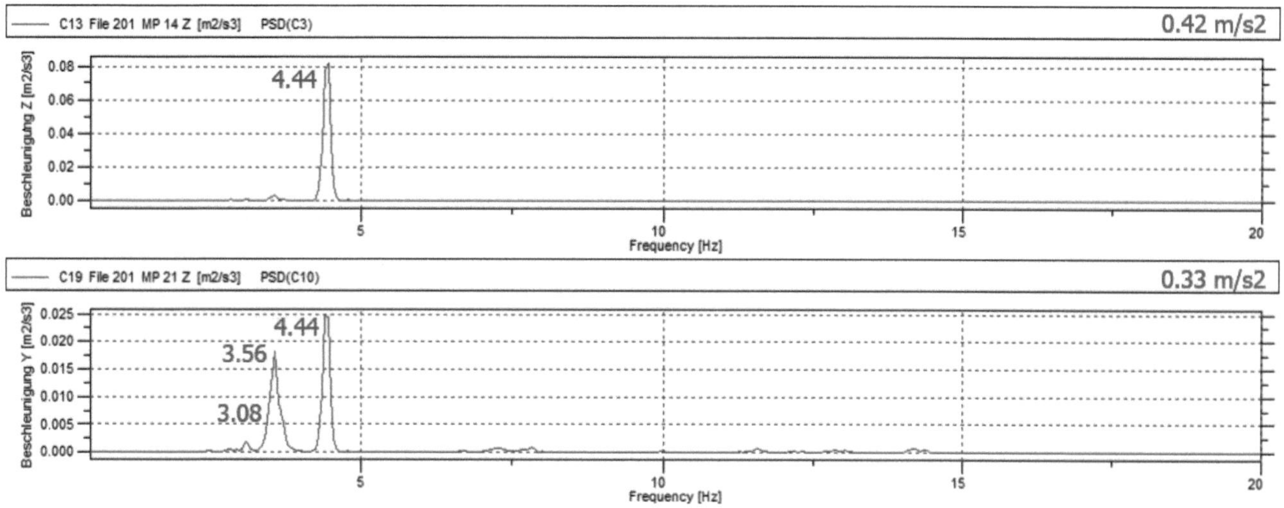


Fig. 17: Steel suspended vehicle, left lane, no plank, 20 km/h.

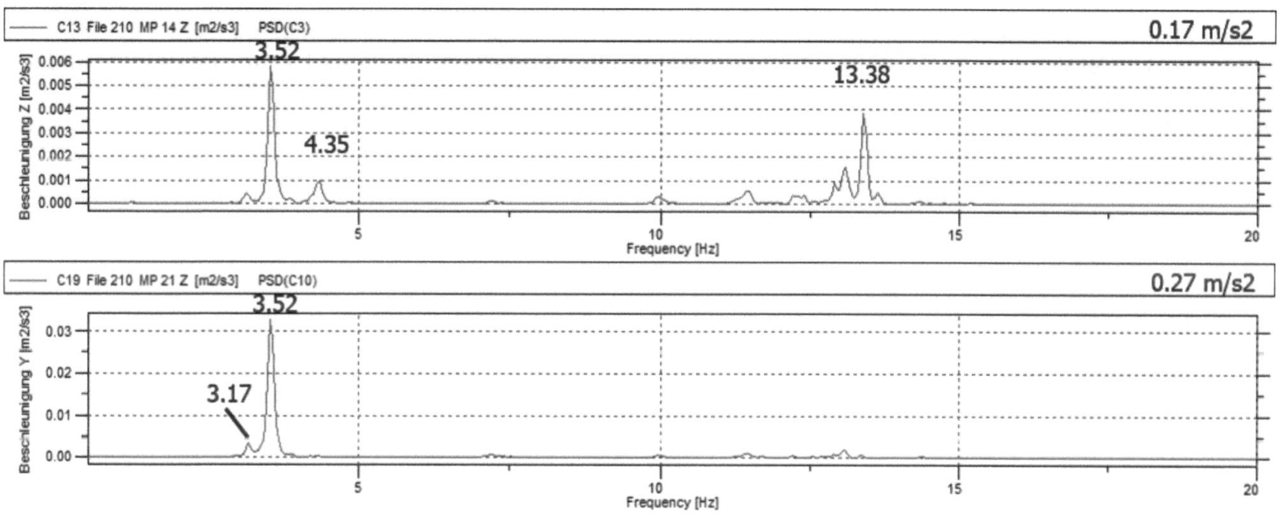


Fig. 18: Steel suspended vehicle, right lane, no plank, 20 km/h.

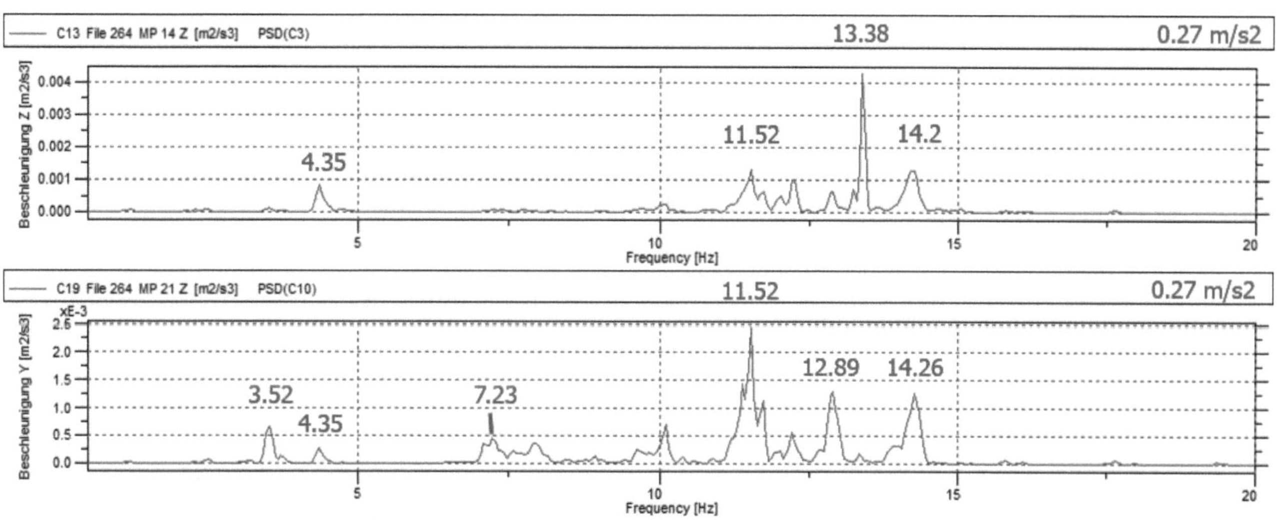


Fig. 19: Air suspended vehicle, left lane, no plank, 20 km/h.

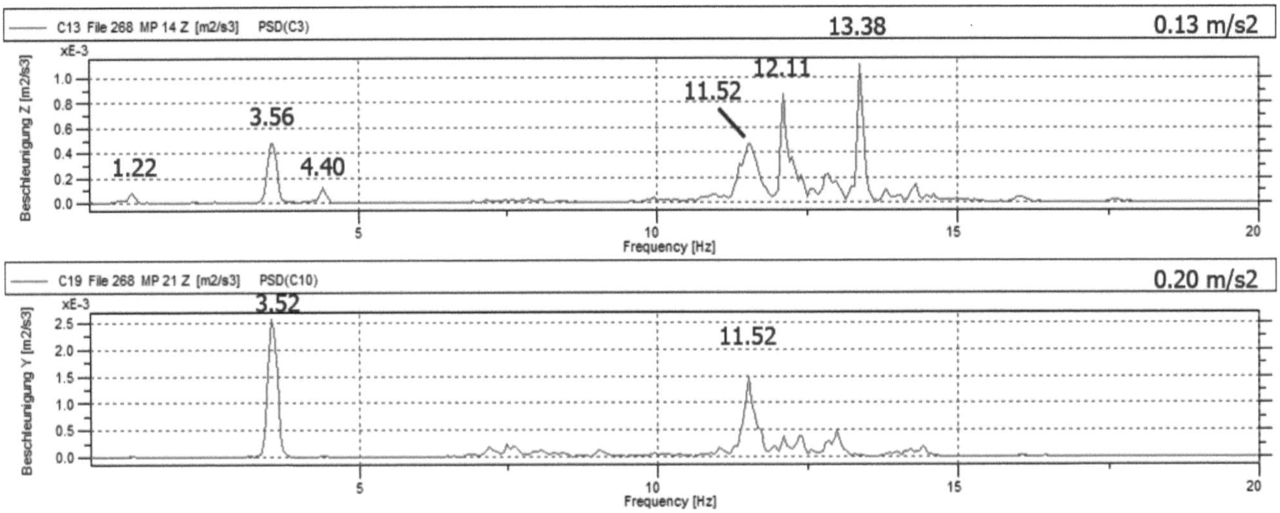


Fig. 20: Air suspended vehicle, right lane, no plank, 20 km/h.

Figures 21 to 26 present the spectra of the vibrations where the maximum accelerations had been measured (Table 2). It can easily be seen that bridge modes in the axle hop range are responsible for these and that the maximum accelerations are reached for air suspended vehicles at relatively high speed. (The maximum speed to be reached was 40 km/h. This was due to the approaches being quite steep gravel ramps.) The intensity of the vehicle's dynamic wheel loads in the axle hop range will significantly increase with deteriorating pavement surface evenness.

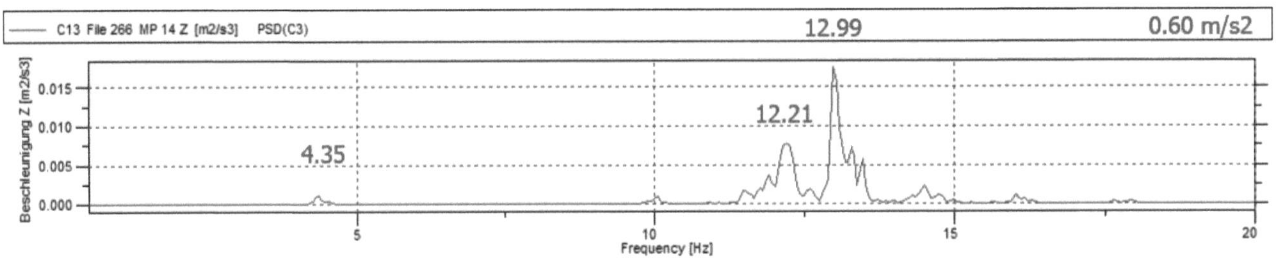


Fig. 21: Spectrum for the peak acceleration value for MP 14, left lane (air, 35 km/h)

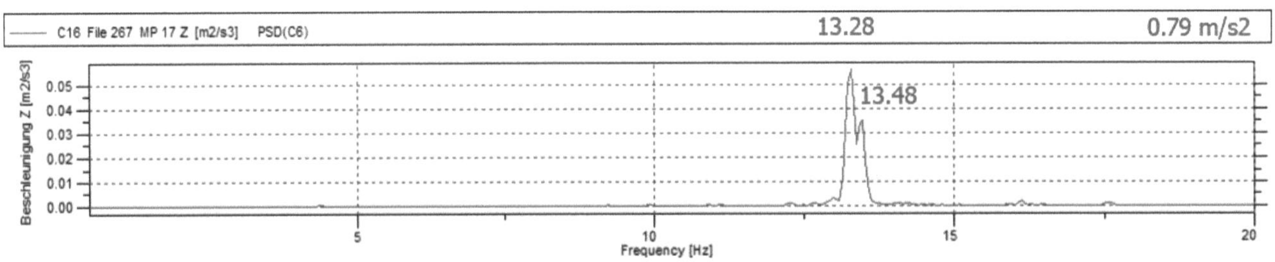


Fig. 22: Spectrum for the peak acceleration value for MP 17, left lane (air, 40 km/h)

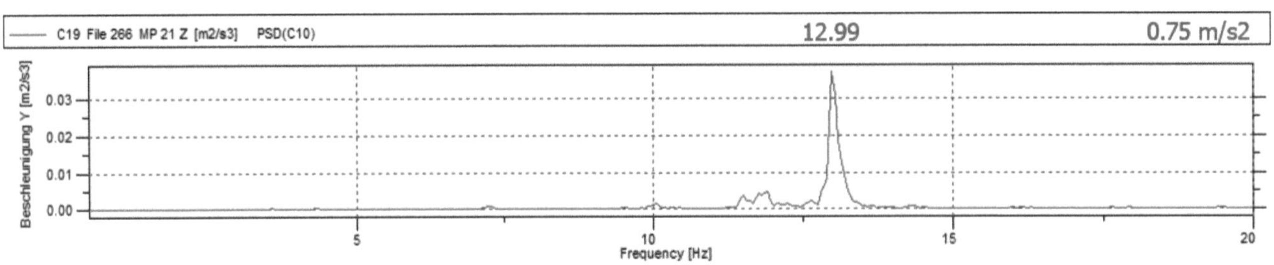


Fig. 23: Spectrum for the peak acceleration value for MP 21, left lane (air, 35 km/h)

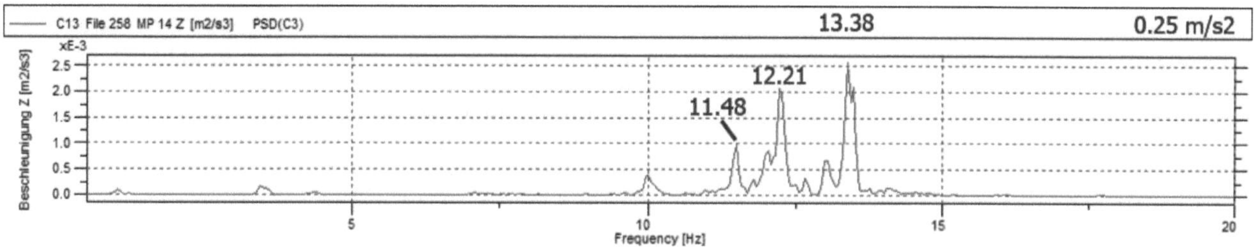


Fig. 24: Spectrum for the peak acceleration value for MP 14, right lane (air + bus, 25 km/h)

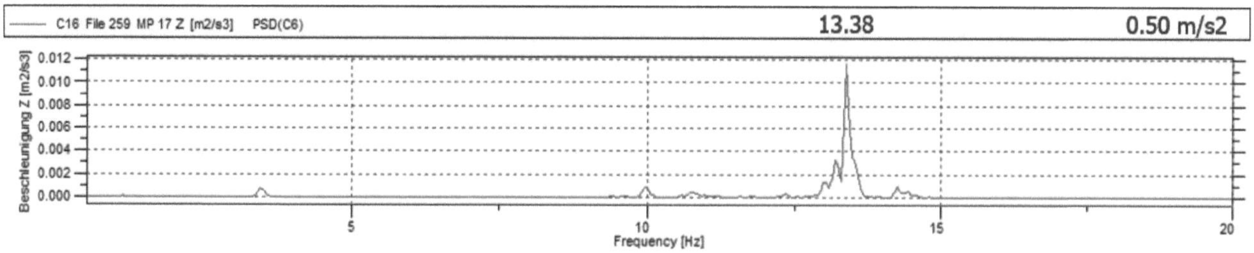


Fig. 25: Spectrum for the peak acceleration value for MP 17, right lane (air + bus, 25 km/h)

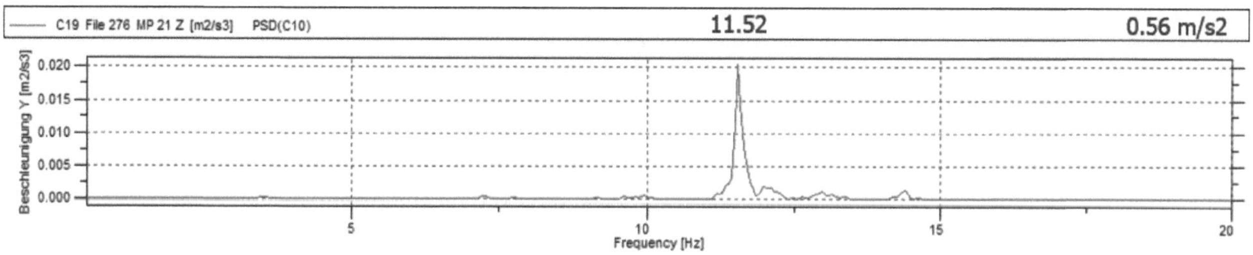


Fig. 26: Spectrum for the peak acceleration level for MP 21, right lane (bus, 30 km/h)

## TUNED MASS DAMPERS (TMD's)

From the discussion above it becomes evident that the effort to minimize traffic excited bridge vibrations using TMD's would have been quite important. Unlike a beam-type bridge with the dynamic behavior being governed by the main girder fundamental bending mode(s) it would not have sufficed here to tune TMD's to one single natural mode. There are at least three modes in the range f = 11...14 Hz which would have been to be worked on. And: The standard thinking of engineers looking for a nice place to hide TMD's would have failed here completely. Placing TMD's in the boxes offered by the steel main girders would not have solved the problem at hand. The locations to put TMD's would be at the free cantilever end where they can easily be seen. May be a solution as chosen for the footbridge crossing the railway tracks at the Cambridge UK main railway station could have been copied here (many small cylindrically shaped TMD's hidden inside "railings"). It is however fortunate that the measured acceleration levels did not clearly cross accepted limits here. The question however remains on how the bridge will behave when finished in December 2009.

## SUMMARY AND CONCLUSIONS

The new Langensand Bridge in Lucerne carries five road traffic lanes in its middle part plus two pedestrian/cyclist lanes on two 6 m long cantilevers over 11 railway tracks close to the Lucerne Main Station. An ambient system identification test on the firstly completed Pilatus side half of the bridge revealed that this steel/concrete composite structure exhibits an unusually high number of nicely distinctive natural modes with frequencies between f = 1.27 Hz for the first and f = 34.04 Hz for the 21st mode. The parameters chosen for this test, a) measurement point grid density, b) degrees of freedom to be measured in the different points, c) sampling rate, and, d) length of the time window per setup were obviously chosen quite reasonably to getting insight into the bridge dynamic natural characteristics having been looked for. It is also the time and place here to tell the community: A pilot test is a very nice thing and helps to prevent tactical errors. The step from good data to rubbish is quite small.

The first couple of modes show a predominantly global shape but for f > 10 Hz, the modes are predominantly local modes of the cantilever. This is not really astonishing when considering the cantilever to being a thin steel slab with cross girders but it is really astonishing for a highway bridge. The dynamic behavior of a "usual" beam-type highway bridge is governed by the first couple of global bending modes of the main girder(s). If necessary, these are to be damped easily with one or two TMD's located at the anti-nodal points of the respective mode shape. It would have been quite a significant task to damp the new Langensand Bridge natural modes exhibiting such a complex shape like modes No. 10 to 12. Fortunately enough, the TMD's design would not have had to deal with the first natural modal mass of this bridge of some 500 t (metric tons mass) or 250 t for half of the bridge, but with the (small) modal mass and (low) damping values related to the cantilever bending modes.

It will be interesting to note how the bridge behaves after completion in December 2009. Problems might arise with the pavement surface evenness becoming bad in the future.

## ACKNOWLEDGEMENTS

The project presented here has been commissioned to rci dynamics by The City of Lucerne Public Works Departement. The approval of this publication is thankfully acknowledged.

## REFERENCES

[1] Cantieni, R., Krebs, W., Heywood, R.: OECD IR6 DIVINE Project, Dynamic Interaction Between Vehicle and Infrastructure Experiment, Element 6, Bridge Research, Final Report. EMPA Test Report No. 153'031, Duebendorf, 2000.

[2] Vibration Problems in Structures – Practical Guidelines, ed. Bachmann, H. Birkhaeuser Verlag Basel, Boston, Berlin, ISBN 3-7643-5148-9, 1995.

[3] Bachmann, H., Ammann, W., Schwingungsprobleme bei Bauwerken. Structural Engineering Documents 3d. IABSE, International Association for Bridge and Structural Engineering, Zurich, 1987. (English version available as Structural Engineering Documents 3e)

Proceedings of the IMAC-XXVIII
February 1–4, 2010, Jacksonville, Florida USA


# Development of a Tamar Bridge Finite Element Model

R. J. Westgate, J. M. W. Brownjohn

The University of Sheffield

Department of Civil and Structural Engineering, Sheffield, United Kingdom, S1 3JD

cip08rjw@sheffield.ac.uk

## ABSTRACT

As part of an ongoing long-term monitoring research project, a finite element model has been developed to model the behavior of the Tamar Bridge in Plymouth, UK. The structure is a suspension bridge which underwent significant remedial and expansion work, including the addition of stay cables to the support structure. This paper presents the production of the model, with holistic static and modal responses of the two cable structural systems acting together.

Keywords: structural health monitoring, hybrid, environmental, finite element, non-linear

## 1 INTRODUCTION

Finite element (FE) analyses are used in conjunction with long-term monitoring of structures, to predict static and dynamic responses under various loads. In the context of Structural Health Monitoring (SHM) research, FE models are often used to simulate conditions that the researcher may wish to emulate with data recorded from the real-life structure.

The structural systems of suspension bridges are complicated compared to other bridges, since they rely on the equilibrium of the deck with the tensions in the cables. The behaviour of the bridge is non-linear, due to the non-compressive strength of cables and hangers, and the movements at the bearings.

Suspensions bridges that are a hybrid of suspension and stay cable support systems may be created as a result of improvements to support increased traffic levels. The number of papers concerning hybrid suspension bridges is limited, and only touch upon the theory. A more detailed approach to model these two systems together, which individually are self-configuring and non-linear processes, would be a welcome contribution to the theoretical modelling of tension structures.

This paper presents a methodology for creating a FE model for this type of suspension bridge, and how the static deflection of the bridge was optimised to get as close to zero displacement as possible. This paper follows a different method compared to other published works [1] and considers the cable forces are all contributing simultaneously. A simple parametric sensitivity study has also been presented, to demonstrate how the components of the hybrid suspension bridge contributes to its natural frequencies.

## 2 BRIDGE DESCRIPTION AND HISTORY

The Tamar Bridge shown in figure 1 carries the A38 trunk road across the River Tamar from Saltash in Cornwall to Plymouth in Devon, and is financed by its toll income for traffic usage. It was first opened in 1961, with a design by Mott Hay and Anderson, as a bridge supported by suspension cables and a truss girder. It has a main span of 335m and symmetrical side spans of 114m. Accompanied by anchorage and support spans, it achieves a span of

Figure 1: Photograph of the Tamar Suspension Bridge, Plymouth, UK

642m. The main towers are constructed out of reinforced concrete, seated on caisson foundations and are 73m tall, with the deck suspended halfway up. The truss is 15.2m wide and 4.9m deep, consisting of rectangular hollow sections, and originally supported a concrete deck. The suspension bridge cables are 0.38m thick, consisting of 31 steel locked cable ropes of 60mm diameter. Vertical locked coil hangers are at 9.2m centers and are 50mm in diameter.

During its life it has supported vehicles loads in excess of its design capacity. The bridge subsequently underwent a strengthening and widening (upgrading) scheme that was completed in December 2001. This involved adding a lane each side of its truss via cantilevers, and replacing the concrete deck with an orthotropic steel deck with shear boxes. Additionally, there was an installation of sixteen 100mm diameter stay cables, spanning from the main towers to either the truss or the side towers. The bridge is continuous at the Plymouth tower, whilst a lateral thrust girder forms an expansion joint at the Saltash tower.

## 3 FINITE ELEMENT MODELING

A high resolution 3D FE model was developed for the studies of a long term health monitoring project, intended as a numerical representation of the suspension bridge, and a research tool for investigating its performance under varying environmental conditions, such as ambient temperature and wind.

The program of choice was ANSYS. Truss elements were designed as beam elements with six degrees of freedom at each node, shell elements were used for the plates on the deck and the towers, and cables were modelled as tension-only spar elements.

Boundary conditions and movement joints are a necessary evil for the analyst; they have to imitate the behaviour of the actual structure correctly, whilst at a resolution agreeable for the whole model. Vertical hinges connecting the truss to a seat on each tower were treated as beam elements with rotational resistance released in the span-wise direction, and the lateral thrust girder was implemented in a similar way. The bridge model is fully fixed at the bases of the towers, and at the road-adjoined side-spans.

## 4 INTERNAL FORCE EQUILIBRIUM

The following section documents the various steps followed to ensure that the forces acting within the suspension and stay cables would provide a neutral profile under dead loading conditions. Modal solutions from the FE analyses are compared to previously obtained results [2] by their frequencies and order of mode shapes, as another measure for the model's accuracy.

## 4.1 ACCEPTABLE PROFILE SHAPE

The first step was to ensure that the finite element model was distorting into a reasonable shape when vertical loads were applied to it. This involved iterating the strains in the suspension cables until the displacements in the deck were minimised. For a suspension bridge, this involves adjusting the main-span and side-span suspension cables, and the stay cables, at every iterative step, to ensure that the deflections in the spans and towers remained small.

The secant iteration method was used to find the cable tension at dead load to reach a suitable profile, since a derivative of the target function is not a requirement within its formula. This is useful when the combination of two variables in the structural system makes linear interpretation impossible, and the equation defining a suitable structural stiffness is indeterminate.

The bi-variable form is:

$$\begin{pmatrix} x_{n+1} \\ y_{n+1} \end{pmatrix} = \begin{pmatrix} x_n \\ y_n \end{pmatrix} - \begin{bmatrix} \frac{F(x_n, y_n) - F(x_{n-1}, y_{n-1})}{x_n - x_{n-1}} & \frac{F(x_n, y_n) - F(x_{n-1}, y_{n-1})}{y_n - y_{n-1}} \\ \frac{G(x_n, y_n) - G(x_{n-1}, y_{n-1})}{x_n - x_{n-1}} & \frac{G(x_n, y_n) - G(x_{n-1}, y_{n-1})}{y_n - y_{n-1}} \end{bmatrix}^{-1} \begin{pmatrix} F(x_n, y_n) \\ G(x_n, y_n) \end{pmatrix} \tag{4.1}$$

where $x$ and $y$ are variables, $F(x, y)$ and $G(x, y)$ are functions of the variables and $n$ is the present iterative step number.

The first pass minimised the vertical deflection on the main span, and the span-wise deflection of the tips of the towers, using the forces in the main and side span suspension as the variables. This setup provided an 'M' shaped sinusoidal deflected shape on the main span, as shown in figure 2, which are a result of the stay cable forces. The side spans had inadequately large vertical deflections, and a simple study showed that this behaviour still existed when there was zero initial forces in the cables. This precluded a reconsideration of the side-cable profile.

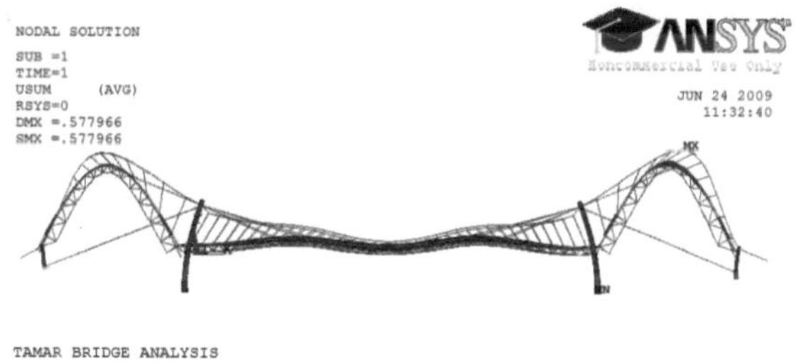


Figure 2: Initial deflected shape.

The gradient of the side-span suspension cables was made at an 11 degree angle to the horizontal at the side towers, which was taken from a past construction paper of the Tamar Bridge by Anderson [3]. The reformed profile produced deflections of 0.1m, which were acceptable. Due to the 'M' deflected shape the minimised outcome was modified to be the mean result of the mid-span and quarterly deflections, along with the deflections of the towers. Least squares method was attempted, but this lead to divergence in the iteration process.

## 4.2 CABLE FORCE ASSEMBLIES

The forces applied in the cables were adjusted to provide a reasonable performance of the bridge. This was achieved by analysing the deflections and modal properties of the suspension bridge, in the structural configurations it was in at 1961 and 2001. These two scenarios were used to check that values of the cable forces could be translated to both stages in the bridge's life comfortably. Modal properties from 1961 were taken from a paper by Williams [4], written before the expansion work. Results from 2001 are in a paper by Brownjohn [2].

Two candidates for possible assemblies were determined:

1. Zeroed mass and initial forces in 2001 elements, and added a concrete deck with iteration of suspension cable strains.

- Followed by non-zeroing previous elements, and iterated stay-cable forces.

2. Used stay cable forces found from site tests, and iterated main suspension cable only.

These arrangements were applied in the finite element model, and static and modal solutions were obtained. The results are presented in tables 1 and 2.

Table 1: Test arrangements for the Tamar Bridge - Deflections.

| Year | Arrangement | Applied Forces($\times 10^4$kN) | | | Displacement (m) | |
|---|---|---|---|---|---|---|
| | | Main susp. cable | Side susp. cable | Stay cable (quarter span) | Mid-span | Side-span |
| 1961 | 1 | 2.4787 | 2.6048 | 0 | -0.0008 | -0.0060 |
| | 2 | 2.0296 | 2.0597 | 2.529 | -0.4239 | -0.2849 |
| 2001 | 1 | 2.4787 | 2.6048 | 0 | 0.3104 | 0.2716 |
| | 2 | 2.0296 | 2.0597 | 2.529 | -0.0397 | 0.0739 |

Table 2: Final arrangements for the Tamar Bridge - Frequencies.

| Year | Arrangement | Modal (Hz) | | | | | |
|---|---|---|---|---|---|---|---|
| | | f1 | Shape | f2 | Shape | f3 | Shape |
| 1961 | Test[4] | - | LS1 | 0.32 | VS1 | 0.37 | VA1 |
| | 1 | 0.2495 | LS1 | 0.3561 | VS1 | 0.3836 | VA1 |
| | 2 | 0.2493 | LS1 | 0.3558 | VS1 | 0.3830 | VA1 |
| 2001 | Test[2] | 0.3930 | VS1 | 0.4570 | LS1 | 0.5946 | VA1 |
| | 1 | 0.3938 | VS1 | 0.4682 | LS1 | 0.5289 | VA1 |
| | 2 | 0.3909 | VS1 | 0.4677 | LS1 | 0.4045 | VA1 |

where LS1 means *first symmetric lateral mode shape*, VS1 *first symmetric vertical mode shape*, and VA1 *first asymmetric vertical mode shape*.

For both arrangements, the frequencies were so close to the previous reported tests, and would be subject to later alterations at the updating stage regardless, that they were not treated as the deciding features.

Although arrangement 1 had the most suitable deflections in the 1961 version of the model, it seemed likely that over the 40 years in service the suspension cables had suffered degradation, and the structural stiffness would be less than that calculated for the 1961 configuration.

The final choice was based upon minimum displacements at the 2001 stage of its life, so arrangement 2 was adopted. This conclusion means that the stay cable forces would be indeterminate from the model due to deterioration from the original arrangement.

## 5 FINAL CONFIGURATION

Figure 3 presents the final assembly chosen, and tables 3 and 4 present the static and modal results prior to model updating.

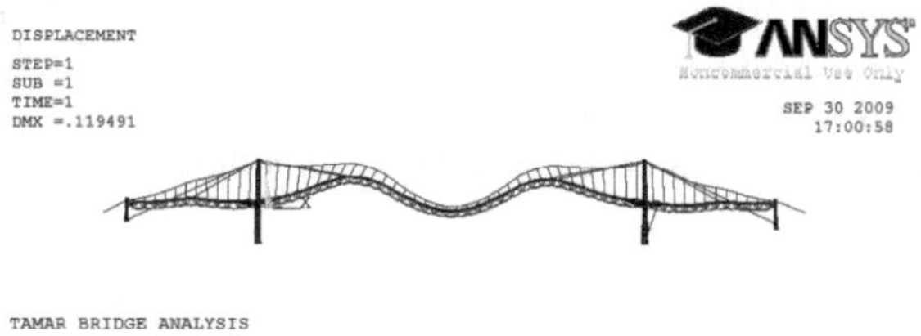


Figure 3: Deflected shape of final FE model for Tamar Bridge, under self weight.

Table 3: Deflection Summary for Figure 3.

| Deflection (m) | | | |
|---|---|---|---|
| Main Span Mid-span | Quarter-span Stay conn. | Side Span Mid-span | Tower Def. Horizontal |
| -0.0397 | 0.0739 | 0.0073 | 0.0036 |

Table 4: Global Frequencies Summary for Figure 3.

| Global Frequency Number | Frequency Number | Mode Shape | Frequency (Hz) | Monitored Frequency (Hz) | Error (%) |
|---|---|---|---|---|---|
| 1 | 1 | VS1 | 0.3938 | 0.393 | 0.2 |
| 2 | 2 | LS1 | 0.4682 | 0.457 | 2.0 |
| 3 | 4 | VA1 | 0.5289 | 0.595 | 11.0 |
| 4 | 7 | LS1 | 0.6565 | | |
| 5 | 12 | TS1 | 0.7840 | 0.726 | 8.0 |
| 6 | 15 | VS2 | 0.8376 | | |
| 7 | 16 | SS-A-VS1 | 0.9036 | | |
| 8 | 17 | SS-S-VS1 | 0.9936 | 0.975 | 1.9 |
| 9 | 22 | TRANS1 | 1.2549 | | |
| 10 | 24 | VA2 | 1.3086 | | |
| 11 | 25 | TA1 | 1.3252 | 1.23 | 7.7 |
| 12 | 34 | SS-A-TS1 | 1.7310 | | |
| 13 | 35 | SS-S-TS1 | 1.7698 | | |
| 14 | 36 | LA1 | 1.7924 | | |

where 'L' is a lateral mode shape, 'V' is vertical mode shape, 'T' is a torsional mode shape, 'TRANS' is a longitudinal translation mode, 'S' is symmetric, 'A' is asymmetric, 'SS' is side span and the numbers are their relevant order.

## 6 FURTHER PARAMETRIC STUDIES

The responses of adjusting the stiffnesses of the stiffening structure (truss, orthotropic deck and cantilevers), the suspension cables and the stay cables were analyzed, and their differences compared to an unadjusted model. This helps explain how these elements would effect the performance of the bridge, in particular how the addition of stay cables changes the modal properties. This also benefits what stiffnesses may need adjusting for model updating, and as a rudimentary form of the expected effect on monitored data if the actual structure deteriorated or strengthened.

For this analysis, the Young's modulus of the three sections were scaled to 50%, 80% and 120%. The first few global frequencies, those which excite the whole structure, are compared with the FE frequencies found in table 4.

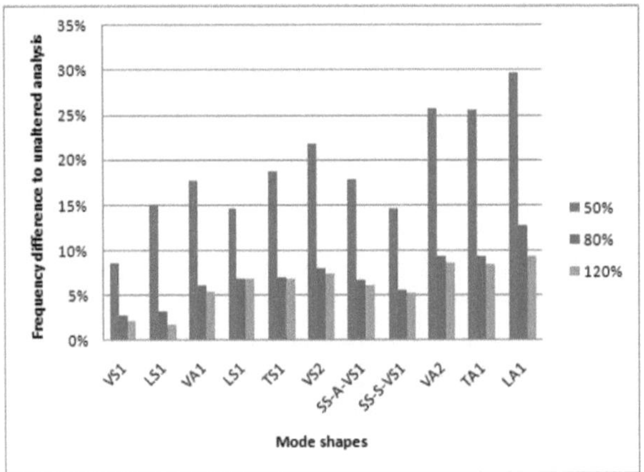


Figure 4: Adjustment of the stiffening structure's Young's modulus.

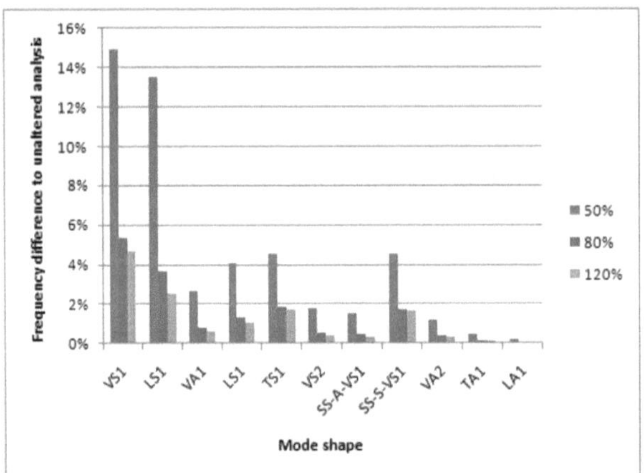


Figure 5: Adjustment of the suspension cable's Young's modulus.

In figures 4 and 5, the bar charts demonstrate the effect adjusting the stiffness of the suspension cable and truss has on the global frequencies of the structure. For all the charts, the scaling of the Young's modulii results in a similar scaling of the differences, with 80% and 120% scaling producing similar sized variations.

Figure 4 demonstrates that adjusting the Young's modulii has a greater effect on later frequencies, showing that the stiffening structure takes an increasing precedence over the modal properties.

As shown in 5, the suspension cable has a more pronounced effect on the first symmetric modes. The first LS1 mode is much more influenced by the suspension cable and the second, which bears further interaction from the stiffening structure. The chart also demonstrates the suspension cable's limited involvement on asymmetric modes, possibly as a result of no additional horizontal tension in the cable for oscillations. Further demonstration of this is in investigations of a smaller suspension bridge [5], where the behaviour of symmetric modes depends on the treatment of the cable, whilst anti-symmetric modes show little change.

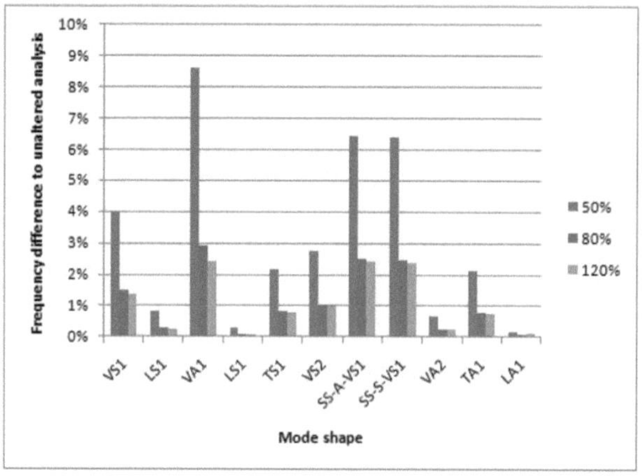


Figure 6: Adjustment of the stay cable's Young's modulus.

Figure 6 demonstrates how adjusting the stiffness of the additional stay cables alters the global frequencies.

It is the author's opinion that their impact on the global performance depends on where on the stiffening structure the stay cables are connected, and their number. For the Tamar Bridge, the first asymmetric mode and the first symmetric side-span modes are pronounced on the bar chart, since the stay cables are connected at a quarter of the main span, and halfway across the side-spans. The same cannot be said for the other vertical modes, where the ends of the cables are only slightly excited.

Another interesting feature is that the stay cables have a negligible effect on lateral modes, with only a very slight modification of the frequencies. This may indicate that the inclusion of the stay cables benefit vertical mode shapes only.

## 7 FUTURE WORK

The next step for refining the model through the geometrical and property tuning of the structure, to match more onerous criteria such as MAC values. There has been some progress in implementing thermal loads onto the model, to demonstrate static and modal changes with ambient temperature, alongside long-term monitored results the group has collected from the Tamar Bridge. The model will be expanded to simulate wind loads on the structure, and identify the bridge's performance changes.

## 8 SUMMARY

This paper presents the methodology used to create the finite element model for a hybrid suspension-stay cabled bridge. This was applied to the Tamar Suspension Bridge, connecting Cornwall to Devon. Non-linear analysis of the suspension cables was performed using the secant method of iteration, whilst additional stay cables used nominal tensions found from monitoring data. The final model has frequencies and mode shapes close to those measured from the real structure.

## REFERENCES

[1] **Ren, W. X., Blandford, G. E.** and **Harik, I. E.**, *Roebling Suspension Bridge. I: Finite-element model and free vibration response*, Journal of Bridge Engineering, Vol. 9, No. 2, pp. 110–118, March-April 2004.

[2] **Brownjohn, J. M. W.** and **Carden, P.**, *Real-time operation modal analysis of Tamar Bridge, 26th International Modal Analysis Conference (IMAC XXVI)*, Orlando, February 2008.

[3] **Anderson, J. K.**, *Tamar Bridge, ICE Proceedings*, pp. 337–360, August 1965.

[4] **Williams, C.**, *Vibration monitoring of large structures*, Experimental Techniques, Vol. 8, No. 12, pp. 29–32, 1984.

[5] **Brownjohn, J. M. W.**, *Vibration charcteristics of a suspension footbridge*, Journal of Sound and Vibration, Vol. 202, No. 1, pp. 29–46, 1997.

Proceedings of the IMAC-XXVIII
February 1–4, 2010, Jacksonville, Florida USA


# Modelling environmental effects on the dynamic characteristics of the Tamar suspension bridge

Elizabeth Cross[1)], Keith Worden[1)], Ki Young Koo[2)] and James MW. Brownjohn[2)]
[1)] Dynamics Research Group, Dept of Mechanical Engineering, University of Sheffield, Sheffield, UK
[2)] Vibration Engineering Section, Dept of Civil & Structural Engineering, University of Sheffield, Sheffield, UK

e.j.cross@sheffield.ac.uk

## ABSTRACT

For structural health monitoring purposes, it has become extremely important to understand, model, and compensate for the effect of environmental variations on the dynamic characteristics of structures under their ambient operating conditions. In this paper, acceleration measurement data from the Tamar Suspension Bridge, UK, processed by the data-driven stochastic subspace identification method is investigated, and the environmental variations in the natural frequencies modelled with respect to the temperature and the wind loadings. Two different approaches are considered; 1) Principal Component Analysis (PCA) which effectively identifies a linear variation pattern induced by environmental effects and 2) meta-modelling, a purely mathematical way to find an input-output relationship of a system based on various combinations of polynomial functions. This study found that the temperature was the biggest contributor to the environmental variations of the bridge's natural frequencies.

## INTRODUCTION

The effect of environmental and operational variations on the dynamic characteristics of civil structures is one of the main problems preventing structural health monitoring (SHM) technologies being transferred from the laboratory to real life structures. Often damage detection techniques developed in the laboratory are unable to function outside the uniform conditions they have been developed in; features chosen for their sensitivity to damage are often also sensitive to variations in the dynamic response of the structure caused by environmental and operational conditions not present in the laboratory. This area of research has received a good amount of attention over the last few years and different approaches to this problem are well summarised in Sohn's review on the effects of environmental and operational variability on SHM [1].

This paper studies the effects of environmental conditions, specifically temperature and wind speed, on the natural frequencies of the Tamar suspension bridge. Previously, several studies have focused on the dependence of modal frequencies of bridges on temperature variation [2-4], and indeed temperature is considered by most to be the dominant environmental factor causing variation in modal properties. Further investigations have been made into the influence of wind loadings on long span bridges [5]. Specifically, [5] found that the natural frequencies of the Hakucho suspension bridge in Japan decreased with increased wind speeds. Temperature, however, was not taken into account. The effect of heavy traffic loading on a long span suspension bridge was studied in [6], traffic weight was found to have little or no effect on the natural frequencies of the long span bridge.

The effect of environmental and operational conditions such as traffic loading and humidity have been neglected in this study, as the bridge in question is not considered to be subjected to heavy traffic loading, nor is the humidity considered to be a contributing factor due to the bridge's location. The aim of this paper is to principally identify the major factors driving the observed daily fluctuations in the bridge deck's natural frequency.

### The Tamar Suspension Bridge

The Tamar Bridge (Figure 1) is a 643m long suspension bridge which stands across the Tamar River connecting Saltash in Cornwall with the city of Plymouth in Devon. Since its construction in 1961 it has provided a vital transport link between the two locations. The bridge is of conventional design with symmetrical geometry, the main span is 335m long and the two side spans are each 114m long. The reinforced concrete towers reach a height of 73m and sit on caisson foundations founded on rock. The main suspension cables are 350mm in diameter, each consist of 31 locked coil wire ropes and carry vertical locked coil hangers at 9.1m intervals. The stiffening truss is 5.5 metres deep and is composed of welded hollow boxes. In 2001 the bridge was strengthened and widened to conform with EU directives. Specifically, eighteen new 100mm diameter locked-coil cables were installed and stressed to supplement the original suspension system, the composite main deck was replaced by a three-lane orthotropic steel deck, and single lane cantilevers were added on each side of the truss.

Figure 1. The Tamar Suspension Bridge, Plymouth, UK.

Several monitoring systems are in place at the Tamar Bridge currently. The Vibration Engineering Section at the University of Sheffield began monitoring the dynamic response of the bridge deck and cables in 2007. This monitoring system includes eight cable accelerometers, three accelerometers installed on the bridge deck, and three extensometers used to track relative motion between the deck sections and the tower. For the purpose of this study modal parameters were extracted from the accelerometer signals using a data-driven approach based on the Stochastic Subspace Identification (SSI) procedure [7]. Natural frequencies and damping parameters were extracted by the SSI routine at ten minute intervals throughout each day.

An additional monitoring system installed by Fugro Structural Monitoring, records cable loads, structural and environmental temperatures, and wind speeds and their directions. Of interest here are the temperature and wind speed monitoring; deck, cable and air temperatures are monitored continuously by ten platinum resistance thermometers, wind speed and direction are measured both at the top of the Saltash tower, and at 5m above deck level.

Readers are referred to [8] or [9] for more in-depth detail of instrumentation and monitoring processes in place at the Tamar bridge.

This paper considers two days worth of data detailing the deck natural frequencies, the deck temperature and the wind speed measured 5m above the deck. The two consecutive days are studied in order to determine the drivers of the variation in the natural frequencies of the deck. Firstly principal component analysis is carried out to attempt to better understand the underlying trends in the natural frequency data. Secondly meta-models are fitted to the response data using temperature and wind speed inputs.

## ANALYSIS TECHNIQUES

### Principal Component Analysis (PCA)

The first method of analysis used here is principal component analysis. The theory of this multivariate analysis technique is well documented (see for example [10]), and will only be briefly outlined here. Principal component analysis takes a multivariate data set and projects it on to a new set of variables, or 'principal components,' which are linear combinations of the old variables. Of these new variables, the first principal component will account for the biggest proportion of the variance in the data set that can be described by a single axis, the second principal component will account for the second biggest proportion of the variance in the data set, and so on. If the original number of variables is some number *p*, up to *p* new variables may be formed. Now, if the first *n* of these principal components represent a significant amount of the variance, it is fair to say that the data can be suitably represented solely by these *n* principal components without loss of any real information. Principal component analysis, therefore, works to reduce the dimensionality of the dataset, which can considerably ease analysis of datasets of high dimensionality.

### Meta Modelling

The term 'meta-model' is used loosely here to describe the process of modelling a system's dynamic response in terms of the presumed inputs to the system. Here, specifically, the bridge deck's natural frequencies are modelled by various quadratic functions dependent on the deck temperature and wind speed. The objective of these models is, for now, to simply give an indication of the influence of these environmental conditions on the natural frequencies of the bridge deck, and to ascertain whether the deck temperature or the wind speed has a more dominant influence. Since all input data is normalised prior to use, this indication comes simply from identifying the largest parameters in the quadratic model.

## FIRST DAY ANALYSIS

The first day chosen for analysis was the 16$^{th}$ December 2008. The variations of the deck's first five natural frequencies, along with the temperature and wind speed are shown in Figure 2. By inspection the temperature and wind speed appear to be correlated to some degree. On further investigation, the covariance of the two normalised time series is equal to 0.744, which confirms that the temperature and the wind speed are reasonably well correlated. This could make it more difficult for the model to separate out the influence of temperature on the natural frequencies from that of the wind speed.

As an initial investigation into the variation of the modal frequencies a principal component analysis was carried out on the first five deck natural frequencies for the chosen day. The first two principal components were found to account for 82% of the variance in the data (60% in the first, 22% in the second). This suggests that there are two latent variables driving the natural frequencies, which we attribute to temperature and wind speed. Figure 3 shows the data projected onto the first two principal components. Principal component analysis retains the temporal order of the data, this enables one to label the new data projections according to what part of the day they occurred in. Figure 3(a) shows the data labelled according to whether the data point occurred in the warmer or colder part of the day. Figure 3(b) shows the data labelled according to whether the data point occurred during the first half of the day where the wind speeds were lower, or the latter part of the day where the wind speeds were higher.

Figure 3(a) demonstrates that data points from lower temperatures can be easily separated from data points from higher temperatures, which suggests the natural frequencies dependence on temperature. The separation of data points from high and low wind speeds is not as clear, and suggests that the wind speed may have less of an influence than temperature on the deck's dynamic response. To further investigate these dependencies meta-models are fitted to the natural frequency data.

Different levels of model complexity have been trialled, ranging from a linear fit to a quadratic fit that includes temperature and wind speed dynamics. Linear fits were found to be inadequate to recreate the variation of the natural frequencies. A quadratic fit using both temperature and wind speed inputs was able to recreate the general trend of each natural frequency, but with large mean squared errors (MSE), as an example the fit to the first natural frequency for the first day is shown in Figure 4.

Much-improved fits to natural frequency data are found by using quadratic meta-models in which time lagged temperature and wind speed data are also included as inputs. It is not unexpected that the dynamic behaviour of the bridge should not only depend on the instantaneous environmental conditions, but those before it. Figure 5 shows a much improved meta-model fit to the first natural frequency, where the frequency at time $t$ is modelled with respect to temperature and wind speed at time $t$ and at time $t-1$ (as well as combinations of

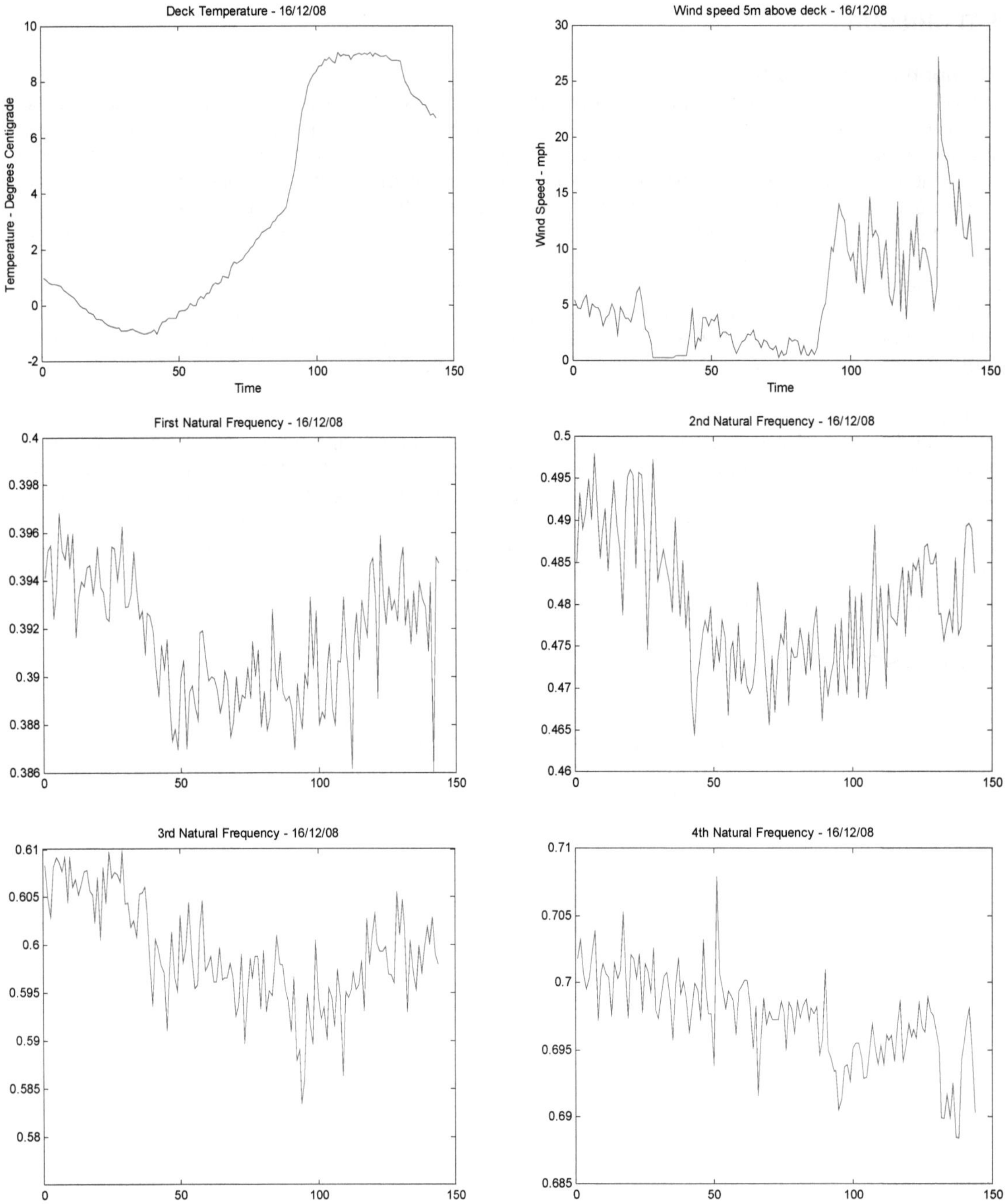


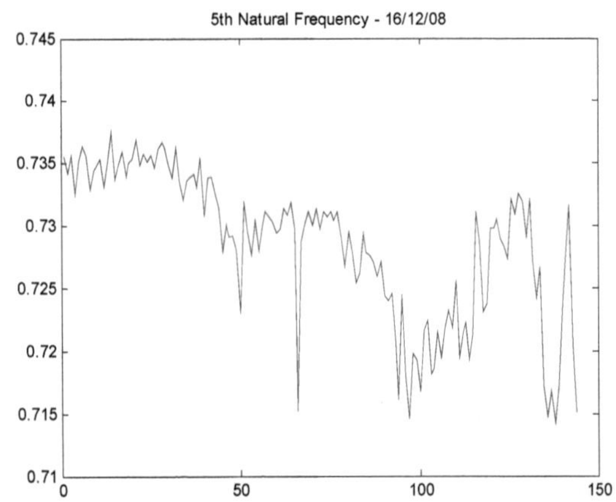


Figure 2: Temperature, wind speed and natural frequencies – 16/12/08

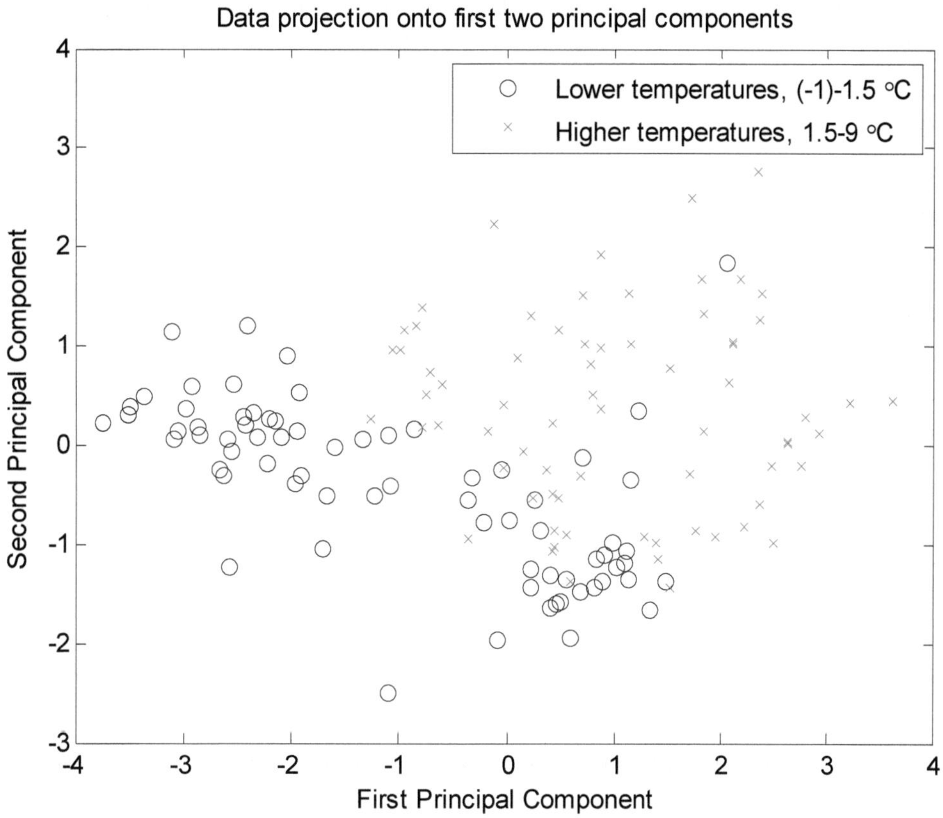


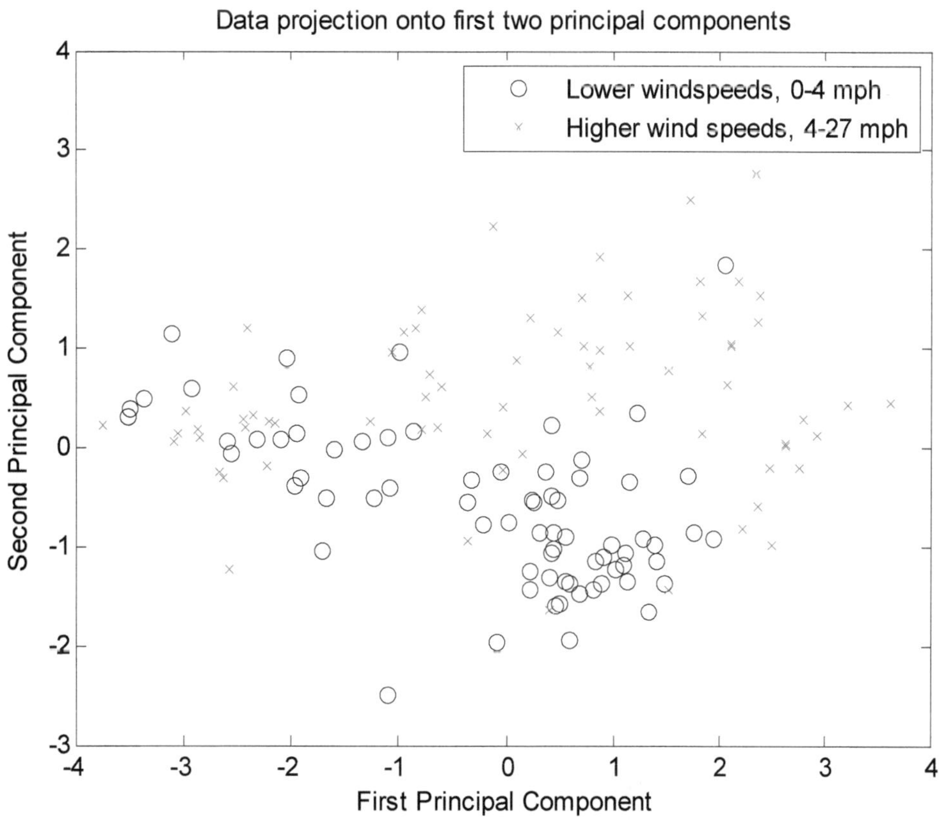


Above: Figure 3(a): Data projection on first two principal components, sorted by temperature.
Below: Figure 3(b): Data projection on first two principal components, sorted by wind speed.

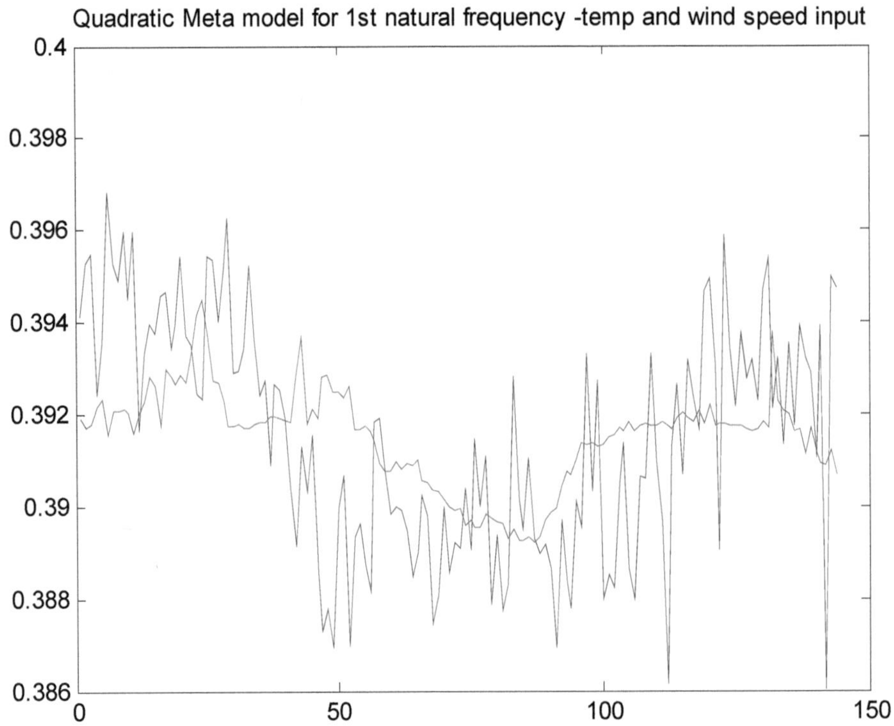


Figure 4: Quadratic temperature and wind speed meta-model fit for the first deck natural frequency.

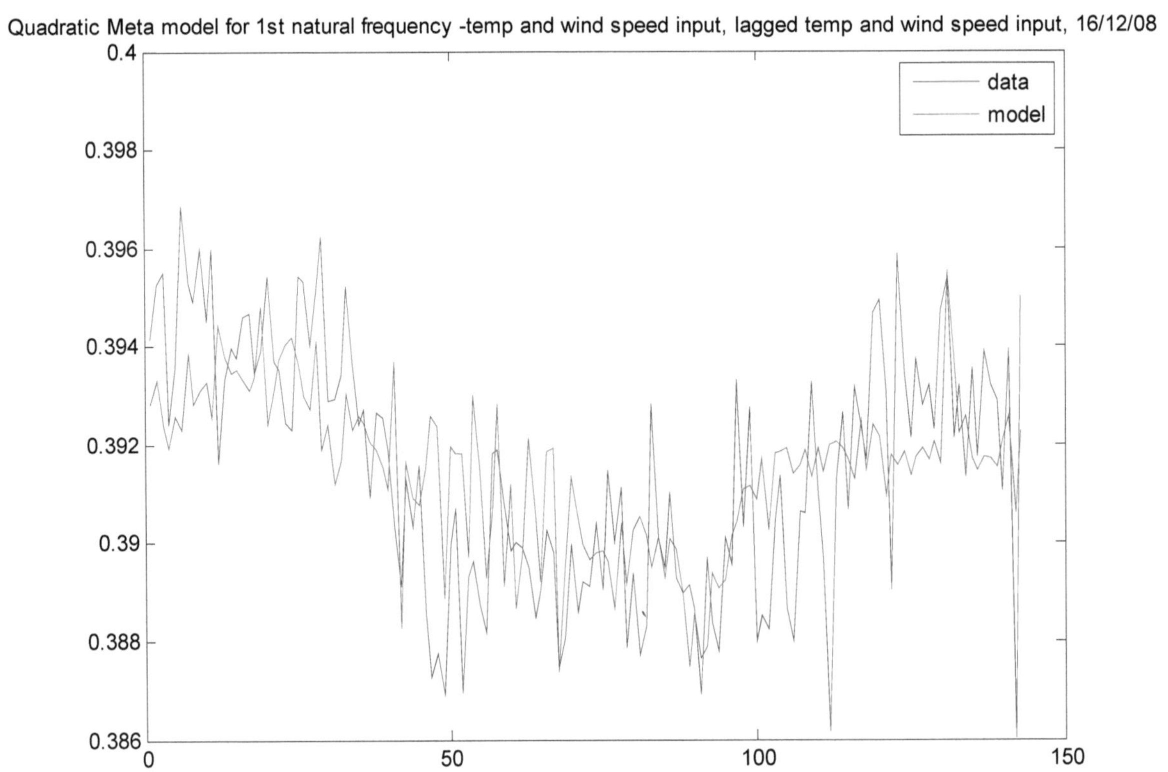


Figure 5: Quadratic temperature and wind speed meta-model fit for the first deck natural frequency including lagged temperature and wind speed data.

these parameters). As the model recreates the frequency so well, it suggests that the underlying relationship is:

$$f_1(t) = 0.3915 - 0.0134\,T(t-1) + 0.0130\,T(t) - 0.0130\,T(t)T(t-1) + 0.0073\,T(t-1)^2 + 0.0059\,T(t)^2 + 0.0005\,T(t)V(t-1) - 0.0005\,T(t-1)V(t-1) + 0.0004\,V(t) - 0.0003\,T(t)V(t) - 0.0003\,V(t-1) + 0.0002\,T(t-1)V(t) + \varepsilon \quad (1)$$

where $f_i$ (t) is the $i$th natural frequency at time $t$, $T(t)$ is the temperature at time $t$, V(t), the wind speed at time $t$, and ε represents a noise process. The parameters occur in order of their significance and less significant parameters have been omitted. To ascertain the importance of either temperature or wind speed one only need look at the corresponding coefficient value in equation (1). It is obvious that temperature has a much more dominant influence on the natural frequency than wind speed, it is also clear that the frequency is also majorly influenced by the lagged temperature input. The meta-model fits for the other four natural frequencies follow similar patterns, their models are represented by equations (2-4).

$$f_2(t) = 0.4781 - 0.0631\,T(t)T(t-1) - 0.0455\,T(t-1) + 0.0442\,T(t) + 0.0336\,T(t-1)^2 + 0.0297\,T(t)^2 + 0.0014\,V(t) - 0.0009\,T(t-1)V(t-1) + 0.0009\,T(t)V(t-1) + 0.0003\,V(t-1) - 0.0001\,T(t)V(t) + 0.0001\,T(t-1)V(t) + \varepsilon \quad (2)$$

$$f_3(t) = 0.5994 - 0.0181\,T(t-1) + 0.0168\,T(t) - 0.0057\,T(t)T(t-1) + 0.0046\,T(t-1)^2 + 0.0018\,T(t)V(t-1) - 0.0017\,T(t-1)V(t-1) + 0.0012\,T(t)^2 + 0.0007\,V(t) + 0.0004\,V(t-1) - 0.0003\,T(t)V(t) + 0.0001\,T(t-1)V(t) + \varepsilon \quad (3)$$

$$f_4(t) = 0.6980 - 0.0013\,T(t) + 0.0010\,T(t-1)^2 + 0.0009\,T(t)W(t-1) - 0.0009\,T(t-1)W(t-1) - 0.0007\,T(t)T(t-1) + 0.0005\,V(t-1) + 0.0004\,T(t-1) + 0.0003\,V(t) - 0.0003\,T(t-1)V(t) + 0.0002\,T(t)V(t) - 0.0002\,T(t)^2 - 0.00041\,V(t)V(t-1) + \varepsilon \quad (4)$$

$$f_5(t) = 0.7318 - 0.0232\,T(t)T(t-1) + 0.0122\,T(t-1)^2 + 0.0112\,T(t)^2 - 0.092\,T(t-1) + 0.0075\,T(t) - 0.0015T(t-1)V(t) + 0.0015\,T(t)V(t) - 0.0004\,T(t)V(t-1) + 0.0004\,T(t-1)V(t-1) + 0.0002\,V(t) + 0.0001\,V(t-1) + \varepsilon \quad (5)$$

As before, temperature is the most dominant influence on the natural frequency variations. We can further confirm this assumption by comparing meta-model fits that use only temperature inputs with those using only wind speed inputs. Figure 6 makes this comparison and clearly shows that the temperature meta-model far out-performs the wind speed model.

## SECOND DAY ANALYSIS

The second day analysed here is 17th December 2008. Figure 7 shows how the temperature, wind speed and first five natural frequencies of the deck vary throughout the 24 hours. Wind speed and temperature are not correlated as before (covariance = -0.004).

As before, the natural frequency data is first analysed with a principal component analysis. The first two principal components account for 83% of the variance in the data, the first principal component accounts for 54% of the variance, whilst the second accounts for 29%. This again suggests that two latent variables are driving the variations of the natural frequencies. Figures 8(a) and (b) show plots of the natural frequency data projected onto the first two principal components. Figure 8(a) shows the data projections sorted according to temperature, Figure 8(b) shows the data projections sorted according to wind speed. Data points occurring at higher temperatures are easily separable from those occurring at lower temperatures, suggesting a clear temperature dependency. Again, the data points sorted according to wind speed are less easy to separate, suggesting a weaker relationship between the natural frequencies and wind speed. This is further investigated through meta-model fitting as before.

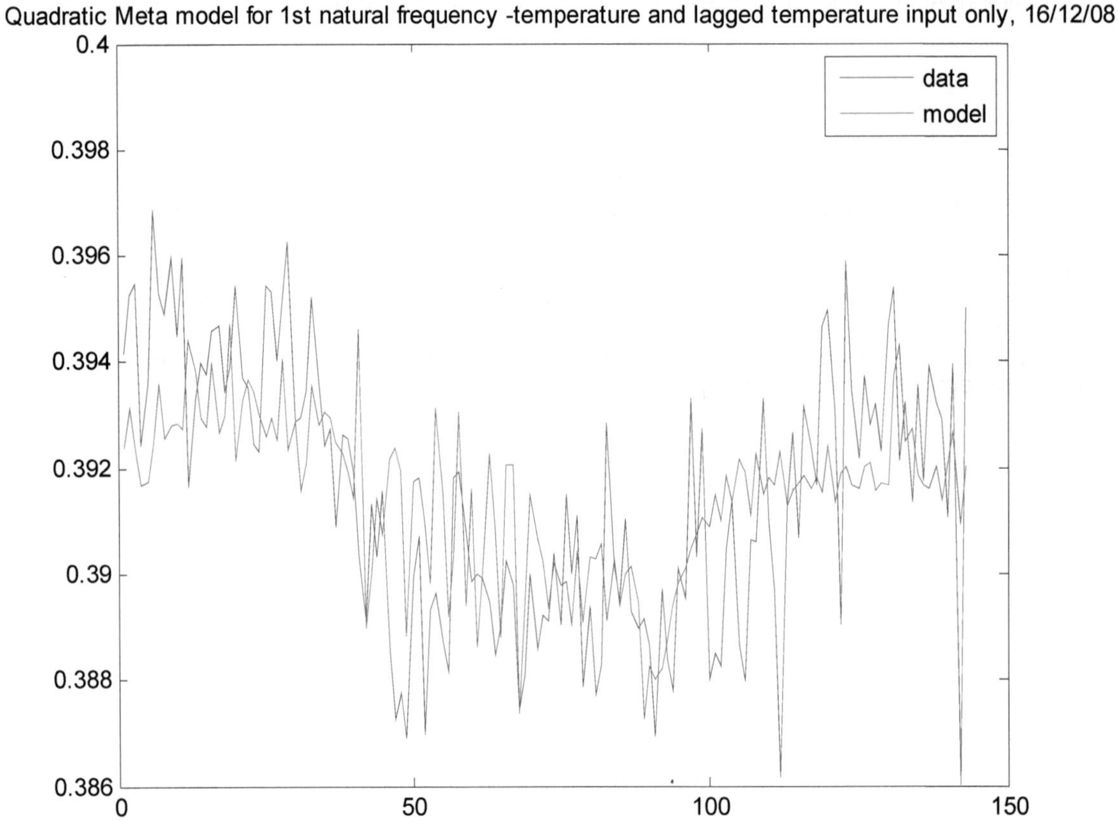


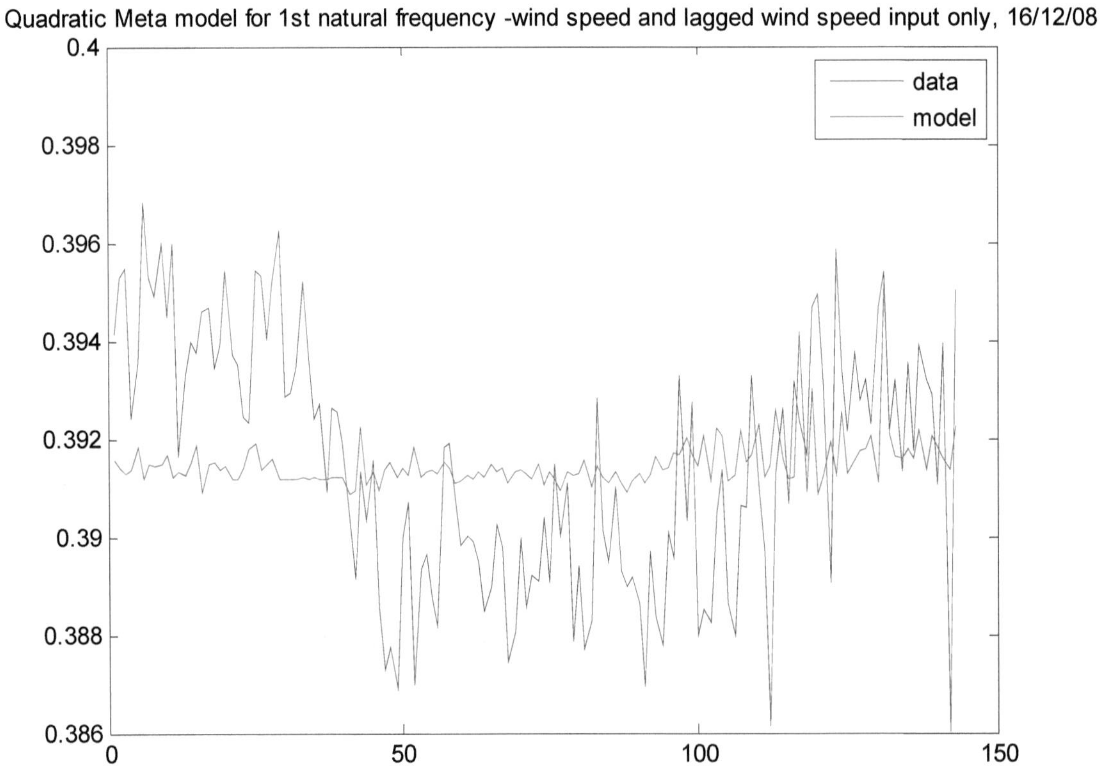


Above: Figure 6(a): Meta-model fit using only temperature inputs.
Below: Figure 6(b): Meta-model fit using only wind speed inputs.

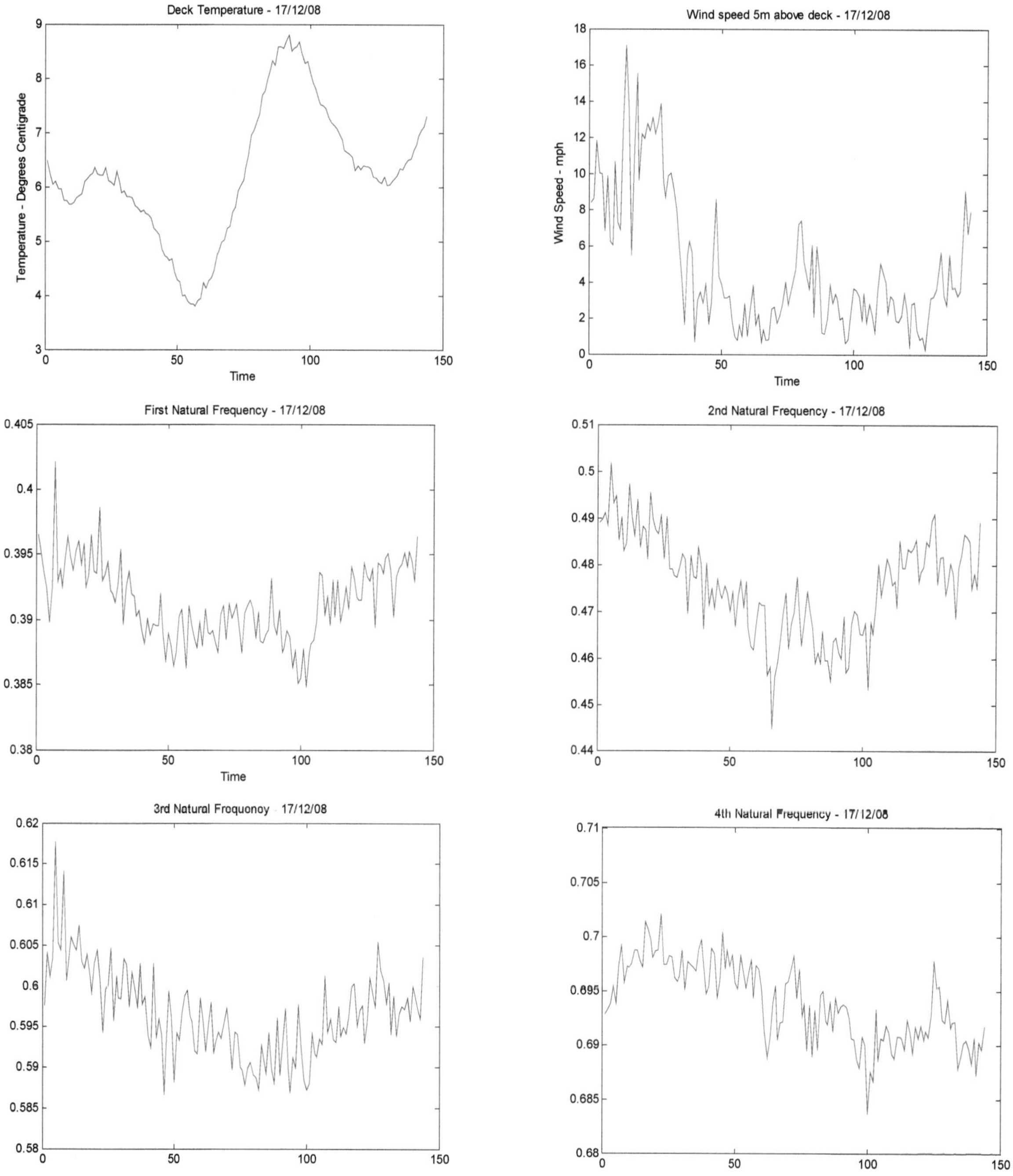


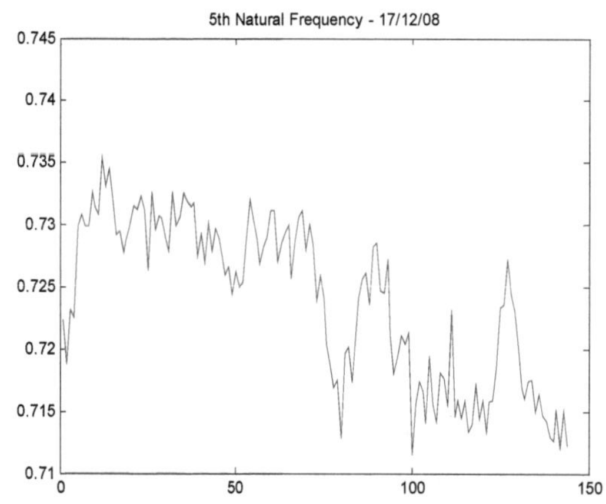


Figure 7: Temperature, wind speed and natural frequencies – 17/12/08

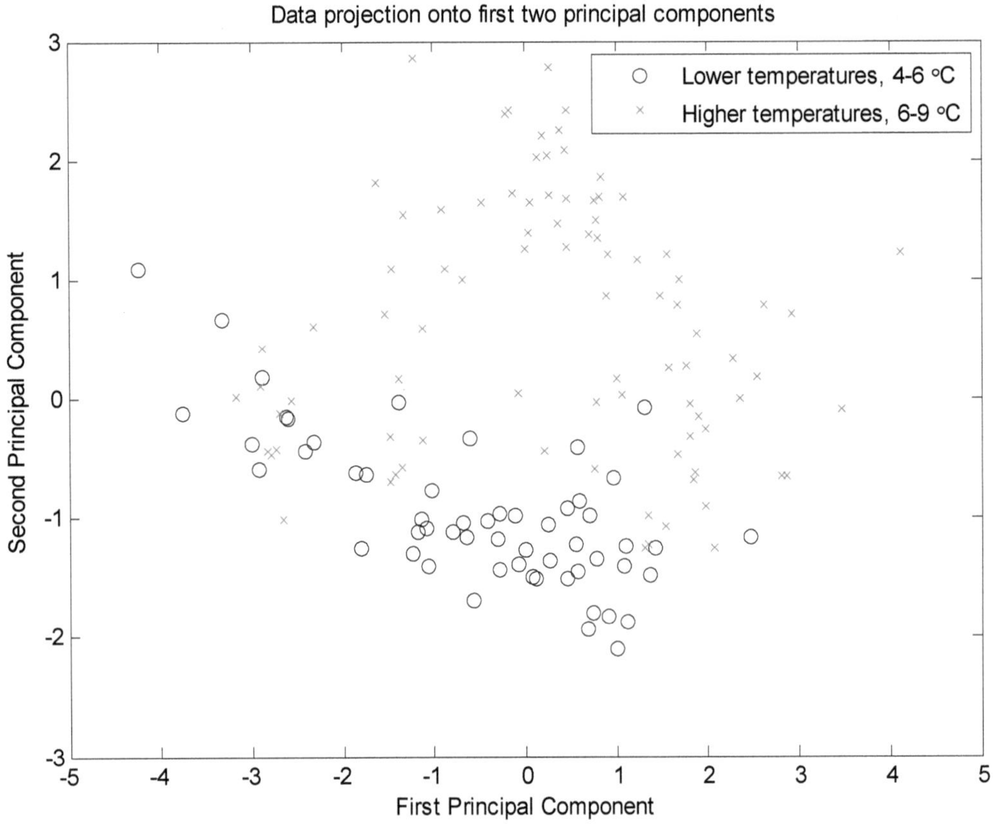


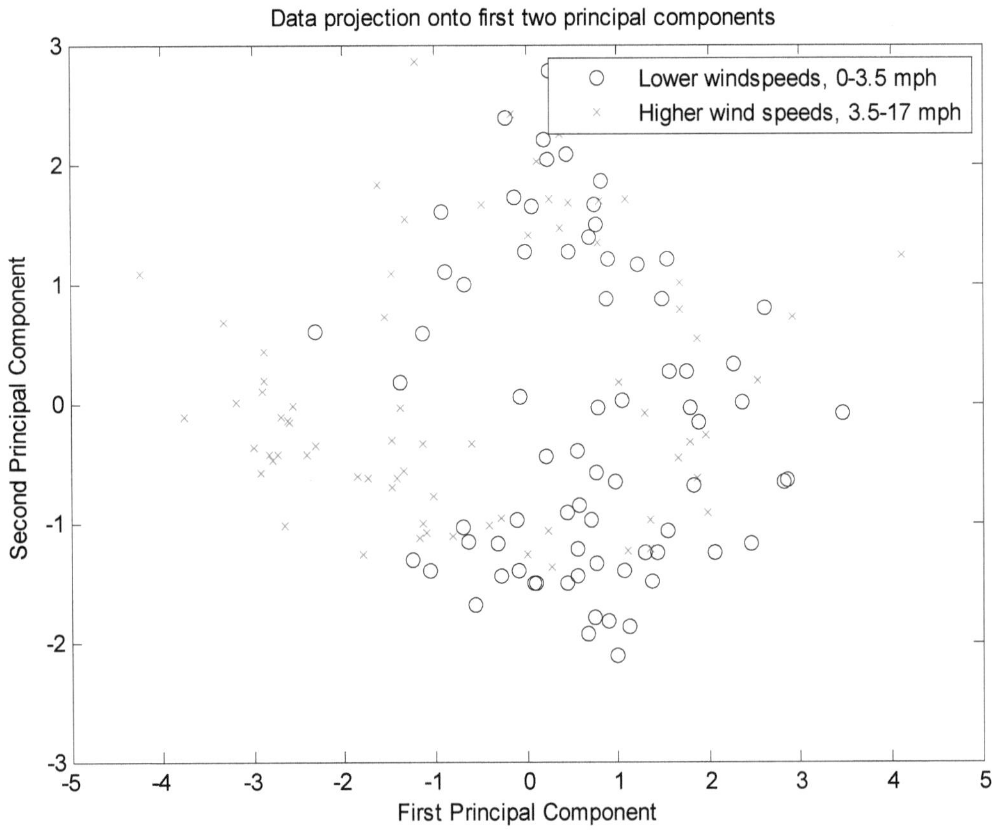


Above: Figure 8(a): Data projection on first two principal components, sorted by temperature.
Below: Figure 8(b): Data projection on first two principal components, sorted by wind speed.

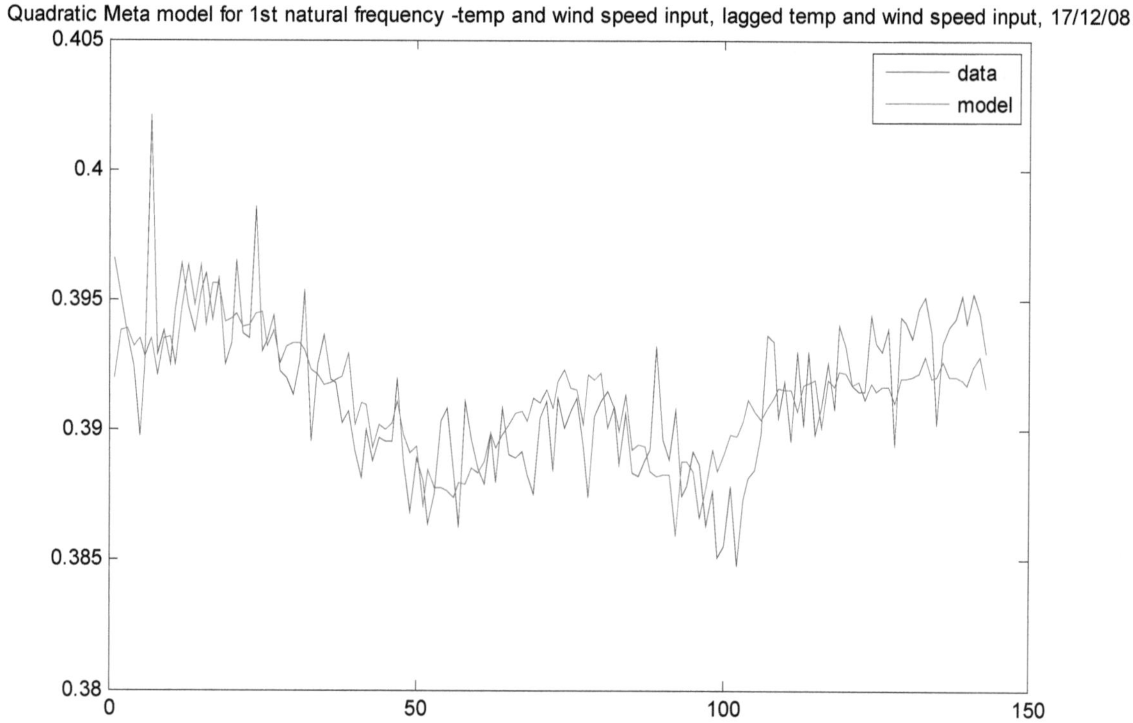


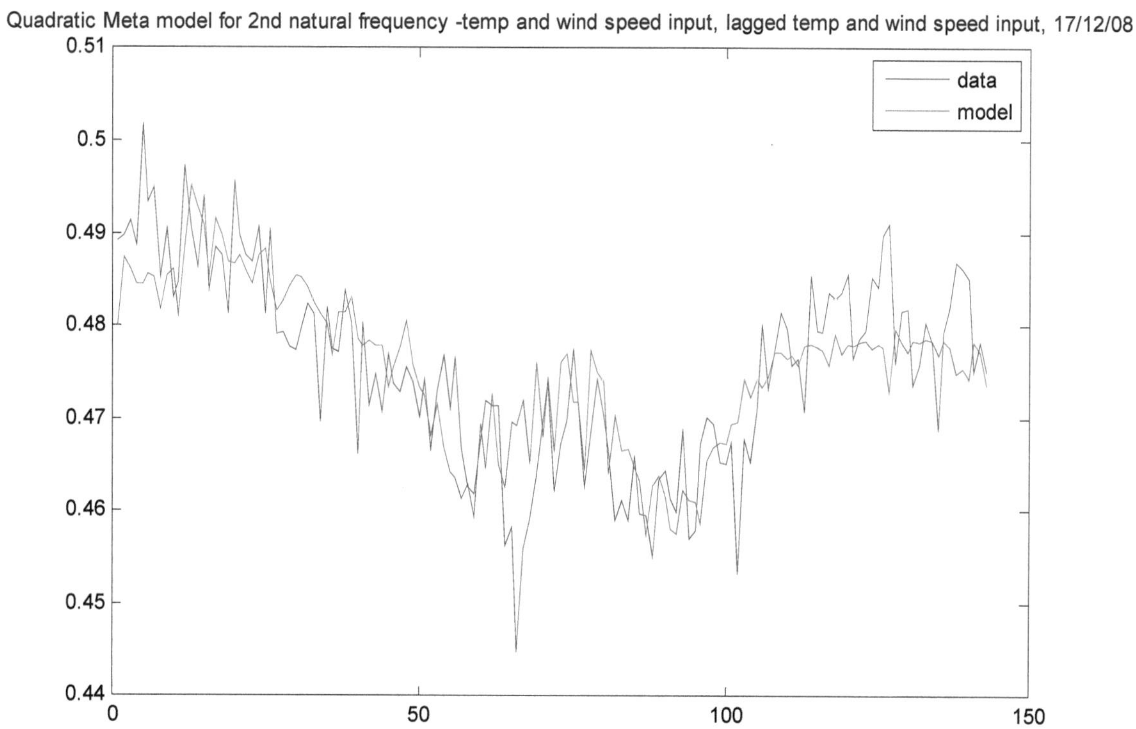


Above: Figure 9(a): Meta-model fit to first natural frequency, temperature, wind speed and lagged temperature and wind speed inputs. 17/12/08
Below: Figure 9(b): Meta-model fit to second natural frequency, temperature, wind speed and lagged temperature and wind speed inputs. 17/12/08

The suitability of different meta-model fits will not be re-investigated here, instead meta-models that include temperature and wind speed inputs as well as lagged temperature and wind speed inputs will be used. Figure 9 shows such meta-models fitted to the first two deck natural frequencies. These meta-models fit the data trends pleasingly well. Ignoring insignificant parameters, the first two natural frequencies can be expressed in terms of temperature and wind speed as:

$$\begin{aligned} f_1(t) = 0.3649 &+ 0.0335\,T(t)T(t-1) - 0.0172\,T(t)^2 - 0.017\,T(t-1)^2 + 0.0063\,T(t-1) + 0.0022\,T(t) \\ &+ 0.0007\,T(t)V(t-1) - 0.0007\,T(t-1)V(t-1) - 0.0003\,T(t)V(t) + 0.0003\,T(t-1)V(t) \\ &+ 0.0003\,V(t) - 0.0002\,V(t-1) + \varepsilon \end{aligned} \tag{6}$$

$$\begin{aligned} f_2(t) = 0.3812 &+ 0.1212\,T(t)T(t-1) + 0.0663\,T(t) - 0.0643\,T(t)^2 - 0.0593\,T(t-1)^2 - 0.0358\,T(t-1) \\ &+ 0.0050\,V(t) - 0.0045\,T(t)V(t) + 0.0038\,T(t-1)V(t) + 0.0026\,T(t)V(t-1) \\ &- 0.0023\,T(t-1)V(t-1) - 0.0019\,V(t-1) + \varepsilon \end{aligned} \tag{7}$$

where $f_i(t)$ is the $i$th natural frequency at time $t$, $T(t)$ is the temperature at time $t$, V(t), the wind speed at time $t$, and $\varepsilon$ a noise process. Again the temperature variation dominates the expressions for the natural frequencies, and as before the expressions are also highly dependent on the dynamics of the temperature.

## SUMMARY

This paper has explored the effect of temperature and wind speed on the first five natural frequencies of the Tamar Bridge. Simple quadratic models of the form

$$f_i(t) = \propto T(t) + \beta V(t) + \gamma T(t)^2 + aV(t)^2 + bT(t)V(t) \tag{8}$$

where $f_i(t)$ is the $i$th natural frequency at time $t$, $T(t)$ is the temperature at time $t$, V(t), the wind speed at time $t$, and $\alpha, \beta, \gamma, a, b$ constants, were able to recreate the general trend of the natural frequencies. More complex models including parameters dependent on time-lagged temperature and wind speed data (equations 1-7) were required to model the natural frequencies to a decent level of accuracy. Studying these fitted models it is clear that temperature has a dominant effect over wind speed on the variation of the natural frequencies. It is also clear that, for measurements taken at ten minute intervals, at a given time $t$, the natural frequency will depend on the environmental conditions at that time $t$, but also on the environmental conditions at the previous time step.

## REFERENCES


[1] Sohn, H., Effects of environmental and operational variability on structural health monitoring, *Philosophical Transactions of the Royal Society A: Mathematical, Physical and Engineering Sciences* **365** pp.539-560, 2007.

[2] Farrar, C., Doebling, S., Cornwell, P., and Straser, E., Variability of modal parameters measured on the Alamosa Canyon Bridge. *Proceedings of the 15th International modal analysis conference (IMAC XV)*, 1997.

[3] Peeters, B., De Roeck, G., One year monitoring of the Z24 bridge: environmental influences versus damage events. *Proceedings of the 18th International modal analysis conference (IMAC XVIII)*, 2000.

[4] Ni, Y.Q., Hua, X.G., Fan, K.Q.,and Ko, J.M., Correlating modal properties with temperature using long-term monitoring data and support vector machine technique. *Engineering Structures*, 27, pp.1762-1773, 2005.

[5] Mahmoud, M., Abe, M. and Fujino, Y., Analysis of suspension bridge by ambient vibration measurement using time domain method and its application to health monitoring. *Proceedings of the 19th International modal analysis conference (IMAC XIX)*, 2001.

[6] Kim, C. Y., Jung, D.S., Kim, N.S., Kwon, S.D., and Feng, M. Q., Effect of vehicle weight on natural frequencies of bridges measured from traffic induced vibration. *Earthquake Engineering and Vibration Engineering,* 2(1), pp 109-115, 2003.

[7] Peeters, P. and De Roeck, G., Reference-Based Stochastic Subspace Identification for Output-Only Modal Analysis. *Mechanical System & Signal Processing*, 13(6), pp. 855-878, 1999.

[8] Carden, P. and Brownjohn, J., Tracking the effects of changing environmental conditions on the modal parameters of the Tamar bridge. *Proceedings of SHMII- 3*, 2007.

[9] Brownjohn, J. And Carden P., Real-time operation modal analysis of Tamar Bridge, *Proceedings of the 26th International modal analysis conference (IMAC XXVI)*, 2008.

[10] Sharma, S., *Applied Multivariate Techniques*, John Wiley and Son, 1996.

Proceedings of the IMAC-XXVIII
February 1–4, 2010, Jacksonville, Florida USA


# Structural Health Monitoring of Bridges in British Columbia, Canada

Carlos E. Ventura

*Professor of Civil Engineering, The University of British Columbia, Department of Civil Engineering, Vancouver, Canada*

Sharlie Huffman

*Bridge Seismic Engineer, British Columbia Ministry of Transportation and Infrastructure, Victoria, Canada*

## ABSTRACT

The BC Ministry of Transportation and Infrastructure and the University of British Columbia have recently embarked on a program to instrument key structures to provide confirmation of seismic capacity, assist in focusing retrofit efforts, detect damage from any cause and provide rapid damage assessment of those structures following a seismic event. The data from this instrumentation will be capable of remote configuration and will automatically upload via the internet. As part of this collaboration effective damage detection algorithms that will provide reliable intelligence close to real time have been developed and implemented. This paper provides an overview of present status of the bridge instrumentation program in BC and illustrates the current approach to monitor bridges.

## 1 INTRODUCTION

The west coast of BC lies in Canada's highest seismic zone under threat of three different types of large, highly destructive earthquakes. The British Columbia Ministry of Transportation and Infrastructure (BCMoTI) is responsible for 400 km of provincial Disaster Response Routes. The loss of any portion of one of these routes could significantly impact emergency response efforts and negatively affect public well being. The Ministry maintains 900 structures in the highest seismic zones, many of which are vulnerable to extensive damage in even a moderate quake and potential collapse in a major earthquake. The loss of the use of several structures would not only have immediate impact on public well being and the ability of emergency vehicles to respond effectively, but would also cripple the economic recovery of the region. The effects would be felt across the nation and for many years into the future.

The better the information on which areas, structures and facilities are most vulnerable, the better planning and preparation can be done. By identifying those structures and facilities most susceptible to seismic forces, decision-makers can do effective risk management. Fast, accurate field intelligence immediately following an earthquake can ensure the most effective deployment of vital services and mitigate damage to the built environment.

In recent years, the BCMoTI has been instrumenting bridges in collaborating with the Earthquake Engineering Research Facility (EERF) of the University of British Columbia (UBC). Presently, five structures have between six and eighteen accelerometers and two structures have a limited number of strain gauges. Three of these structures have recorded motions from earthquakes in the last decade. Only two structures upload data to an internet site.

The BCMoTI and UBC have embarked on a program called **Smart Infrastructure Monitoring System (SIMS)** to instrument key structures to provide confirmation of seismic capacity, assist in focusing retrofit efforts, detect damage from any cause and provide rapid damage assessment of those structures following a seismic event. The data from this instrumentation will be capable of remote configuration and will automatically upload via the internet. As part of this collaboration effective damage detection algorithms that will provide reliable intelligence close to real time have been developed and are now being implemented.

## 2 GOALS AND OBJECTIVES OF THE SIMS PROJECT

***The goals of this Project are to:***

1. develop and implement a real-time seismic structural response system to enable rapid deployment and prioritized inspections of the Ministry's structures; and
2. develop and implement a health monitoring program to address the need for safe and cost-effective operation of structures in BC.

This Project will help transform the current practice of inspecting and evaluating all structures after an earthquake to a more rational and effective one that makes effective use of state-of-the-art sensing technology with fast and efficient techniques for data analysis and interpretation. Inspections can then be focused and prioritized to maximize the effectiveness of scarce resources. This Project will further provide ongoing structural condition monitoring for impact or deterioration to enable timely inspection and intervention.

***The objectives of this Project are to:***

1. develop and implement a cost-effective, reliable Structural Health Monitoring (SHM) technology that makes effective use of sensors and broad band digital communications for remote monitoring of structures subjected to dynamic loads;
2. identify and implement effective algorithms for system identification, model updating and damage detection suitable for remote monitoring of structures subjected to seismic forces and condition changes due to deterioration or impact; and
3. develop an integrated decision support system that incorporates geographical information and information about ground shaking and structural performance of a portfolio of remotely monitored structures distributed throughout the province. This system will include decision-making tools that will expedite the process of prioritization, risk assessment and damage evaluation to assist decision-makers with post earthquake response and recovery options.

The state-of-the-art technology to be developed by the EERF will be used to detect, analyze and localize damage to structures and transmit the data in real time via the Internet and display in animated and static web pages as appropriate for use by the Province and UBC. The alert systems and public access web pages will display real time seismic data from the BC Strong Motion Network to provide input to assessments for non-instrumented bridges. They will also provide other agencies, emergency responders and engineers with instantaneous situational awareness. Either structural deterioration or impact damage that results in a pre-set level of change in readings of the instruments will trigger an alert as well as triggering a full diagnostic evaluation automatically. The structural condition information and displays will be accessed only by designated parties. The distribution list for alerts will be based on the type of alert and source of damage. Seismic damage will have a wider distribution but condition and impact alerts from a single structure will go only to selected institutions.

The data collected will be analyzed at the EERF with off-site backup, and the results will be used by the Ministry to make decisions about the operability of the monitored structures after a significant seismic event.

## 3 PROJECT SCHEDULE AND INTERNATIONAL COLLABORATIONS

The monitoring project started in the summer of 2009 and should be fully operational by the spring of 2010. During this period of time three important milestones need to be completed: 1) Evaluation and selection of damage detection algorithms; 2) Development of software and acquisition of hardware to implement a real-time Internet-base monitoring system; and 3) Commissioning of the system and remote access to existing instrumented structures.

Because of the complexity of the this project, collaboration has been formally established with institutions and companies with different types of expertise. Interaction with suppliers of monitoring equipment for bridges has also been established. The key collaborators are:

1. Pacific Geoscience Centre of Geological Survey of Canada in Sidney, British Columbia. Dr. Andreas Rosenberger oversees the strong motion instrumentation network in British Columbia and will facilitate access to this network so that concurrent information of ground motion data and bridge motion data can be obtained through the monitoring system.
2. TVP Engineering of Vancouver Ltd, British Columbia is assisting in the selection and implementation of the various damage detection algorithms that will be implemented as part of the SIMS project. Dr. Martin Turek of TVP is also assisting in the instrumentation of the bridges that will be part of the network of instrumented structures.
3. A partnership of Structural Vibration Solutions A/S (SVS) of Aalborg, Denmark, Dynamic Design Solutions (DDS) of Leuven, Belgium and the Institut National de Recherche en Informatique et en Automatique (INRIA) – Bretagne Atlantique, Rennes Cedex, France is responsible for developing the key software components that will be the core of the structural health monitoring project and provides a seamless interface between the tools developed by each of these organizations. Dr. Palle Andersen of SVS is leading this part of the project and is working in close collaboration with Dr. Eddy Dascotte of DDS and Dr. Laurent Mevel of INRIA. The graphical interfaces are being developed by Henrik Vollesen of SVS.

## 4 COMPONENTS OF SIMS

This section provides a general description of the main components of monitoring system. The functional components of SIMS will perform structural assessments in two fashions: a) monitoring before and after an event (see Fig. 1), and b) evaluation of performance of the instrumented bridges immediately after a significant earthquake (see Fig. 2). Customized graphical interfaces will permit a fast and effective identification of the structures that have recorded strong motion data or that are behaving in an uncharacteristic manner. Figures 3 through 7 show samples of information screens that will be part of the graphical interface system.

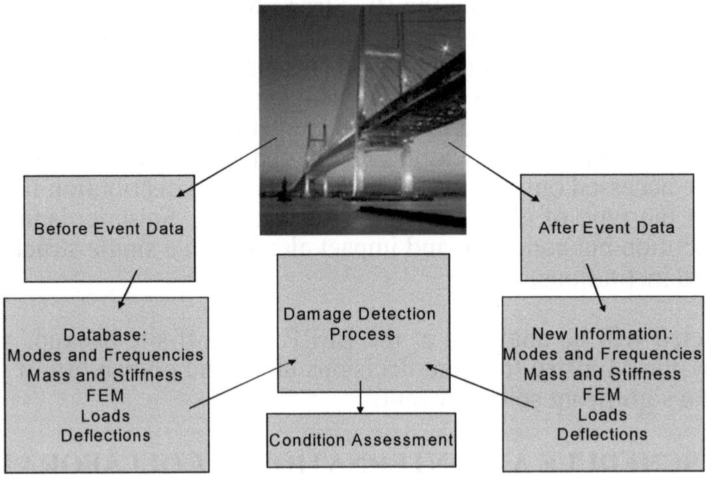


**Figure 1. Structural assessment pre and post event**

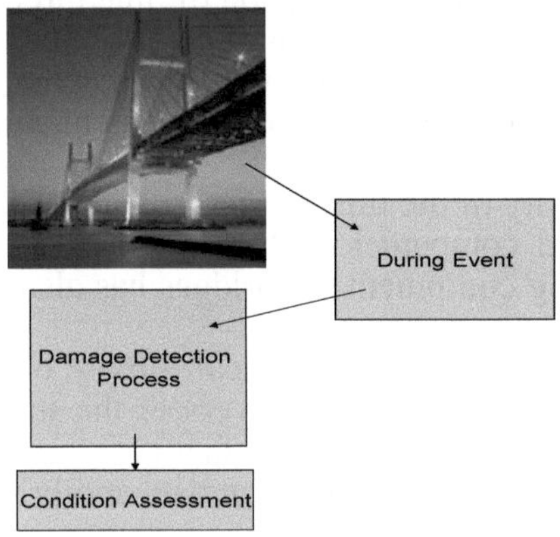


**Figure 2. Structural assessment during event**

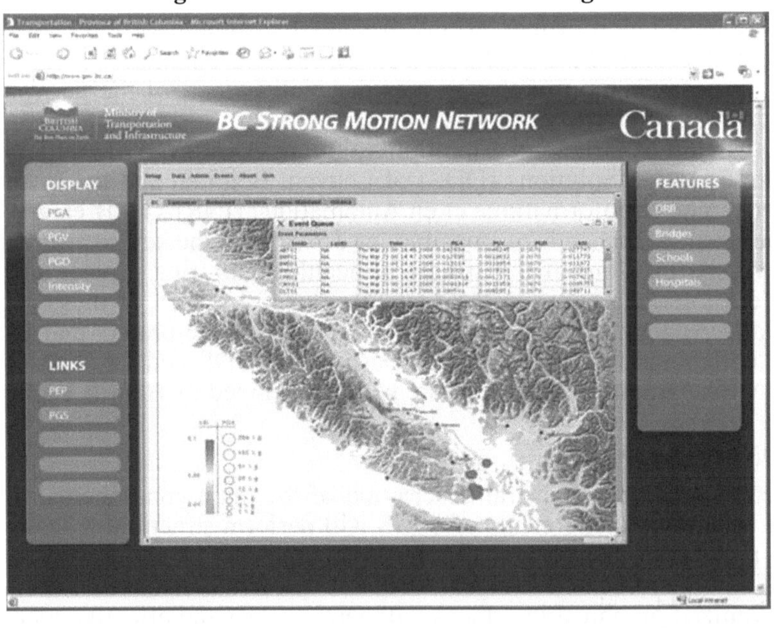


**Figure 3. Main screen of SMIS showing a the levels of shaking caused by a simulated Magnitude 6 earthquake near Victoria, BC**

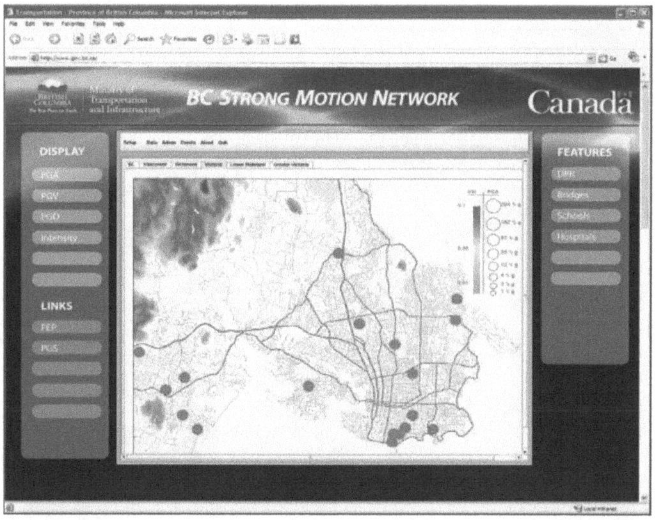


**Figure 4. Zooming into Victoria to identify stations that have recorded the earthquake**

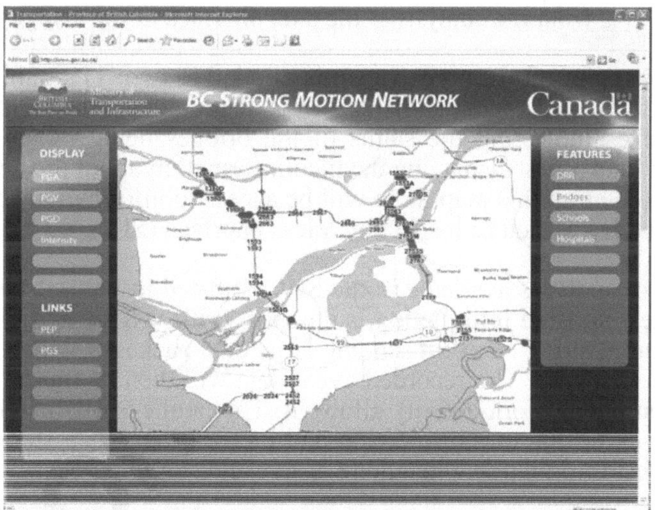


**Figure 5. After clicking the features for bridges, the map shows the bridges in the region that have recorded motions**

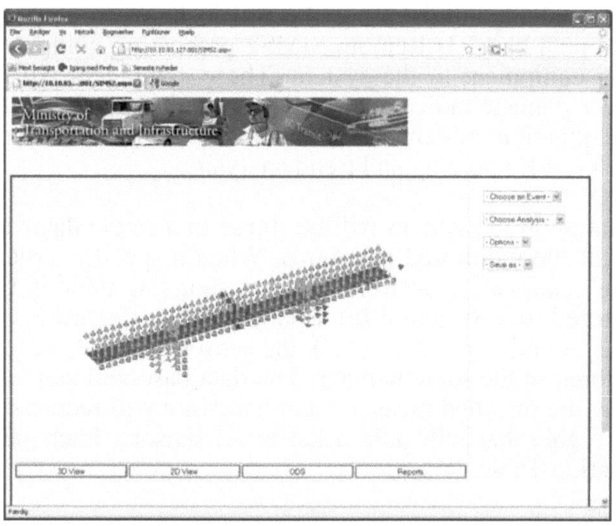


**Figure 6. Sample of instrumentation at a bridge that has recorded the earthquake**

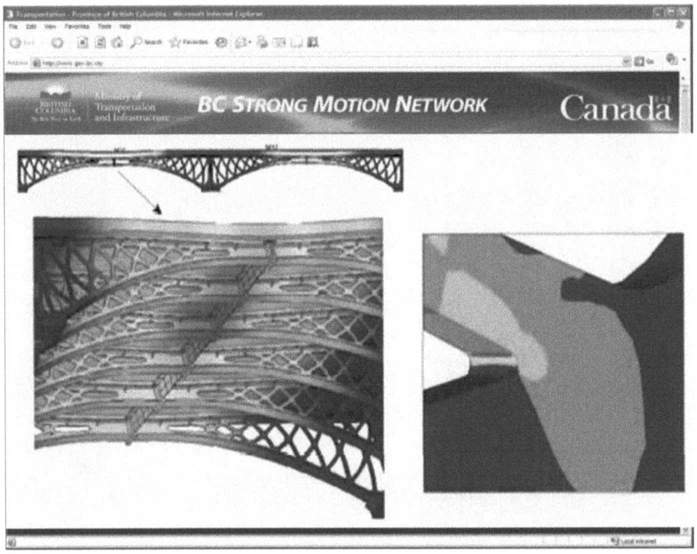


**Figure 7. Example of screen showing the results of the damage assessment obtained from the analysis of the data obtained at the bridge**

From a user point of view the system is divided in two: 1) a public available Shake Map View of a selected region showing levels of ground shaking and with general performance indexes indicating the status of the bridges being monitored, and 2) a restricted Shake Map View displaying detailed information about the health of the bridges being monitored. From this map there will be access to information about the individual structures through a Structure Inspection View interface (SIV).

The types of analyses available in the SIV are listed in Table 1 below. The speed of getting the latest updated information is the most critical issue and the system will take this into account. During an event sending large data packages will be avoided so as not to overload the network. The recorded event data will then be transmitted to the central data base when the network activity allows it. The public available GUI will ultimately have its own server, internet connection and its own local data storage that receives a copy of the latest information when there is an event.

The SIMS system is a Supervisory Control and Data Acquisition (SCADA) system consisting of the following subsystems: 1) Public available and restricted web-based Graphical User Interfaces capable of displaying relevant information. 2) One SIMS1 supervisory computer administrating the traffic between sensors, remote processing computers, and data base. 3) Remote SIMS2 processing computers connecting the on-site monitoring equipment to the system. These computers will analyze the measurements to produce damage indicators and OMA results and should be as close as possible to the data acquisition system. 4) One or more SIMS3 computers used for post-processing such as FE updating and trend analysis.

All information in the system will be stored in a single database. The data base will keep a list of all the monitored structures. When a specific structure is selected in the Shake Map the details are read from the corresponding table in the data base. The information is stored in a Structure Information Table. Regardless of the location of the sensors (on the ground or on structures), the sensors and data acquisition parameters are handled and stored in the same manner. The data base will store all the settings that influence the way the installed measurement hardware will record data. The will be a Data Acquisition Table that will hold a list to all sensors. Each sensor is described in a Sensor Information Table.

The way to set up triggers and scheduled measurements are administrated by a Trigger Information Table and a Measurement Information Table. Each recorded event is stored in an individual record in the data base, and the analyses made on the measurements are stored in Analysis Information Tables. The recorded data will be stored as binary large objects. The backup of all the data and information will as such be based on synchronization of two or more databases located at different sites.

**Table 1 Types of analyses available in SIV**

| Type | Analysis Display Options | Description | Avail. In View |
|---|---|---|---|
| Health Status | Damage indication from event data, INRIA and OMA damage indication on post event data, damage indication using updated FE. | Indicate the areas of the structure where health is deteriorated using contour colors. | 3D, Report |
| Reference FE | Displacements, stresses, mode shapes, MAC diagram. | Display results of the reference FE model. | 3D, Report |
| Updated FE | Displacements, stresses, mode shapes, MAC diagram. | Display results of the updated FE model. | 3D, Report |
| FE Correlation Updated versus Reference Model | Displacements, stresses, mode shapes, MAC diagram. | Compare the updated FE model with the reference model. | 3D, Report |
| FE trend analysis | Mode shapes, natural frequencies, stress and displacement for selected DOF's. | Perform a historical trend analysis from reference state to currently selected event. Optionally, perform a linear prediction into the future. | 2D, Report |
| Reference OMA | Mode shapes, MAC diagram, Processed Spectral Data, Stabilization Diagrams. | Display results of the reference OMA analysis. | 3D, 2D, Report |
| Updated OMA | Mode shapes, MAC diagram, Processed Spectral Data, Stabilization Diagrams. | Display results of each of the OMA analyses performed on the post event measurements. | 3D, 2D, Report |
| OMA trend analysis | Mode shapes, natural frequencies, damping ratios | Perform a historical trend analysis from reference state to currently selected event. Optionally, perform a linear prediction into the future. | 2D, Report |
| Damage Indicator trend analysis | Damage Index Method (DIM), drift, INRIA | Perform a historical trend analysis from reference state to currently selected event. Optionally, perform a linear prediction into the future. | 2D, Report |

## 5 SUMMARY AND CONCLUSIONS

The technology to be implemented by the Ministry and UBC will be used to: i) detect, analyze and localize damage to structures; ii) transmit the data regarding these structures in real time via the internet; iii) display in animated and static web pages the data as appropriate for use by the Ministry and UBC. The alert systems and public access web pages will display real time seismic data from the BC Strong Motion Network to provide input for assessments by the Ministry of non-instrumented bridges. These systems may also provide other agencies, emergency responders and engineers with situational awareness.

Emerging technologies during the last decade allow us to monitor the "structural health" of any type of structures via the Internet. The research community has developed tools and algorithms that can be used now reliably to determine the state of health of a structure during its normal operating conditions and immediately after a significant event.

This Project will help transform the current practice of inspecting and evaluating all structures after an earthquake to a more rational and effective one that makes effective use of sensing technology with fast and efficient techniques for data analysis and interpretation. Inspections by the Ministry can then be focussed and prioritized to maximize the effectiveness of scarce resources.

## ACKNOWLEDGMENTS

The authors would like to acknowledge financial support for this project provided mainly by the Bridge Branch of the British Columbia Ministry of Transportation and Infrastructure, and partially by the Natural Science and Engineering Research Council of Canada (NSERC). The support of Mr. Felix Yao of the EERF at the University of British Columbia is acknowledged with thanks.

Proceedings of the IMAC-XXVIII
February 1–4, 2010, Jacksonville, Florida USA


# Structural Assessment of Damaged Bridges Using Ambient Vibration Testing

Carlos E. Ventura

*Professor of Civil Engineering, The University of British Columbia, Department of Civil Engineering, Vancouver, Canada*

Juan C. Carvajal

*Research Assistant, The University of British Columbia, Department of Civil Engineering, Vancouver, Canada*

## ABSTRACT

This paper presents a simple method for preliminary structural assessment of bridges located in the province of British Columbia, Canada, using ambient vibration testing. The method is applied to bridges that have been hit by trucks or showed deterioration in some elements caused by bad construction and aging. The Earthquake Engineering Research Facility at The University of British Columbia, UBC-EERF, carried out a testing campaign of these bridges in December 2008. The bridges were tested along the deck taking measurements at 100 locations approximately. The damage criteria was based on identification of modal properties (mode shapes, frequencies and damping) and evaluation of maximum vertical displacements of the deck under normal traffic load conditions. The results show that the proposed approach can be used for preliminary evaluation of the structural integrity of the type of bridges investigated.

## 1 INTRODUCTION

An on-line structural health monitoring system for a bridge requires several major components, which can be grouped into three categories: instrumentation, data processing, and data transmission. Traditionally, for seismic applications the instrumentation measures accelerations and the data can be transmitted to a remote location via the Internet or by Telemetry. At the remote location the data can be analyzed to determine the response of the bridge, and can also be used to determine the intensity and characteristics of the ground shaking in the vicinity of the bridge. One of the benefits of using vibration measurements is that information about the global behaviour of the structure can be obtained with relatively few sensor locations. However, vibration measurements offer challenges in data processing, particularly for damage detection due to noise, environmental conditions and small changes in structure response due to damage [1]. In recent years significant research has been devoted to developing algorithms for detecting damage. In general, these algorithms compare information from the structure in an undamaged state, to a subsequent case in which the possibility of damage exists.

Damage detection algorithms available in the literature have been developed and applied by universities around the world, and some consulting firms are starting to use them in an exploratory basis. In addition, the performance of these algorithms has not been established on a universal basis; as a result, each must be evaluated on a case by case basis. Research at the University of British Columbia [2] has been conducted to develop a method to evaluate the algorithms based on calibrated finite element models. The method calibrates the model of an existing structure using ambient vibration results obtained from that structure.

A key aspect of any algorithm is the evaluation of its robustness. The term robustness for an algorithm is meant to describe the ability of the algorithm to identify damage in a variety of situations, which can include: a) A variety of damage types and levels, b) A variety of contamination of data, c) Various amounts of instrumentation, d) Varying levels of excitation of the structure; e) Varying environmental conditions, and f) A variety of load types, i.e. wind, seismic, collision etc. In addition to the work on evaluation of the algorithms, it has been demonstrated that combining several algorithms together improves the robustness of the overall process [3], therefore improving robustness of the SHM system. Many damage detection algorithms that have been developed are from a mechanical engineering origin, and evolved into civil engineering applications. There are several review papers in the literature, such as those by Los Alamos National Laboratory [4, 5]. However, none of the review papers cover all of the methods that are either available or under development.

Typically for bridge damage detection several algorithms are cited. Four of these are: a) Change in mode shape method [6], b) Change in mode shape curvature method [7], c) The damage index method [8], and d) Change in flexibility method [9]. The first three methods require only consistently normalized mode shape vectors in the undamaged and damaged conditions, while the last method requires the resonant frequencies in addition to the mode shape vectors in the undamaged and damaged conditions. None of these methods considers environmental effects, which tends to limit accuracy and robustness. Several other algorithms are available in the literature and have a variety of mathematical structures, and each offers benefits and limitations. One such example is the statistical model-base approach developed at IRISA in France [10 and 11].

The evaluation and calibration of the damage detection algorithms requires the availability of vibration data from a bridge before and after damage has occurred. At the present time no such data is available for bridges in British Columbia and, therefore, it is not possible to test and calibrate these algorithms using "real" data. Because of this limitation it was decided instead to investigate the use of ambient vibration tests of damaged structures to determine what type of information could be obtained to correlate the test data with the actual damage of the structures investigated. To this end, the Earthquake Research Engineering Facility (EERF) at UBC Department of Civil Engineering conducted ambient vibration tests at three bridges in the Cranbrook region of British Columbia. These bridges have experienced some level of damage and the ambient vibration tests results are used here to show what type of information can be inferred that could be correlated to the actual damage at each bridge. The following sections discuss the results from the analyses conducted for each bridge.

## 2 BRIDGES TESTED AND RESULTS

Based on current bridge design practice in Canada, a five-point criterion to define an acceptable dynamic performance of the bridge was established:

1. Frequency of the First vertical mode $F_{1V} > 2.0$Hz
2. Damping of the First vertical mode $\xi_{1V} < 2.0\%$
3. Symmetry in modes shapes of vibration
4. No presence of any unexpected mode shapes
5. Vertical deflection $\Delta v < 15.0$mm. The maximum deflection is calculated for a bridge with $F_{1V} = 5.2$Hz and without sidewalks.

This criterion was implemented to assess the condition of the bridges descry below.

### GOAT FELL BRIDGE

Goat Fell Bridge is located in the south-eastern side of British Columbia, 75 km West of Cranbrook. The superstructure of this 49.3m-long, 9.1m-wide, three-span bridge consists of a concrete deck slab supported on four continuous longitudinal reinforced concrete I girders of variable depth. The support conditions of the bridge are Roller bearings on both abutments. This bridge shows high level of deterioration caused by a poor quality of construction. The deterioration of this bridge consists of concrete spalling of the girders at the North end and several thin cracks in the girders located close to the intermediate supports. The West girder has a thick shear crack located at the North span of the bridge (see Figs. 1 and 2). No temporary repair has been done on this girder yet, however, a steel column has been placed under this girder to work as a temporary support in case the girder deflects more than 2.5 cm. The temporary steel column is not working yet as a support because the cracked girder is still able to take load without large deflections.

The deck was tested at 25 locations along each one of the longitudinal concrete I girders. This resulted in 100 testing locations. The duration of each ambient vibration measurement was 4 minutes and the sampling rate was 500 sps. Five mode shapes of vibration were identified for this bridge within the frequency range of 2.4Hz to 13.2Hz. The frequencies of vibration and damping of the first longitudinal, first transverse, first vertical, first torsional, and first plate mode are 2.4Hz and 6.8%, 4.3Hz and 3.2%, 5.2Hz and 2.8%, 6.9Hz and 2.1%, and 13.2Hz and 1.9%, respectively. The first longitudinal mode shape is associated with a rigid body motion caused by the roller bearing support conditions. The maximum vertical dynamic displacement estimated under normal traffic conditions was 2.0 mm.

According to this, the evaluation of the dynamic performance of the bridge is:

1. $F_{1V} = 5.2Hz > 2.0Hz$ →Yes
2. $\xi_{1V} = 3.2\% < 2.0\%$ →No
3. Symmetry of the modes shapes →No
4. No presence of any unexpected mode shapes →Yes
5. $\Delta v = 2.0mm < 15.0mm$ →Yes

The anomalies associated with "damage" determined from the dynamic behaviour of the bridge are the high damping ratio of the first vertical mode and the asymmetry of the first torsional and first plate mode. The analysis of the operational deflected shape of the East girder indicates symmetry between the South and the North spans. On the contrary, the West (cracked) girder did not show this symmetry. Its deflected shape indicated that the displacement at the location of the crack in the North span is 50% bigger than the displacement at the same location in the South span. Another anomaly found in the data from this bridge is that only a few mode shapes could be identified with the ambient vibration data, whereas in a "normal" bridge a good number of mode shapes are generally identified with confidence. This may be due to the deterioration of the bridge and the damage of the West girder.

**Figure 1 Overall view of Goat Fell bridge and damage details**

**Figure 2 Temporary support at Goat Fell bridge**

**Dutch Creek Bridge**

Dutch Creek Bridge is located in the southeastern part of British Columbia at 1.0 hr North of Cranbrook. The superstructure of this 37.5m-long, 9.8m-wide bridge consists of a steel Warren truss at each side of the deck. The concrete deck slab is supported on six longitudinal and seven transverse steel I girders. The support conditions of the bridge are a Pinned bearing on the North abutment and a Roller bearing on the South abutment. This bridge was hit at the portal header beam at the North end of the bridge. The observed damage was a total loss of the header beam and some adjacent elements, and loss of vertical alignment of the trusses at the North end (Fig. 3). The temporary repair consisted of replacement of the damaged members. However, the trusses still show the loss of the vertical alignment at the North end of the bridge.

The deck was tested at 20 locations along each one of the longitudinal steel I girder, which gives a total of 120 testing locations. The duration of each ambient vibration measurement was 4 minutes and the sampling rate was 500 sps. Thirteen mode shapes of vibration were identified for this bridge within the frequency range of 4.3Hz to 38.4Hz. The frequencies of vibration and damping of the first vertical, first transverse, and first torsional mode are 4.3Hz and 2.8%, 6.1Hz and 1.4%, and 7.6Hz and 1.0%, respectively. The maximum vertical dynamic displacement estimated under normal traffic conditions was 1.5 mm.

The evaluation of the dynamic performance of the bridge is:

1. $F_{1V} = 4.3Hz > 2.0Hz$ →Yes
2. $\xi_{1V} = 2.8\% < 2.0\%$ →No
3. Symmetry of the modes shapes →Yes
4. No presence of any unexpected mode shapes →Yes
5. $\Delta v = 1.5mm < 10.0mm$ →Yes

The anomaly associated with damage found in the dynamic behaviour of the bridge is the high damping ratio of the first vertical mode. The ambient vibration testing did not identify any unexpected mode of vibration that should have appeared because of the damage of the trusses. This is due to the fact that the bridge has already been repaired and both trusses keep the loss of vertical alignment almost in a symmetrical way.

It is unknown if the fundamental frequency of this bridge changed after the damage and the repair. This would have been a good reference to identify permanent damage of the trusses. However, a comparison of the first vertical frequency of Dutch Creek Bridge (4.3Hz) with the first vertical frequency of the Loop Bridge (4.4Hz) suggests that the Dutch Creek Bridge could have lowered its fundamental frequency after the damage and the repair. This is due to the fact that both bridges have similar structural configuration but the Dutch Creek Bridge is 20% shorter, which means that a higher fundamental frequency of vibration would be expected for the Dutch Creek Bridge ($F_1 > 4.3Hz$).

**Figure 3 Overall view of Dutch bridge and damage to the truss**

**Loop Bridge**

Loop Bridge is located in the southeastern part of British Columbia over Michel Creek at 1.25 hrs East of Cranbrook. The superstructure of this 46.9m-long, 8.7m-wide bridge consists of a steel Pratt truss at each side of the deck. The concrete deck slab is supported on five longitudinal and nine transverse steel I girders. The support conditions of the bridge are a Pinned bearing on the West abutment and a Roller bearing on the East abutment. This bridge was hit by a truck at the central column of the North truss. The damage of the column consists of loss of vertical alignment and wrapping of the cross section. The column has not been replaced yet. The temporary repair consists of a horizontal steel member added between the damaged column and the closest diagonal steel member (Fig. 4).

The deck was tested at 25 locations along each one of the longitudinal steel I girder, which gives a total of 125 testing locations. The duration of each ambient vibration measurement was 4 minutes and the sampling rate was 500 sps. Nineteen mode shapes of vibration were identified for this bridge within the frequency range of 2.9Hz to 48.6Hz. The frequencies of vibration and damping of the first vertical, first transverse, and first torsional mode are 4.4Hz and 2.8%, 5.4Hz and 1.9%, and 7.1Hz and 1.8%, respectively. An unexpected mode of vibration was identified at 2.9Hz with damping ratio of 2.4%. This is a coupled mode that has the first transverse and first torsional mode components. The maximum vertical dynamic displacement estimated under normal traffic conditions was 2.0 mm.

According to the proposed criterion, the evaluation of the performance of the bridge is:

1. $F_{1V} = 4.4\text{Hz} > 2.0\text{Hz}$ →Yes
2. $\xi_{1V} = 2.8\% < 2.0\%$ →No
3. Symmetry of the modes shapes →Yes
4. No presence of any unexpected mode shapes →No
5. $\Delta v = 2.0\text{mm} < 15.0\text{mm}$ →Yes

The anomalies associated with damage found in the dynamic behaviour of the bridge are the high damping ratio of the first vertical mode and the presence of the unexpected mode of vibration at 2.9Hz. The origin of this mode of vibration is due to a change in the stiffness of the North truss caused by the damage of the vertical column. Under this condition, the stiffnesses of the North and South trusses are different. This explains why the unexpected mode of vibration has transverse and torsional components of vibration.

**Figure 4 Overall view of Loop bridge and partial repair to damaged truss**

## 3 CONCLUSIONS

The benefit of using vibration measurements for a structural health monitoring program is that information about the global behaviour of the structure can be obtained with relatively few sensor locations. However, vibration measurements offer challenges in data processing, particularly for damage detection due to noise, environmental conditions and small changes in structure response due to damage. This study shows that even in the absence of "pre-damage condition" test data, it is possible to obtain some preliminary information about the condition of the type of bridge investigated in this study, based on its dynamic properties obtained from ambient tests results after damage has occurred. The three case studies presented in this paper show a possible way of making use of this data for preliminary damage estimation.

## ACKNOWLEDGMENTS

The authors would like to acknowledge Ms. Sharlie Huffman, Senior Bridge Engineer from the Ministry of Transportation and Infrastructure of British Columbia for her support and interest in conduction the project described here. Support provided for the field work by Felix Yao and Jose Centeno of the EERF at the University of British Columbia is acknowledged with thanks.

## REFERENCES

1. Moon, F.L., Aktan, A.E., (2006) "Impacts of Epistemic (Bias) Uncertainty on Structural Identification of Constructed (Civil) Systems", The Shock and Vibration Digest, September 2006, pp. 399-420
2. Turek, M., Ventura, C., "Calibrated Simulations for the Design of SHM Systems", Proc. of IMAC XXX, Orlando, Fla, Feb 2007, Paper No. 249
3. Turek, M., Ventura, C., "Development of a Combined Damage Detection Methodology", Proc. Of IMAC XXV, Orlando, FLA, February 2007, Paper 248
4. Farrar, C.R., Sohn, H., "Condition/Damage Monitoring Methodologies", Consortium of Organizations for Strong-Motion Observation Systems – Workshop on Structural Instrumentation, Emeryville, CA., November 2001
5. Farrar, C.R., Doebling, S.W., "An Overview of Modal-Based Damage Identification Methods", DAMAS Conference, Sheffield, UK, June 1997
6. Zhou, Z., Wegner, L.D. and Sparling, B.F., 'Vibration-based damage detection on a prestressed concrete girder', Proc. of the 5th CSCE Structural Specialty Conference, Saskatoon, June 2004 (CSCE , Montreal, 2004) Paper ST-085.
7. Pandey, A.K., Biswas, M. and Samman, M.M., 'Damage detection from changes in curvature mode shapes', J. Sound Vibration 145(2)(1991) 321-332.
8. Stubbs, N., Kim, J.T. and Farrar, C.R., 'Field verification of a nondestructive damage localization and severity estimation algorithm', Proc. of the 13th International Modal

Analysis Conference, Nashville, TN, February 1995 (Society for Experimental Mechanics, Bethel, CT, 1995) 210-218.

9. Pandey, A.K., and Biswas, M., 'Damage detection in structures using changes in flexibility', J. Sound Vibration 169(1)(1994) 3-17.
10. Balmes, E., Basseville, M., Mevel, L., Nasser, H., Zhou, W., (2008) 'Statistical model-based damage localization: A combined subspace-based and substructuring approach' Structural Control and Health Monitoring, 15:857-875
11. Andersen, P., Basseville, M., Brincker, R.,Mevel, L., Ventura, C.E., and Zhou, W. (2007) 'Seismic Damage Assessment in Structures Using Stochastic Subspace-based Algorithms' ECCOMAS Thematic Conference on Computational Methods in Structural Dynamics and Earthquake Engineering M. Papadrakakis, D.C. Charmpis, N.D. Lagaros, Y. Tsompanakis (eds.) Rethymno, Crete, Greece, 13–16 June.

Proceedings of the IMAC-XXVIII
February 1–4, 2010, Jacksonville, Florida USA


# Model Updating of the Ironworkers Memorial Second Narrows Bridge, Vancouver, Canada

Martin Turek
*TVP Engineering Ltd., Vancouver, Canada*

Carlos E. Ventura
*The University of British Columbia, Department of Civil Engineering, Vancouver, Canada*

Eddy Dascotte
*Dynamic Design Solutions, Leuven, Belgium*

The Ironworkers Memorial Second Narrows Crossing is a critical transportation link in the Lower Mainland Region of British Columbia, Canada, which is a zone of high seismic risk. The IMSNC is being instrumented as part of a real time seismic and health monitoring system, which will also include many other bridges around the region. As part of the system, a finite element model has been developed and updated using previously obtained ambient vibration measurements. This paper focuses on updating of one approach truss section of the bridge. The model was updated in several steps using both manual and automated techniques. Updating included stiffness of a set of seismic isolators that were added to the bridge in a retrofit in the mid 1990's. The complete updated set of models will be used in fatigue and damage studies as a part of the monitoring system.

## 1 INTRODUCTION

The Ironworkers Memorial Second Narrows Crossing (IMSNC) is the second bridge constructed at the Second Narrows of Burrard Inlet in Vancouver, British Columbia, Canada. It is an essential transportation corridor along the Trans-Canada highway (Route 1), connecting the City of Vancouver to the District of North Vancouver, the City of North Vancouver, and West Vancouver.

Vancouver is located in a zone of high seismic hazard in Western Canada. New evidence about potential earthquake sources, magnitudes and seismic activity in the area indicated the bridge would not withstand the design earthquake as per current code with the seismic design of the 1950's. The bridge was then retrofitted in 1994 by strengthening some elements and changing some of the original expansion and pin-bearings to isolation bearings.

The Ministry of Transportation and Infrastructure (MTI) and the University of British Columbia (UBC) have initiated a structural health monitoring program of bridges in the province of British Columbia. The purpose of the program is to detect damage of the structures using field data and state-of-the-art damage detection algorithms. The IMSNC is one of those bridges that will be instrumented. As part of the system, a finite element model has been developed and updated using previously obtained ambient vibration measurements [Ventura et al., 2009].

This paper presents the results of an updating study as performed on a portion of model of the IMSNC. The updating was done in two phases; the first to obtain a working stiffness value for a set of seismic isolation bearings installed in several locations on the bridge and the second to update several modes of a typical approach truss. The updated model will be used in fatigue and damage studies as a part of the monitoring system.

## 2 DESCRIPTION OF THE IMSNC

The Ironworkers Memorial Second Narrows Crossing is a 1292m-long composite structure. It carries six lanes of traffic (three lanes for each direction) and a 0.6m-diameter gas pipeline. The general configuration of the bridge and bent designation is illustrated in Figure 1. This paper focuses on a typical approach truss (between Section 10 and 11) only. For detailed information on the entire bridge see [Ventura et al. 2009].

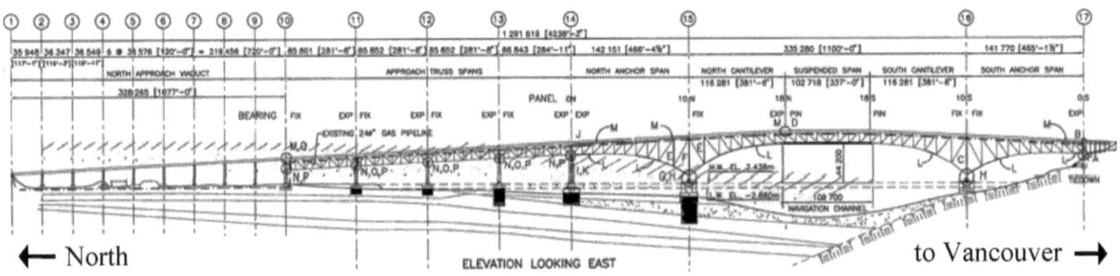


Figure 1: General configuration and bent designation of the IMSNC.

There are four 86.0m-long steel approach truss spans starting on the North shore of the crossing and extending out into the inlet. Each span has two 13.0m-depth custom steel warren trusses separated 14.6m apart. Both trusses are connected by horizontal and vertical bracing. The 0.2m-thick RC deck slab is supported on thirteen 0.8m-depth steel I-girders in the longitudinal direction and on a 1.7m-depth x 22.0m-long steel I-floor beam located transversally every 10.5m along the truss. The longitudinal girders are also connected transversally every 3.5m by a 0.5m-depth steel I-girders. The warren trusses are supported on isolation bearings protected with bumpers. The deck has transverse expansion joints at the ends of each span, so there is no structural continuity among them. The transition between two approaches and between the southernmost approach and the main cantilever section of the bridge is shown in Figure 2.

a) Deck floor and steel warren trusses

b) Bent 14: south end of the approach truss spans

Figure 2: Approach truss spans.

## 3 REAL-TIME SEISMIC MONITORING SYSTEM

A real-time monitoring system (RTMS) is currently being installed on the IMSNC. The monitoring system will part of a province wide network, with its central hub located at the University of British Columbia (UBC) in Vancouver, Canada. The monitoring system will provide real-time information regarding structural performance and safety, primarily for seismic, but also applicable for a variety of load types. The monitoring system will be implemented in two phases: first to install the on-site hardware and second to implement a customized software and data processing system unique to the MTI/UBC network.

The general purpose of the system is to monitor the structural health of the bridge for seismic, impact and deterioration effects. This considers two loading levels: severe infrequent events, such as seismic and impact/collision; and frequent long-term effects, such as wind, traffic, etc. The system instrumentation will consist of:

- Vibration measurements utilizing uni- and tri- axial accelerometers
- Strong motion measurements off the structure, including free-field and down-hole accelerometers
- Strain measurements on the deck floor beams and major truss elements
- Temperature measurements at several locations, both the North and South ends
- Wind speed measurements at midspan

The data will be collected at a central data recorder, in which a certain amount of on-site data processing will occur. Then processed and raw data will be sent to UBC for further processing and storage. The monitoring system has approximately 100 channels; however due to the configuration of isolation bearings and expansion joints, the bridge is essentially split into 10 smaller independent structures. This means that in reality there is an average of about 10 channels per structure.

## 4 SUMMARY OF AVT RESULTS

A comprehensive ambient vibration test program of the IMSNC has been planned in several phases, the first of which was carried out in January 2009. This phase included testing of the entire bridge along the deck. For full details of the testing including test locations please refer to [Ventura et al., 2009]. Future phases include detailed testing of the trusses and piers.

Identification of modal vibration properties of the IMSNC was performed using ARTeMIS Extractor [SVS, 2009]. Several system identification techniques are available in the software including Stochastic Subspace Identification (SSI) and Frequency Domain Decomposition (FDD). Typically data was processed using the automated identification feature in the software, with the SSI-CVA algorithm. The analysis as described here focused only on the approach trusses, and the data was processed in the 0-5Hz bandwidth.

Two different tests were performed on the approach trusses. First the *simplified test*, measuring three orthogonal directions at two locations at midspan of the approach (one on either side of the bridge); the simplified test was repeated on each of the four approaches. Second the *detailed test*, measuring three orthogonal directions at 36 locations along the approach, 18 on either side. The detailed test was only performed on one of the approaches.

Four modes below 5Hz were identified in each of the four simplified tests, listed in Table 1. The modes include a horizontal mode in each orthogonal direction (modes 1 and 2), which are considered as rigid body modes. This rigid body behavior is caused by the isolation bearings at the four corners, which are flexible relative to the truss. From the detailed test, only the last two modes were clearly identified. This could be explained by the fact that the simplified tests were recorded for twice the duration of the detailed test; the reduced dataset length does not allow for the proper identification of the lowest frequency modes. Consequently updating using the first two modes will be based on frequency only; the third and fourth modes will compare frequency and mode shapes.

Table 1: Ambient vibration results.

| Mode | Freq. [Hz] | Descr. |
|---|---|---|
| 1 | 1.34 | Transverse rigid body |
| 2 | 1.56 | Longitudinal, rigid body |
| 3 | 2.38 | 1st Vertical |
| 4 | 3.06 | 1st Torsional |

## 5 FINITE ELEMENT MODEL

A finite element model of the IMSNC has been developed, partly from a previous model that was used in the 1994 retrofit of the bridge; the original model was created in SFRAME [SOFTEK, 2009] format. Subsequently the bridge was remodeled into SAP2000 [CSI, 2009], and then input to FEMTools [DDS, 2008] for the updating. SAP2000 is an industry standard finite element analysis software, which will be used for structural analysis of the bridge. FEMTools is a multi-function tool that has capabilities for validation and updating of FE models.

The FE model consists of a series of frame elements with added masses; in general the elements are typical W sections for braces, beams and stringers, and custom rectangular elements with equivalent properties to represent the built-up sections found in the truss. The seismic isolation bearings are modeled as link (spring) elements with axial stiffness only. The links are then placed in each of the orthogonal horizontal directions at each of the four supports. A single approach truss section as described in this paper is shown in Figure 3a. A typical isolator of the type used at IMSNC is shown in Figure 3b.

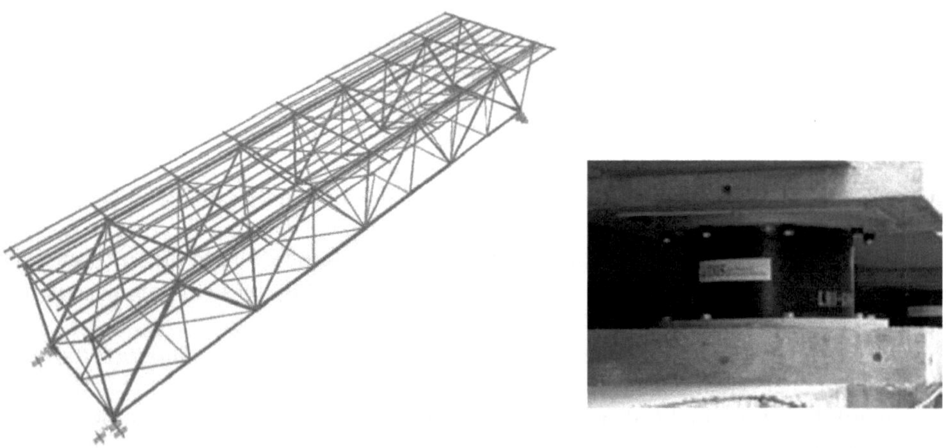

Figure 3: a) FE model of typical approach truss span b) typical seismic isolation bearing

To establish the translational stiffness (K) of the bearings, it was first assumed that the second mode from the AVT results could be treated as a SDOF system. This is based on the idea that the truss is very stiff longitudinally compared to the bearings, and so the stiffness (in mode 2) comes only from the bearings and the truss acts as the mass. The bearing stiffness is then tuned as to match the frequency of the FE Model second mode to the AVT result (1.56Hz). The modes of the original pinned model are given shown in Table 2; the transverse mode is shown at 1.78Hz with no longitudinal mode observed below 5Hz. To obtain the starting value of K for the bearings, first the translational restraint in the transverse direction was removed. The value of K was then manually tuned until a reasonable match of frequency was obtained; this resulted in the preliminary value of 40MN/m, with a second mode frequency of 1.64Hz. Then the longitudinal-direction restraints were removed, and using the same K value for the links in that direction the rest of the modes were computed (shown in Table 2).

Table 2: SAP2000 model results

| Mode | AVT | Descr. | SAP2000 (Pinned model) | SAP2000 (Iso. model) | Difference (Iso. vs AVT) |
|---|---|---|---|---|---|
| | Freq. [Hz] | | Freq. [Hz] | Freq. [Hz] | [%] |
| **1** | 1.34 | Trans, rb | 1.78 | 1.23 | -8.94 |
| **2** | 1.56 | Long, rb | N/A | 1.64 | 4.88 |
| **3** | 2.38 | Vert | 2.99 | 2.57 | 7.39 |
| **4** | 3.06 | Tors | 3.72 | 3.40 | 10.00 |

*trans only, no rigid body

## 6 FINITE ELEMENT MODEL UPDATING

The model updating presented in this paper has two stages. The first stage is to determine a reasonable working stiffness for the seismic isolation bearings; the second is to do a general overall updating of the truss. The preliminary updating of the bearing stiffness was presented in the last section, with results shown in Table 2.

The software FEMTools (FT) was used for both automated and manual updating of the truss. The initial SAP2000 model described in the last section was imported into FEMTools using a code translator developed at the University of British Columbia. The initial FT model modes are shown in Table 3. There are very small differences between FT and SAP, but as shown in the table the FT initial frequencies are identical to the SAP results. The first updating step is to automatically change the stiffness of the longitudinal bearings to obtain a match in the second frequency; the updated stiffness will then be used for the transverse direction bearings as well. The results are shown in the Table 3 as Update 1 (UD1). One consequence of the second mode frequency match is that the first mode also decreases in frequency (expected since K is updated in both horizontal directions). The updated stiffness value, which will be used for all the bearings on the bridge, is 35.2MN/m (a parameter change of -12%).

Table 3: Updating results, bearing stiffness.

| Mode | AVT | FT (initial) | Difference | FT (UD1) | Difference |
|---|---|---|---|---|---|
| | Freq. [Hz] | Freq. [Hz] | [%] | Freq. [Hz] | [%] |
| **1** | 1.34 | 1.23 | -8.24 | 1.19 | -11.08 |
| **2** | 1.56 | 1.64 | 5.56 | 1.55 | -0.16 |
| **3** | 2.38 | 2.57 | 10.22 | 2.57 | 10.02 |
| **4** | 3.06 | 3.38 | 7.79 | 3.36 | 7.17 |

The second updating step addresses the difference in first mode frequency (-11%). Assuming that the bearings have the same stiffness in all directions and that the mass of the truss contributes the same in both horizontal directions, then the difference in frequency between modes 1 and 2 must be a result of contributions of the truss transverse bending stiffness. The truss model elements were separated into sets for the updating: top and bottom chords, east and west sides; top and bottom diagonals, north and south groups. A normalized sensitivity analysis was performed on element sets using FEMTools. The analysis was done comparing the first mode frequencies as the response for matching, and using modulus of elasticity, E, and moment of inertias, Iy and Iz as potential parameters for updating. The resulting sensitivity matrix is shown in Figure 4. It is given here as a typical example and further sensitivity matrices will not be shown.

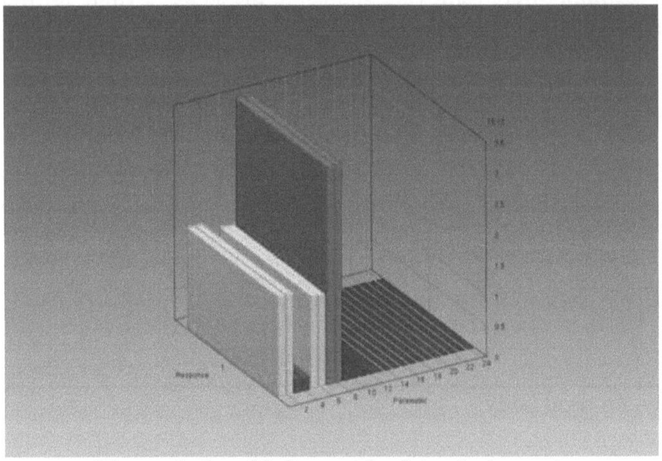

Figure 4: Sensitivity matrix, lateral elements

From the analysis it was found that the transverse mode is most sensitive to changes in the top diagonal properties. A trial updating was performed using modulus of elasticity, E, of top diagonals as the parameter, and its value was allowed to change without restriction. A very close match could be obtained (FEA to 1.30Hz) with a large parameter change (+630%). The implication is that the concrete deck slab, which was previously ignored in the model except as contributing only as mass, has a potentially significant effect on lateral stiffness. Therefore a series of additional lateral braces were added at the deck level, given a unit area, negligible mass and an initial E value of 2 x$10^5$ MPa. The model was then manually updated by increasing E; it was found that there was a maximum obtainable frequency of 1.315 Hz (meaning further increases of E had no effect on frequency). This limit occurred at an E value of 2.5 x $10^5$ MPa. The updated modes from the additional braces are shown in Table 4, as Update 2 (UD2). A side effect of adding the additional braces is that there is an increase in the frequencies of modes 3 and 4 (as expected).

Table 4: Updating results, lateral (2) and vertical (3)

| Mode | AVT | FT (UD2) | Difference | FT (UD3) | Difference |
|---|---|---|---|---|---|
| | Freq [Hz] | Freq [Hz] | [%] | Freq [Hz] | [%] |
| **1** | 1.34 | 1.32 | -2.07 | 1.34 | -0.03 |
| **2** | 1.56 | 1.55 | -0.15 | 1.56 | 0.09 |
| **3** | 2.38 | 2.60 | 11.19 | 2.39 | 0.42 |
| **4** | 3.06 | 3.64 | 15.81 | 3.36 | 8.93 |

The last stage is to update the 2[nd] and 3[rd] modes. This can be done using the results from the detailed ambient vibration test, which includes modes shapes and MAC values (modal assurance criterion, MAC, is a commonly used indicator to compare mode shapes, given in percent). Table 5 shows a comparison of the 3[rd] and 4[th] modes. From the results of the previous updating step (UD2), it can be seen that the MAC values are already reasonably high, being both over 90%. Both of the frequency values are higher, implying that the truss is vertically too stiff. Therefore it is meaningful to examine the behavior of the vertical chords/diagonals of the model.

A sensitivity analysis was again performed, and from the results it was seen that the moment of inertia properties were the least sensitive; they were removed from the updating. Since the initial MAC values were already high, the updating response chosen to match was the 3rd mode frequency, and the updating parameter was the cross sectional area of the vertical diagonal element set. The area was reduced by 25%, then automatically refined for a total area reduction of 53%. The results are shown in Table 5 referred to as Updating 3 (UD3). The first mode frequency matched nearly perfectly; the second mode error was reduced by half. In addition, for both modes the MAC value increased.

Table 5: Updating results, comparing 3rd and 4th modes with MAC values.

| Mode | AVT | FT (UD2) | Difference | MAC |
|---|---|---|---|---|
| | Freq [Hz] | Freq [Hz] | [%] | [%] |
| 3 | 2.38 | 2.60 | 9.14 | 96.1 |
| 4 | 3.06 | 3.64 | 18.96 | 90.7 |
| **Mode** | **AVT** | **FT (UD3)** | **Difference** | **MAC** |
| 3 | 2.38 | 2.39 | 0.63 | 96.7 |
| 4 | 3.06 | 3.35 | 9.50 | 93.2 |

Even though Updating 3 focussed on the 3rd mode (vertical) there is a change to the 4th mode (torsion) as expected. While there is sizeable improvement there remains some error. A further updating was explored to reduce this error; the primary focus was on the vertical braces located at 5 locations along the truss (these brace against transverse deformation of the truss). Despite several variations of the updating step, no significant improvement in torsional frequency match was made.

## 7 SUMMARY AND CONCLUSIONS

This paper presented the results of a model updating study on a typical approach truss of the Ironworkers Memorial Second Narrows Crossing (bridge), in Vancouver, Canada. The updating study was successful at achieving two primary objectives: to obtain a working stiffness value for the bridge seismic isolator bearings and to obtain a well-correlated dynamic model.

Results of a series of ambient vibration tests were used in the updating; four modes of vibration were obtained in those tests. These included two horizontal rigid-body type modes which were indicative of the isolator bearing stiffness; and two structural modes of the truss including vertical and torsional. The rigid body modes were used to find a working stiffness value for the bearings, identified as 35.2MN/m. The frequency and shapes of all four modes was used to modify the model, and the results of the updating has errors of less than 1% on each of the first three modes; and less than 10% for the fourth mode. The MAC values for the 3rd and 4th mode were both greater than 93%.

The results of the final step of the updating feature a large parameter change; the cross-sectional area of the vertical truss diagonals was reduced by 53%. This is not a realistic physical parameter change; however for a constructed system such as IMSNC there is significant uncertainty in properties of built-up, riveted members that have been in service for nearly 60 years. It is more reasonable to assume that changes to modulus and section properties would be distributed throughout the truss. Further study will be done on the model using:

- The working stiffness value for the isolator bearings
- Data from future ambient vibration tests, including more detail of the truss
- Data from the monitoring system

## ACKNOWLEDGMENTS

The authors would like to acknowledge Ms. Sharlie Huffman, from the Ministry of Transporation and Infrastructure of British Columbia. They would like to acknowledge Mr. Juan-Carlos

Carvajal and the students at the University of British Columbia who performed the testing on the bridge. They would also like to acknowledge Mr. Devin Sauer of the University of British Columbia for his work on updating the SAP2000 translator for Version 12.

REFERENCES

ARTeMIS Extractor Software. Structural Vibration Solutions, Inc. 1999-2003. Denmark.

FEMTools 3.3.1 (32-bit), Dynamic Design Solutions N.V., 1993-2008. Belgium.

SAP2000 Advanced, V12.0.0 Copyright 1976-2008 Computers and Structures Inc., Berkeley, CA

SFrame Version 7.02, Copyright 1995-2006 Softek Services Ltd., Canada

Ventura, C.E., Carvajal, J.C., Centeno, J., Pandey, B., 2009. Identification of modal properties of the Ironworkers Memorial Second Narrows Crossing: testing on deck. Proc. 3rd International Operational Modal Analysis Conference – IOMAC, Ancona, Italy

Proceedings of the IMAC-XXVIII
February 1–4, 2010, Jacksonville, Florida USA


# Pre- and Post-identification Merging for Multi-Setup OMA with Covariance-Driven SSI

M. Döhler[1], E. Reynders[2], F. Magalhães[3], L. Mevel[1], G. De Roeck[2], and Á. Cunha[3]

[1] *INRIA, Centre Rennes - Bretagne Atlantique, Campus de Beaulieu, F-35042 Rennes, France*

[2] *Katholieke Universiteit Leuven, Dept. of Civil Engineering, Kasteelpark Arenberg 40, B-3001 Leuven, Belgium*

[3] *Faculty of Engineering of the University of Porto (FEUP), R. Dr. Roberto Frias, 4200-465 Porto, Portugal*

**Abstract**

In Operational Modal Analysis (OMA) of large structures we often need to process sensor data from multiple nonsimultaneously recorded measurement setups. These setups share some sensors in common, the so-called reference sensors that are fixed for all the measurements, while the other sensors are moved from one setup to the next. To obtain the modal parameters of the investigated structure, it is necessary to process the data of all the measurement setups and normalize it as the unmeasured background excitation of each setup might be different. For this we compare three different approaches in this paper which differ in the order of the data merging, normalization and system identification step: The classical PoSER (identification-normalization-merging), the PoGER (merging-identification-normalization) and the PreGER (normalization-merging-identification). Special care was taken with the PreGER method and its efficiency has been tested with respect to the two other methods. The system identification is done with the SSI-cov/ref method. We apply these methods to the extraction of the modal parameters (natural frequencies, damping ratios and mode shapes) of the Luiz I arch bridge in Porto, Portugal, compare them and evaluate the different methods.

## 1 Introduction

Subspace-based linear system identification methods have been proven efficient for the identification of the eigenstructure of a linear multivariable system in many applications. Our main motivation in this paper is output-only structural identification in vibration mechanics. This problem consists in identifying the modal parameters (natural frequencies, damping ratios and mode shapes) of a structure subject to ambient unmeasured vibrations, by using accelerometer measurements or strain gauges. This is output-only system identification, as the excitation input is unknown and not measured. Examples are, amongst others, offshore structures subject to swell, bridges subject to wind and traffic, etc.

We wish to analyze how the covariance-driven stochastic system identification can be adapted when several successive data sets are recorded, with sensors at different locations in the structure. For doing this, some of the sensors, called the *reference sensors,* are kept fixed, while the others are moved. Like this, we mimic a situation in which lots of sensors are available, while in fact only a few are at hand. However, there is one unpleasant feature of structural identification of structures subject to ambient excitation, namely that excitation is typically turbulent in nature and nonstationary. For example, fluid/structure interaction in offshore structures results in shock effects causing nonstationary excitation, and the same holds for wind and traffic on bridges. Like this, the excitation level can change from setup to setup.

The relevance of merging successive records, and its implementation in the case of nonstationary excitation, are the subject of this paper. Three different approaches are considered which differ in the order they handle the main steps of the modal parameter extraction: the classical PoSER (identification-normalization-merging), the PoGER (merging-identification-normalization) and the PreGER (normalization-merging-identification). They are tested and compared on data measured on the Luiz I bridge in Porto using 26 different measurement setups.

A first analysis of the Luiz I bridge was made in [2] and a preceding comparison of the three methods was performed in [9]. The current paper extends the previous research and focuses on the PreGER approach, since a typo in the description of this method [4] easily leads to a wrong implementation. In Section 3.3 the corrected algorithm for the PreGER approach is presented.

## 2 Reference-based Covariance-driven Stochastic Subspace Identification

### 2.1 Single Setup

We consider a linear multi-variable output-only system described by a discrete-time state space model

$$\begin{cases} X_{k+1} & = & F\,X_k + V_{k+1} \\ Y_k^{(\text{ref})} & = & H^{(\text{ref})}\,X_k \\ Y_k^{(\text{mov})} & = & H^{(\text{mov})}\,X_k \end{cases} \tag{1}$$

with

- $X_k$ the state vector at time instant $k$,
- $Y_k^{(\text{ref})}$ the observed output vector of the reference sensors,
- $Y_k^{(\text{mov})}$ the observed output vector of all the sensors minus the reference sensors (the remaining sensors),
- $H^{(\text{ref})}$ the observation matrix with respect to the reference sensors,
- $H^{(\text{mov})}$ the observation matrix with respect to the remaining sensors,
- $F$ the state transition matrix,
- $V_k$ the unmeasured stationary Gaussian white noise.

Let furthermore

- $Y_k = \begin{pmatrix} Y_k^{(\text{ref})} \\ Y_k^{(\text{mov})} \end{pmatrix}$ all the observed output,
- $H = \begin{pmatrix} H^{(\text{ref})} \\ H^{(\text{mov})} \end{pmatrix}$ the full observation matrix.

The classical reference-based subspace identification of the eigenstructure $(\lambda, \phi_\lambda)$ of the system (1) consists of the following steps: We build the output correlations

$$R_i \stackrel{\text{def}}{=} \mathbf{E}\left(Y_k\, {Y_{k-i}^{(\text{ref})}}^T\right) \tag{2}$$

between all the sensors and the reference sensors. $\mathbf{E}$ denotes the expectation operator. In the next step the Hankel matrix

$$\mathcal{H} \stackrel{\text{def}}{=} \begin{pmatrix} R_0 & R_1 & R_2 & \dots & R_q \\ R_1 & R_2 & R_3 & \dots & \vdots \\ R_2 & R_3 & R_4 & \dots & \vdots \\ \vdots & \vdots & \ddots & \dots & \vdots \\ R_p & \vdots & \ddots & \dots & R_{p+q} \end{pmatrix} \tag{3}$$

is filled. For the Hankel matrix a factorization $\mathcal{H} = \mathcal{O}\,\mathcal{C}$ into matrix of observability and controllability holds with

$$\mathcal{O} \stackrel{\text{def}}{=} \begin{pmatrix} H \\ HF \\ HF^2 \\ \vdots \\ HF^p \end{pmatrix}, \quad \mathcal{C} \stackrel{\text{def}}{=} \begin{pmatrix} G & FG & F^2G & \dots & F^qG \end{pmatrix}, \tag{4}$$

where $G \stackrel{\text{def}}{=} \mathbf{E}\left(X_k Y_k^{(\text{ref})^T}\right)$. This factorization is obtained by a singular value decomposition, where the singular values are cut at a desired model order:

$$\mathcal{H} = \begin{pmatrix} U_1 & U_2 \end{pmatrix} \begin{pmatrix} \Sigma_1 & 0 \\ 0 & \Sigma_2 \end{pmatrix} \begin{pmatrix} V_1 & V_2 \end{pmatrix}^T, \quad \mathcal{O} \stackrel{\text{def}}{=} U_1\, \Sigma_1^{1/2}, \quad \mathcal{C} \stackrel{\text{def}}{=} \Sigma_1^{1/2} V_1^T.$$

For obtaining a stabilization diagram that contains the modal parameters for different model orders, we repeat this procedure and cut at the respective model orders. From the observability matrix $\mathcal{O}$ the matrices $H$ and $F$ can be retrieved to finally obtain the eigenstructure $(\lambda, \phi_\lambda)$ of the system (1). More details about this algorithm can be found in [1, 8].

### 2.2 Multiple Setups

Instead of a single record for the output $(Y_k)$ of the system (1), $N_s$ records

$$\underbrace{\begin{pmatrix} Y_k^{(1,\text{ref})} \\ Y_k^{(1,\text{mov})} \end{pmatrix}}_{\text{Record } 1} \quad \underbrace{\begin{pmatrix} Y_k^{(2,\text{ref})} \\ Y_k^{(2,\text{mov})} \end{pmatrix}}_{\text{Record } 2} \quad \cdots \quad \underbrace{\begin{pmatrix} Y_k^{(N_s,\text{ref})} \\ Y_k^{(N_s,\text{mov})} \end{pmatrix}}_{\text{Record } J} \tag{5}$$

are now available collected successively. Each record $j$ contains data $Y_k^{(j,\text{ref})}$ from a fixed *reference* sensor pool, and data $Y_k^{(j,\text{mov})}$ from a *moving* sensor pool. To each record $j = 1, \ldots, N_s$ corresponds a state-space realization in the form

$$\begin{cases} X_{k+1}^{(j)} & = & F\, X_k^{(j)} + V_{k+1}^{(j)} & \\ Y_k^{(j,\text{ref})} & = & H^{(\text{ref})}\, X_k^{(j)} & \text{(reference pool)} \\ Y_k^{(j,\text{mov})} & = & H^{(j,\text{mov})}\, X_k^{(j)} & \text{(sensor pool n}^o j) \end{cases} \tag{6}$$

with a single state transition matrix $F$. Note that the unmeasured excitation $V^{(j)}$ can be different for each setup $j$ as the environmental conditions can slightly change between the measurements. However, during each setup $j$ the noise $V^{(j)}$ is assumed to be stationary. Note also that the observation matrix $H^{(\text{ref})}$ is independent of the specific measurement setup if the reference sensors are the same throughout all measurements $j = 1, \ldots, N_s$.

For each setup $j$ we obtain a "local" Hankel matrix $\mathcal{H}^{(j)}$ according to equations (2)-(3) that is filled with correlations between all the sensor data and data from the reference sensors of this setup. The question is now how to adapt the subspace identification from Section 2.1 to

- merge the data from the multiple setups $j = 1, \ldots, N_s$ to obtain global eigenvalues and mode shapes, and to
- normalize or re-scale the data from the multiple setups as the background excitation may differ from setup to setup.

In the following section we present three merging strategies that are different in the order of the merging, normalization and system identification step.

## 3 Merging strategies

### 3.1 The PoSER approach

In classical *Experimental Modal Analysis* (EMA), where both the forces that are applied to the structure and the response (acceleration, velocity, displacement, strain, ...) in some *Degrees Of Freedom* (DOFs) are measured, merging data obtained in different measurement setups before extracting the modal parameters is common practice, certainly when measured Frequency Response Functions (FRFs) are used for the identification. A typical example is a roving hammer test.

In output-only or *Operational Modal Analysis* (OMA) on the other hand, processing each setup separately and then merging the modal data is the common practice. Since in a standard OMA test the mode shapes can not be mass-normalized, it is necessary to have measured DOFs that are common to each setup, the so-called reference DOFs. Before merging the partial mode shapes, obtained in each setup, they are re-scaled in a least-squares

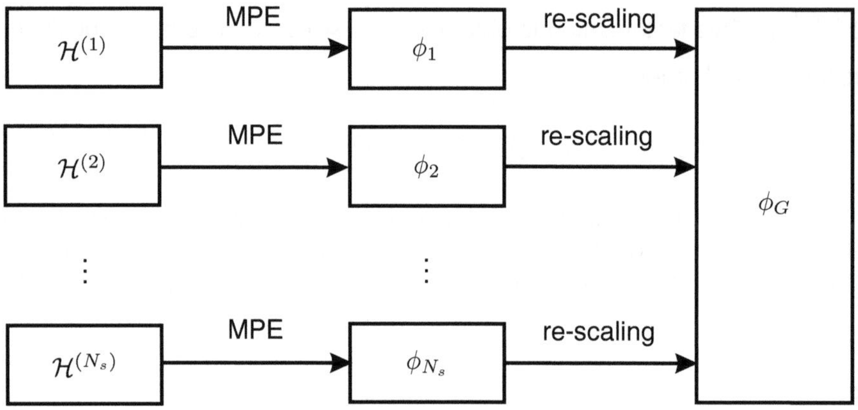


Figure 1: Merging partial mode shape estimates $\phi_j, j = 1, ..., N_s$ into a global mode shape estimate $\phi_G$ in the PoSER approach. MPE stands for Modal Parameter Estimation.

sense, for instance to the reference DOFs of the first setup (see Figure 1). In the remainder of this text, this "classical" approach is called the PoSER (Post Separate Estimation Re-scaling) approach.

The PoSER approach is tiresome, especially when the number of setups is large. Often, not all modes are well excited in all setups, and often mode pairing between different setups is difficult due to closely spaced modes. Furthermore, when a modal test is performed in operational conditions, the noise levels are usually much higher and the stabilization diagrams for the individual setups less clear than in an EMA test. Therefore, the PoSER approach is less attractive in operational modal analysis.

### 3.2 The PoGER approach

Parloo et al. [6, 7] define two other approaches for frequency domain OMA methods, but they are also applicable to time domain methods. One of them is called PoGER (Post Global Estimation Re-scaling). In the PoGER approach, first a nonparametric system model is identified for each setup separately. In the time domain, this nonparametric model consists of the correlations between all measured outputs, see Equation (2).

In a second step, the output correlations obtained from the different setups are stacked on top of each other. Extracting the modal parameters from the resulting correlation function yields global values for the eigenfrequencies and damping ratios. The identified partial mode shapes are stacked on top of each other in a global mode shape. However, due to the non-stationary ambient excitation level and the non-stationary ambient excitation color, it is necessary to re-scale the partial mode shapes in a least-squares sense, for instance to the reference DOFs of the first partial mode shape, just as in the PoSER approach [6, 7] (see Figure 2).

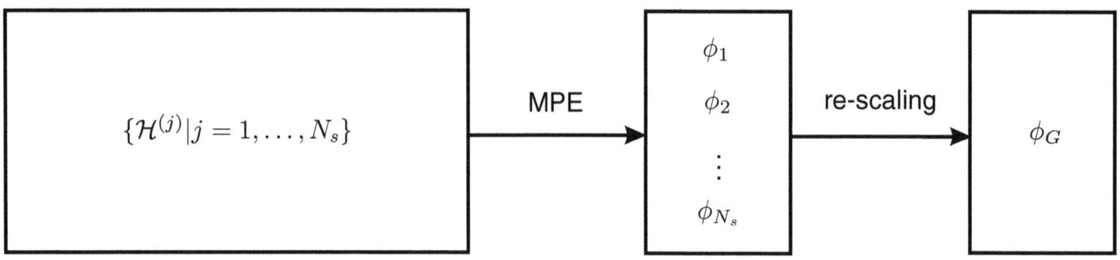


Figure 2: Merging partial mode shape estimates $\phi_j, j = 1, ..., N_s$ into a global mode shape estimate $\phi_G$ in the PoGER approach.

### 3.3 The PreGER approach

Another merging approach described in [4, 5] makes use of the factorizations $\mathcal{H}^{(j)} = \mathcal{O}^{(j)}\mathcal{C}^{(j)}$ of the Hankel matrices of each setup $j = 1, \ldots, N_s$ into matrix of observability and controllability, see Equation (4). In order to merge the data we first take into account the difference in the excitation between setups, information which is only present in the matrix of controllability since the matrix of observability is only dependent of the observation matrix $H^{(j)}$ and state matrix $F$ that are not affected. In the first step, all the Hankel matrices $\mathcal{H}^{(j)}$ are re-scaled with a common matrix of controllability, then the resulting matrices are merged and a global modal parameter estimation is finally done on the merged matrix. In doing so, we make use of the special structure of the Hankel matrices $\mathcal{H}^{(j)}$ that are filled with correlations $R_i^{(j)}$, and their decomposition $R_i^{(j)} = H^{(j)}F^iG^{(j)}$. We call this method PreGER (Pre Global Estimation Re-scaling). It is consistent under nonstationary excitation [5].

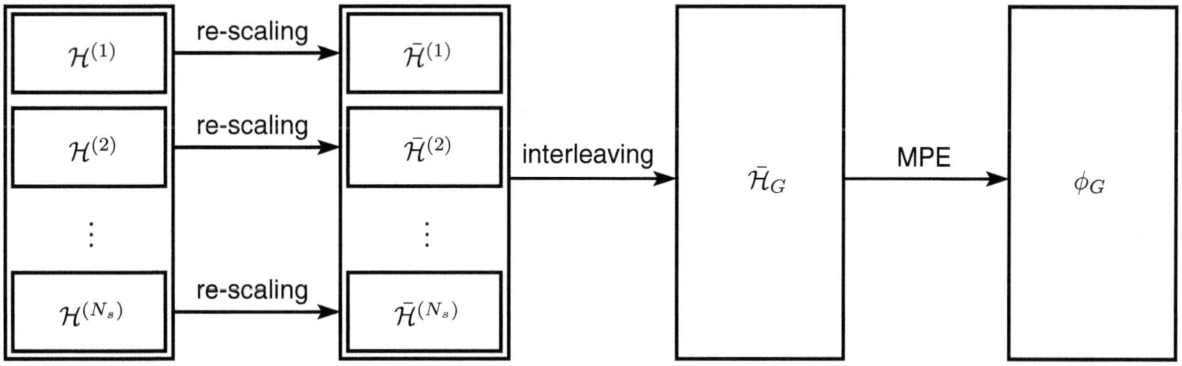


Figure 3: Merging Hankel matrices of each setup to obtain a global Hankel matrix and global mode shape estimate $\phi_G$ in the PreGER approach.

In detail, we separate the Hankel matrices $\mathcal{H}^{(j)}$ into matrices $\mathcal{H}^{(j,\text{ref})}$ and $\mathcal{H}^{(j,\text{mov})}$, where the matrix $\mathcal{H}^{(j,\text{ref})}$ contains only the auto-correlations of the reference sensor data $R_i^{(j,\text{ref})}$ and $\mathcal{H}^{(j,\text{mov})}$ contains the correlations of the moving sensor data vs reference sensor data $R_i^{(j,\text{mov})}$ with

$$R_i^{(j,\text{ref})} = \mathbf{E}\left(Y_k^{(j,\text{ref})}\, Y_{k-i}^{(j,\text{ref})^T}\right) = H^{(\text{ref})}F^iG^{(j)}, \quad R_i^{(j,\text{mov})} = \mathbf{E}\left(Y_k^{(j,\text{mov})}\, Y_{k-i}^{(j,\text{ref})^T}\right) = H^{(j,\text{mov})}F^iG^{(j)}.$$

Furthermore it holds

$$\mathcal{H}^{(j,\text{ref})} = \mathcal{O}^{(\text{ref})}\,\mathcal{C}^{(j)}, \quad \mathcal{H}^{(j,\text{mov})} = \mathcal{O}^{(j,\text{mov})}\,\mathcal{C}^{(j)}.$$

For the normalization we need the matrices $\mathcal{C}^{(j)}$ in the same state basis, that we obtain from a Hankel matrix filled with the re-arranged correlations

$$R_i^{(\text{allref})} \stackrel{\text{def}}{=} H^{(\text{ref})}F^iG^{(\text{all})} \quad \text{with} \quad G^{(\text{all})} \stackrel{\text{def}}{=} \begin{pmatrix} G^{(1)} & G^{(2)} & \ldots & G^{(N_s)} \end{pmatrix}$$

by decomposing it into matrix of observability and controllability at the maximal model order and taking the appropriate columns of the matrix of controllability. Note that this matrix of controllability has the shape

$$\mathcal{C}^{(\text{all})} = \begin{pmatrix} G^{(\text{all})} & FG^{(\text{all})} & \ldots & F^qG^{(\text{all})} \end{pmatrix},$$

so the appropriate columns of every block $F^iC^{(\text{all})}$ have to be taken and juxtaposed to obtain $\mathcal{C}^{(j)}$, in contrary to the indications in [4]. In the next step, we choose one setup $j^* \in \{1, \ldots, N_s\}$ and re-scale the matrices $\mathcal{H}^{(j,\text{mov})}$ to

$$\bar{\mathcal{H}}^{(j,\text{mov})} = \mathcal{H}^{(j,\text{mov})}\, \mathcal{C}^{(j)^T} (\mathcal{C}^{(j)}\, \mathcal{C}^{(j)^T})^{-1}\, \mathcal{C}^{(j^*)}.$$

These matrices now contain the correlations $\bar{R}_i^{(j)} = H^{(j)}F^iG^{(j^*)}$. Finally, we fill the merged Hankel matrix with the re-arranged correlations

$$\bar{R}_i^{(\text{total})} \stackrel{\text{def}}{=} \begin{pmatrix} H^{(\text{ref})^T} & H^{(1)^T} & H^{(2)^T} & \ldots & H^{(N_s)^T} \end{pmatrix}^T F^iG^{(j^*)}$$

and perform the subspace system identification on it to obtain the global modal parameters.

## 4 Output-only Modal Analysis of the Luiz I bridge

### 4.1 Bridge description and ambient vibration test

The Luiz I Bridge (Figure 4) is a 172 m tied metallic arch bridge over the Douro River at the city of Porto, in the north of Portugal, built in 1885 by Eiffel's former engineer Theophile Seyrig.

The bridge is composed by a hinged metallic arch supporting two different decks: one at the top of the arch and another one at the level of the respective supports. The arch spans 172 m between abutments and rises 45.1 m. It has a variable thickness, between 16.7 m near the supports and 7.1 m at midspan, and presents a parabolic geometry both in vertical and plan views. The upper deck is 391.25 m long and 5 m height, and it is supported by the arch, at midspan and at two intermediate sections, by 5 piers and the abutments. The two columns that transfer the vertical loads of the upper deck to the arch are monolithically connected to both structural elements. The connection between the deck and the other columns and the abutments is materialised by roller supports. The lower deck has a height of 3.25 m, is suspended from the arch by four ties and provides a roadway connection between the cities of Porto and Vila Nova de Gaia.

Since 2005, the upper deck of this bridge is used by the light metro of Porto. The installation of the metro line required the development of significant rehabilitation works, involving the modification of the upper deck and the replacement and strengthening of a significant number of structural components.

Figure 4: General view of the Luiz I Bridge.

The Laboratory of Vibrations and Monitoring (VIBEST, www.fe.up.pt/vibest) of the Civil Engineering Department of the Faculty of Engineering of the University of Porto (FEUP) was contracted to perform two ambient vibration tests on the Luiz I bridge: one before and another after the rehabilitation works, with the aim of accurately identifying the most relevant dynamic structural properties and correlate them with the values provided by finite element modeling. The present paper is only focused on the processing of the data collected during the second test. This dynamic test was conducted without disturbing the normal use of the bridge, so the measured accelerations were mainly induced by the wind, by the roadway traffic on the lower deck and by the metro passing over the upper deck. The dynamic response of the bridge was recorded with four seismographs equipped with tri-axial force-balance accelerometers and duly synchronized by GPS. Because of the limited number of available sensors and the need to measure a high number of points, the measurements were carried out in 26 different setups during two days. For the test of the upper deck, two recorders served as references, permanently located at section 15 (Figure 5), at both the upstream and downstream sides, while the other two recorders scanned the bridge deck in 18 consecutive setups, measuring the acceleration along the 3 orthogonal directions, at both the upstream and downstream sides.

During the ambient vibration test of the lower deck, the reference sensors of the upper deck kept their positions and an additional reference was introduced at the downstream side at section 22 (Figure 5), while the remaining seismograph was used as a moving sensor along the upstream side of the lower deck. For each setup, time series of 16 minutes were collected with a sampling frequency of 100 Hz. Further details of the dynamic test can be found in [2], together with a first processing of the data that includes the application of the p-LSCF method using the PoSER approach and of the SSI-COV method using the PoGER strategy.

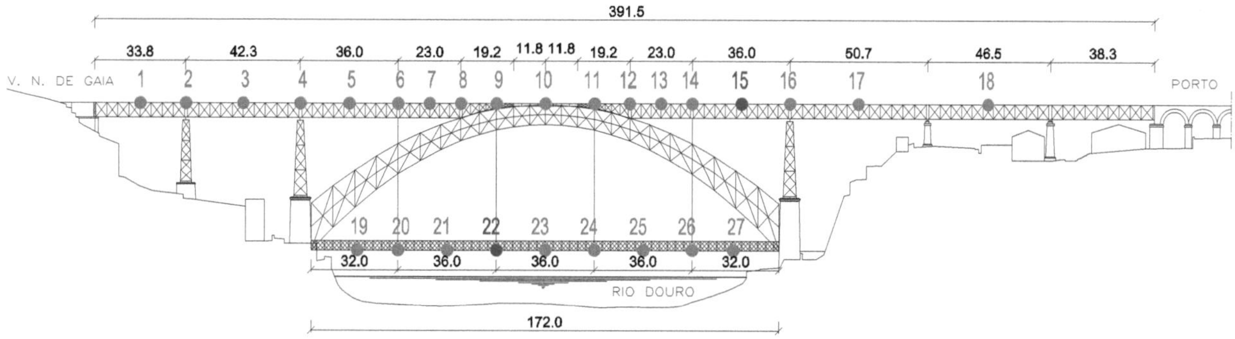


Figure 5: Instrumented sections.

The data to be processed consists of the lateral and vertical accelerations of the bridge, is low-pass filtered at 4 Hz and re-sampled with the sampling frequency 20 Hz.

## 4.2 PoSER approach

First, the classical PoSER approach was applied using the *REFerence-based COVariance-driven Stochastic Subspace Identification* (SSI-cov/ref) [8]. Details of the analysis can be found in [9]. The analysis turned out to be very cumbersome and time-consuming, not only because $26$ different stabilization diagrams needed to be analyzed, but also due to the fact that most of the modes contain a main contribution from either the upper or the lower deck, while the other bridge deck remains relatively undeformed.

The set of fourteen modes that have been extracted from the data in the frequency range of interest, $[0 - 4\text{Hz}]$, is shown in Figure 6. They can be characterized as follows:

- modes $1$ and $7$ are lateral bending modes of the lower deck. Mode $1$ can even be considered as a purely local mode, i.e., it does not appear in any of the stabilization diagrams of the seventeen setups that cover the upper deck only;
- modes $2, 3, 4, 6, 8$ and $9$ are lateral bending modes of the upper deck. Only for mode $2$, there is a clear but minor deformation of the lower deck;
- modes $5$ and $10$ are vertical bending modes were both decks move in phase;
- modes $11$, $12$ and $13$ are vertical bending modes of the lower deck;
- mode $14$ is a local bending mode at the side spans of the Porto side of the bridge (see Figure 5).

Modes $9$ and $10$ make up a double mode. Mode $10$ is a global vertical bending mode, while mode $9$ is a lateral local bending mode of the upper deck. The partial mode shapes of these two modes, that were estimated from the different setups, have been paired as follows. First, the MAC values between both modes, estimated in each setup, and mode $9$, estimated in the first setup, were calculated using the mode shape estimates at reference section $15$ (see Figure 5). Then, the partial mode shape with the highest MAC value was retained and glued to mode $9$. The other partial mode shape was glued to mode $10$. Following this procedure, multiple crossings of the eigenfrequencies of both modes have been observed.

The large vertical bending component of the lower deck in mode $9$ is probably due to the fact that for the lower deck, the double mode has only a bending component. It should be noted that when double modes are present, their mode shapes are not unique, since any linear combination of both mode shapes yields another valid mode shape [3, p. 57].

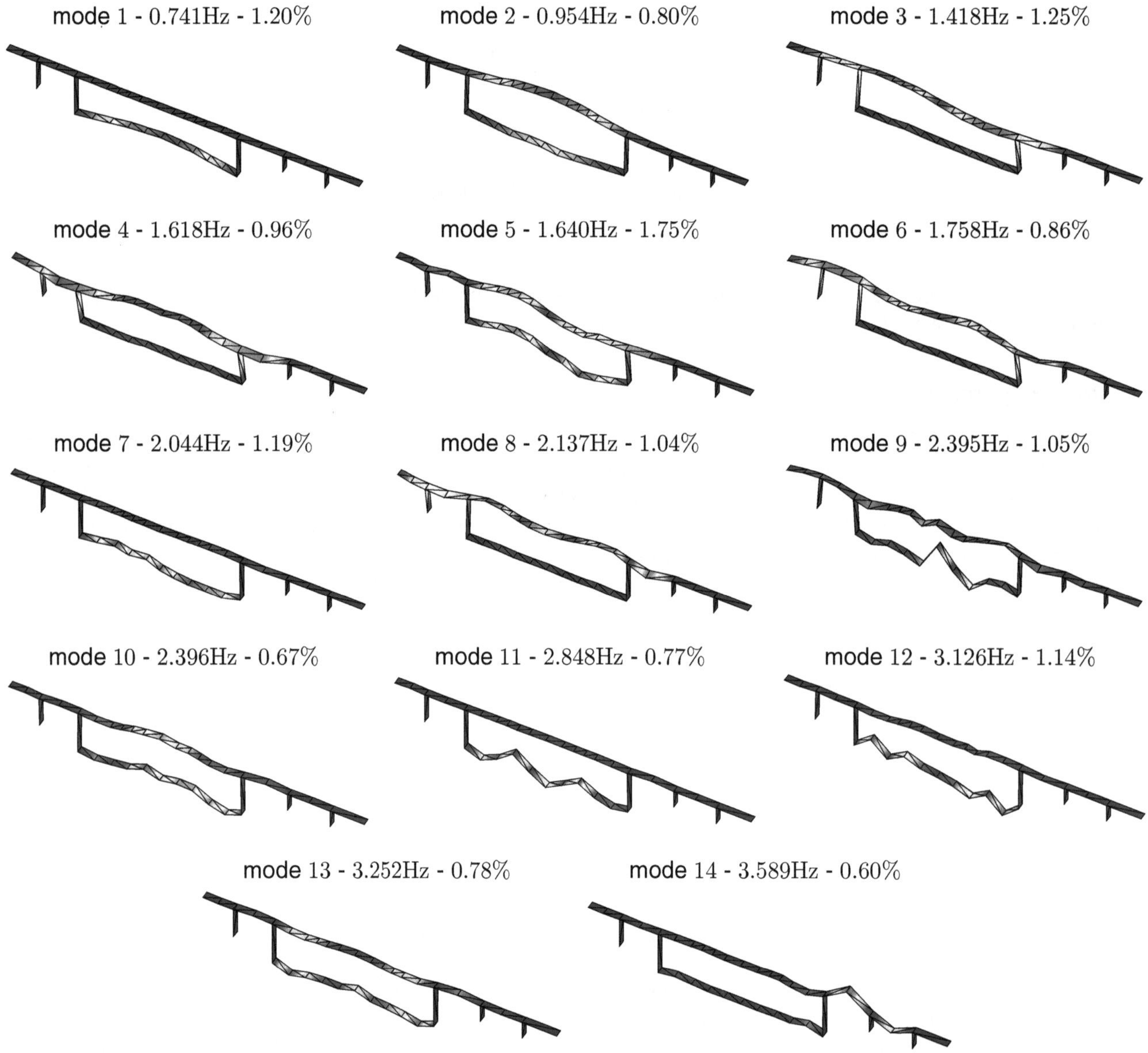


Figure 6: SSI-cov/ref: identified modes with the PoSER approach [9].

### 4.3 PoGER approach

As an alternative to the classical PoSER approach, the PoGER approach was investigated using the SSI-cov/ref method [8]. Details of the analysis can be found in [9]. All setups were analyzed at once. The three signals that are common to all setups were selected as reference signals for the identification.

The PoGER analysis was much faster and much less cumbersome than the PoSER analysis, since only one clear stabilization diagram needed to be analyzed. On the other hand, the number of outputs to be processed at once was 26 times larger than in a separate identification [9]. This increased the computation time and memory usage considerably. As can be seen from Figure 7, the extracted modal parameters (eigenfrequencies, damping ratios and mode shapes) compare very well to the results of the PoSER approach. Only for the mode shapes of the double mode at 2.39Hz, the differences are larger, but as said in section 4.2, the mode shapes are not uniquely determined when multiple modes are present.

From Figure 7, it can also be noted that modes 1 and 7 were not extracted. This is because they almost exclusively involve lateral movement of the lower deck of the bridge, and this movement is obviously not observed by the reference DOFs that are common to all setups, i.e. the vertical and lateral movement at the downstream side of section 15 and the vertical movement at the upstream side of section 15. In order to overcome this problem, the data measured at the upper deck and the lower deck could be processed separately, and both partial mode shape estimates could then be glued together, just as in the PoSER approach.

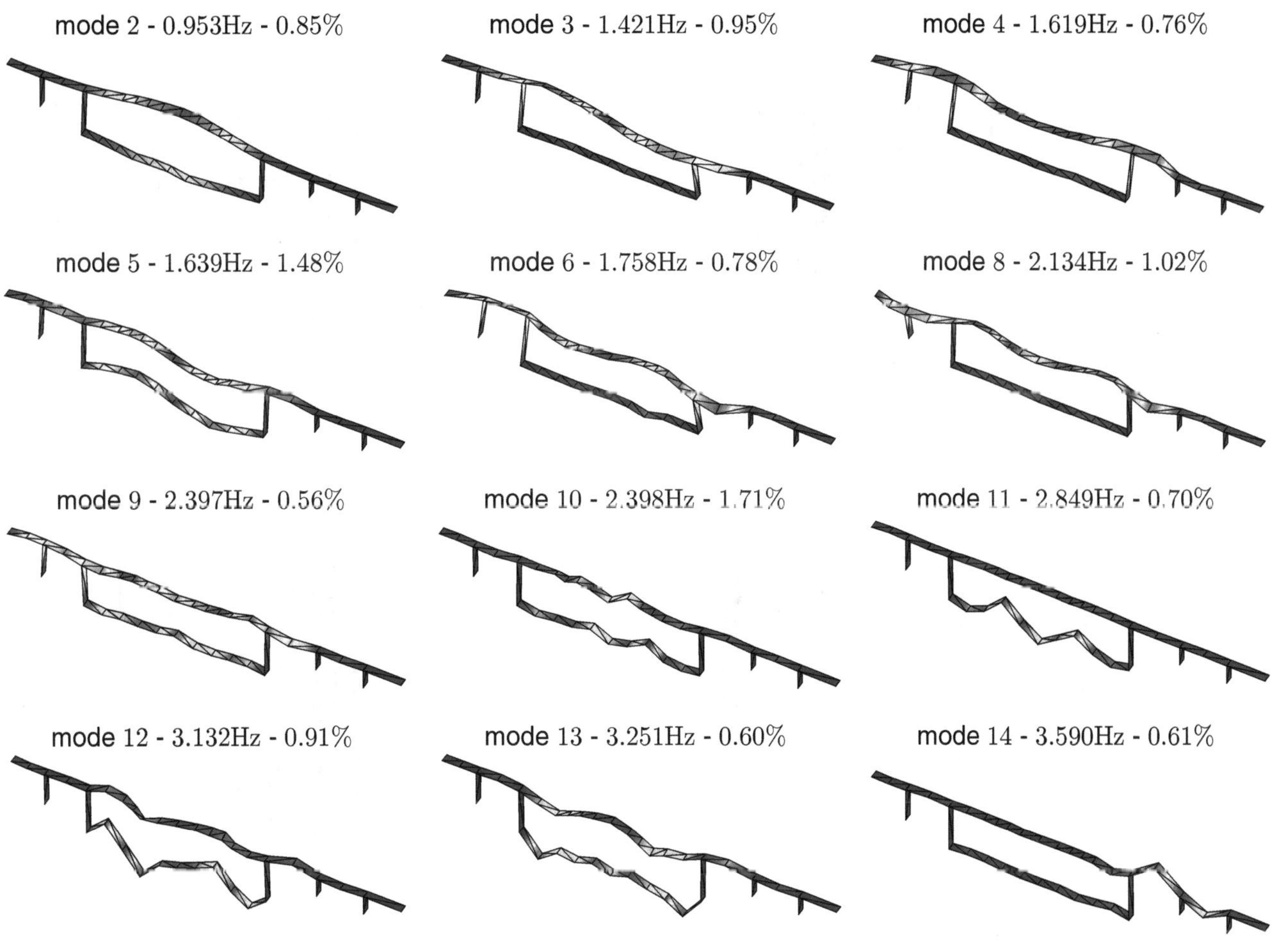


Figure 7: SSI-cov/ref: identified modes with the PoGER approach [9].

### 4.4 PreGER approach

For the analysis with the PreGER method we separated the data into their lateral and vertical components and processed them separately because there are only purely lateral and purely vertical bending modes present in the frequency range of interest. This led to better results than processing the data from both directions together as the mode shape of a lateral bending mode would contain noise in the vertical direction, and vice versa. Apart from this separation, all setups were processed together. For the analysis the half number of correlation lags was 150 ($p = q = 150$ in Equation (3)) and the model order ranged from 1 to 150. The modes were manually chosen in a stable model order band in the resulting stabilization diagrams.

For the lateral components, one reference sensor common to all setups was available on the upper deck, and for the vertical components, two reference sensors were available on the upper deck. With these settings all modes from the PoSER approach could be identified, except mode $1$ and $7$ like in the PoGER approach. These are local lateral bending modes of the lower deck, and the lateral reference sensor only from the upper deck was used. Besides, an additional lateral bending mode was identified at $3.088\text{Hz}$ (mode $11$A).

In order to identify the local lateral bending modes of the lower deck, we continued the analysis only with the setups containing data from the lower deck, including the reference sensor from the lower deck. This led to the identification of modes $1$ and $7$, and an additional local lateral bending mode at $3.963\text{Hz}$ (mode $15$). Another separate analysis of the upper deck (with the reference sensor from there) led to the identification of a local lateral bending mode of the upper deck at $3.393\text{Hz}$ (mode $13$A). None of the modes $1$, $7$, $13$A and $15$ were present on both decks.

The stabilization diagrams with these different settings used for the modal extraction with the PreGER approach can be found in Figure 8.

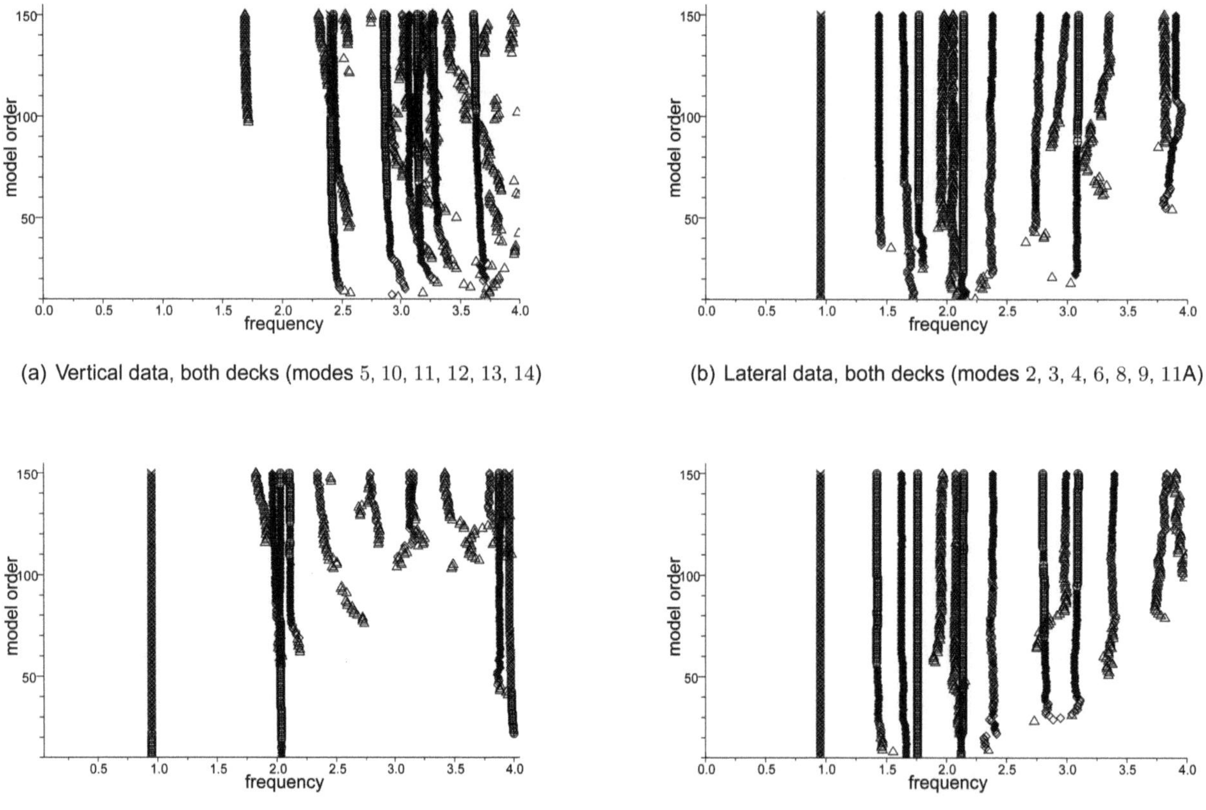


(a) Vertical data, both decks (modes $5$, $10$, $11$, $12$, $13$, $14$)

(b) Lateral data, both decks (modes $2$, $3$, $4$, $6$, $8$, $9$, $11$A)

(c) Lateral data, lower deck (modes $1$, $7$, $15$)

(d) Lateral data, upper deck (mode $13$A)

Figure 8: Stabilization diagrams used to identify the modes in the PreGER approach.

All the identified mode shapes can be seen in Figure 9 and they correspond very well to the results of the PoSER and the PoGER approach, as well as the frequencies and damping ratios. Only the damping ratio of mode $5$ was badly estimated as it has not stabilized yet in the stabilization diagram. Furthermore, the double mode $9$ and $10$

turns out to be the superposition of a lateral and vertical bending mode, as mode 9 is a lateral and mode 10 is a vertical bending mode.

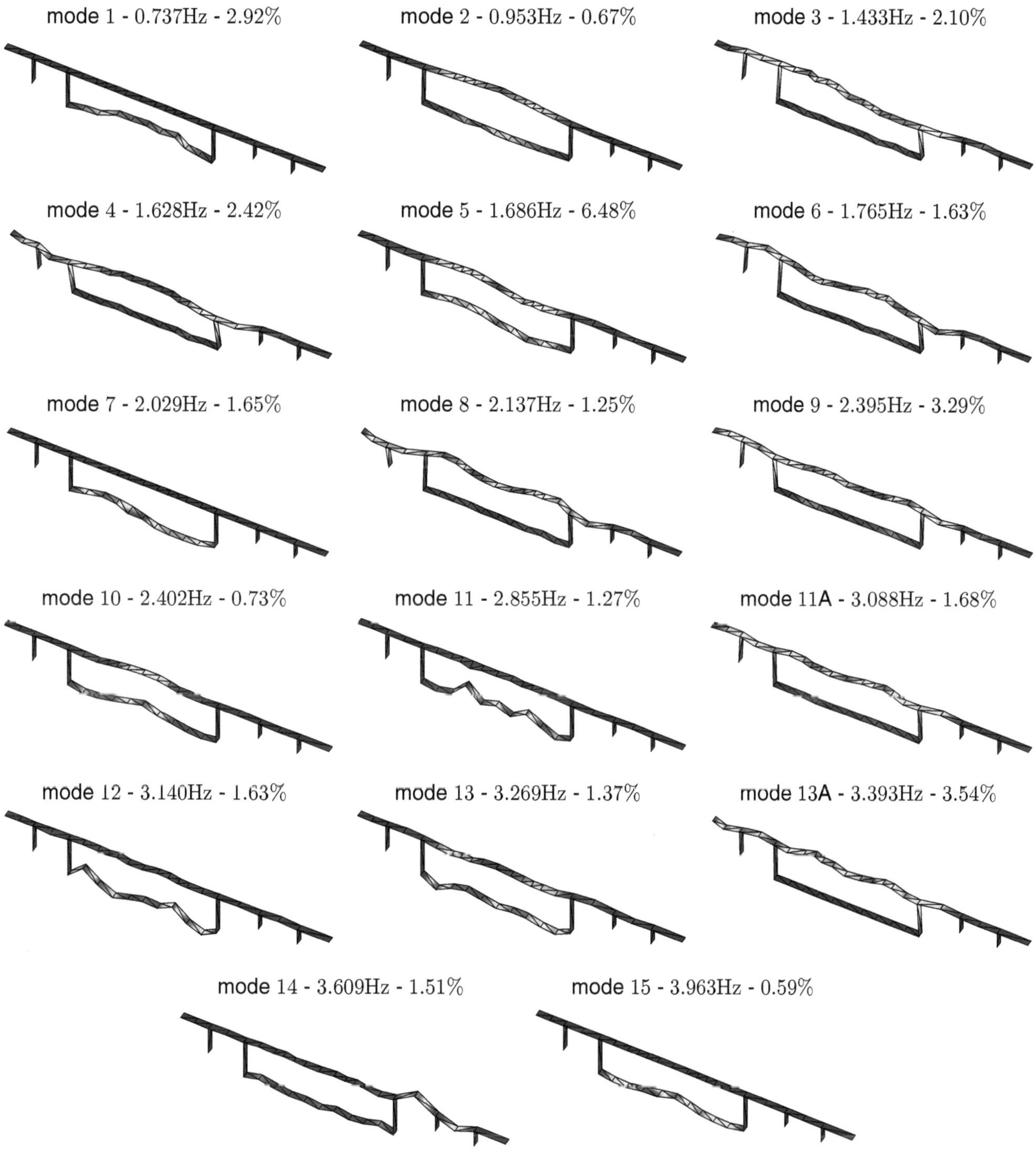


Figure 9: SSI-cov/ref: identified modes with the PreGER approach.

## 5 Summary of results

Tables 1 and 2 provide a comparative overview of the natural frequencies and damping ratios estimated with the different merging approaches (PoSER, PoGER and PreGER) using the covariance-driven SSI-cov/ref method. For a comparison of the mode shapes we use the Modal Assurance Criterion (MAC) in Figure 10. The following observations can be made:

- Modes $1$, $7$ and $15$ were not observed in the PoGER and initial PreGER estimates, due to the lack of a lateral reference DOF at the lower deck that is common to all setups; however they were observed in the PreGER estimates after analyzing only the lower deck.
- Modes $11$A, $13$A and $15$ were not observed in the PoSER and PoGER estimates.
- The differences in natural frequencies for the PoSER approaches and the PoGER approach are negligibly small. Their differences to the PreGER approach are slightly larger, but still less than $1\,\%$.
- For the damping ratios, the differences between PoSER and PoGER estimates are larger but still not significant considering the large standard deviations on the estimates. For some modes however, the damping ratios of the PreGER estimates are considerably higher than the PoSER and PoGER estimates. This might be due to the fact, that the natural frequencies in each setup are slightly different. Then, the resulting frequency for each mode obtained by the PreGER approach is associated to a higher damping ratio, consequence from the merging of overlapping frequencies.
- The MAC values between the mode shapes of the different merging approaches are very high ($> 0.9$) for most of the modes, meaning that the identified mode shapes are very similar. Only the mode shapes of modes $6$ and $11$ show larger differences, and, as expected, the double modes $9$ and $10$.

| mode | 1 | 2 | 3 | 4 | 5 | 6 | 7 | 8 | 9 | 10 | 11 | 12 | 13 | 14 |
|---|---|---|---|---|---|---|---|---|---|---|---|---|---|---|
| PoSER | 0.741 | 0.954 | 1.418 | 1.618 | 1.640 | 1.758 | 2.044 | 2.137 | 2.395 | 2.396 | 2.848 | 3.126 | 3.252 | 3.589 |
| $\sigma$ | 0.008 | 0.002 | 0.005 | 0.005 | 0.012 | 0.010 | 0.035 | 0.008 | 0.014 | 0.014 | 0.011 | 0.016 | 0.008 | 0.009 |
| PoGER | – | 0.953 | 1.421 | 1.619 | 1.639 | 1.758 | – | 2.134 | 2.397 | 2.398 | 2.849 | 3.132 | 3.251 | 3.590 |
| PreGER | 0.737 | 0.953 | 1.433 | 1.628 | 1.686 | 1.765 | 2.029 | 2.137 | 2.378 | 2.402 | 2.855 | 3.140 | 3.269 | 3.609 |

Table 1: An overview of the estimated eigenfrequencies (in $\mathrm{Hz}$) obtained from the different merging strategies using SSI-cov/ref.

| mode | 1 | 2 | 3 | 4 | 5 | 6 | 7 | 8 | 9 | 10 | 11 | 12 | 13 | 14 |
|---|---|---|---|---|---|---|---|---|---|---|---|---|---|---|
| PoSER | 1.20 | 0.80 | 1.25 | 0.96 | 1.75 | 0.86 | 1.19 | 1.04 | 1.05 | 0.67 | 0.77 | 1.14 | 0.78 | 0.60 |
| $\sigma$ | 1.00 | 0.22 | 1.00 | 0.43 | 0.70 | 0.52 | 0.92 | 0.37 | 0.44 | 0.49 | 0.51 | 1.06 | 0.38 | 0.20 |
| PoGER | – | 0.85 | 0.95 | 0.76 | 1.48 | 0.78 | – | 1.02 | 0.56 | 1.71 | 0.70 | 0.91 | 0.60 | 0.61 |
| PreGER | 2.92 | 0.67 | 2.10 | 2.42 | 6.48 | 1.63 | 1.65 | 1.25 | 3.29 | 0.73 | 1.27 | 1.63 | 1.37 | 1.51 |

Table 2: An overview of the estimated damping ratios (in $\%$ of critical) obtained from the different merging strategies using SSI-cov/ref.

## 6 Conclusions

In this paper, we compared different approaches for obtaining the global modal parameters from a multi setup experiment with reference-based covariance-driven SSI, namely

- Post Separate Estimation Re-Scaling (PoSER),
- Post Global Estimation Re-Scaling (PoGER),

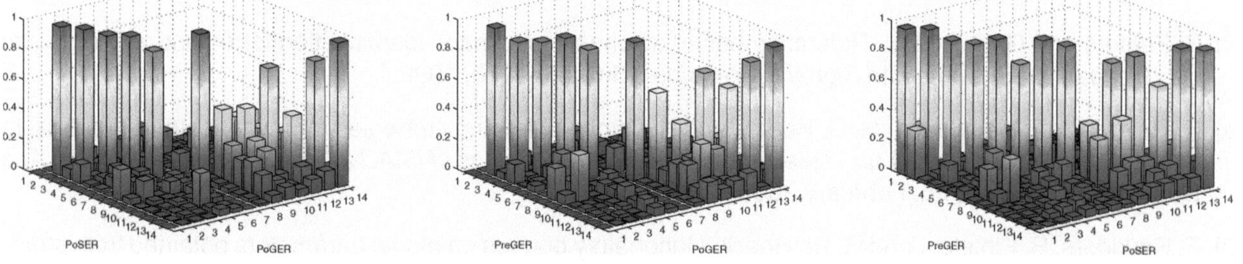


Figure 10: MAC values between the mode shapes from the PoSER and PoGER approach (left), from the PreGER and PoGER approach (middle) and from the PreGER and PoSER approach (right).

- Pre Global Estimation Re-Scaling (PreGER).

The main advance of this paper is the proof of feasibility for the PreGER approach, unlike previous results obtained in [9]. Both PreGER and PoGER approaches offer comparable qualitative results as shown in Table 1, Table 2 and Figure 10. However, some of the estimates, especially the damping ratios in the PreGER approach, show different results. A bigger number of correlations in the used Hankel matrices might improve the results, but in general the damping ratios have quite big statistical uncertainties. This could be further investigated by computing the variances on the modal parameters, e.g., by following a sensitivity-based approach as in [10].

Concerning both methods

- PreGER has a slight memory advantage, because it does not stack reference sensors for all setups in the identification process. Depending on the number of setups, it may be a plus.
- PoGER rescales the mode shapes in the different setups by making correspondence between the different mode shape values on the reference sensors among all the different setups, not taking into account the inherent uncertainties in the identification results. Coupling PoGER with uncertainties computed as in [10] might improve the robustness of PoGER with respect to this.

These comments still have to be verified and it is necessary to continue exploring the applied methodologies on further examples.

## References

[1] A. Benveniste and J.-J. Fuchs. Single sample modal identification of a non-stationary stochastic process. *IEEE Transactions on Automatic Control*, AC-30(1):66–74, 1985.

[2] A. Cunha, F. Magalhães, and E. Caetano. Output-only modal identification of Luiz I bridge before and after rehabilitation. In P. Sas and M. D. Munck, editors, *Proceedings of ISMA2006 International Conference on Noise and Vibration Engineering*, Leuven, Belgium, September 2006.

[3] D. J. Ewins. *Modal testing*. Research Studies Press, Baldock, U.K., 2nd edition, 2000.

[4] L. Mevel, M. Basseville, A. Benveniste, and M. Goursat. Merging sensor data from multiple measurement setups for nonstationary subspace-based modal analysis. *Journal of Sound and Vibration*, 249(4):719–741, 2002.

[5] L. Mevel, A. Benveniste, M. Basseville, and M. Goursat. Blind subspace-based eigenstructure identification under nonstationary excitation using moving sensors. *IEEE Transactions on Signal Processing*, SP-50(1):41–48, 2002.

[6] E. Parloo. *Application of frequency-domain system identification techniques in the field of operational modal analysis*. PhD thesis, Vrije Universiteit Brussel, 2003.

[7] E. Parloo, P. Guillaume, and B. Cauberghe. Maximum likelihood identification of non-stationary operational data. *Journal of Sound and Vibration*, 268(5):971–991, 2003.

[8] B. Peeters and G. D. Roeck. Reference-based stochastic subspace identification for output-only modal analysis. *Mechanical Systems and Signal Processing*, 13(6):855–878, 1999.

[9] E. Reynders, F. Magalhães, G. D. Roeck, and A. Cunha. Merging strategies for multi-setup operational modal analysis: application to the Luiz I steel arch bridge. In *Proceedings of IMAC 27, the International Modal Analysis Conference*, Orlando, FL, February 2009.

[10] E. Reynders, R. Pintelon, and G. De Roeck. Uncertainty bounds on modal parameters obtained from stochastic subspace identification. *Mechanical Systems and Signal Processing*, 22(4):948–969, 2008.

Proceedings of the IMAC-XXVIII
February 1–4, 2010, Jacksonville, Florida USA


# Time-Variant Reliability and Load Rating of a Movable Bridge Using Structural Health Monitoring

Taha Dumlupinar, H. Burak Gokce, F. Necati Catbas
Structures and Systems Research Laboratory, Eng2-116, Civil & Environmental Engineering Department, University of Central Florida, Orlando, FL 32816, USA,
tdumlupi@mail.ucf.edu, hgokce@mail.ucf.edu and catbas@mail.ucf.edu

Dan M. Frangopol
Department of Civil and Environmental Engineering, Center for Advanced Technology for Large Structural Systems (ATLSS Center), Lehigh University
Bethlehem, PA 18015-4729, USA
dan.frangopol@lehigh.edu

## ABSTRACT

After the recent bridge collapses in the US, the engineering community demands critical effects of bridge deterioration over long-term to be investigated and closely monitored. Combined with reliability techniques, structural health monitoring (SHM) can provide objective and accurate assessment of existing condition for safety and serviceability trends based on collected data. Two main approaches for this evaluation are rating and reliability. These approaches will be demonstrated analytically on a finite element model and the SHM data of a movable bridge: Sunrise Boulevard Bridge. Firstly, the reliability index and load rating will be analyzed under truck loading. A moving truck will be simulated on the model, obtaining the response reliability indices and ratings as a function of time. Strains from the model will be evaluated as monitoring data and component reliabilities will be calculated according to assumed limit states using random variables for material properties. Then, the same procedure is repeated but this time using the real-time data collected from the actual bridge. Finally, the results coming from both cases are compared and interpreted. These demonstrations will establish a guideline for applying reliability assessment based on monitoring data.

## NOMENCLATURE

| | | |
|---|---|---|
| C | : | Factored load carrying capacity |
| DC | : | Dead load of structural components |
| DW | : | Dead load of the wearing surface |
| P | : | Dead load concentrated at a single point |
| LL | : | Live load effect |
| IM | : | Impact factor |
| $\gamma$ | : | Safety factors |
| $F_y$ | : | Yield strength |
| $Z_x$ | : | Plastic section modulus |
| RF | : | Rating factor |
| $\beta$ | : | Reliability Index |
| $\mu$ | : | Mean |
| $\sigma$ | : | Standard deviation |
| C | : | F |
| A | : | Statistical random variable |
| B | : | Statistical random variable |

## INTRODUCTION

Bridges are critical links of transportation networks. Any damage or collapse of a bridge not only results in loss of property and human fatalities but also has severe effects on the regional economy. Deterioration of civil infrastructures in North America, Europe and Japan has been well documented and publicized. In United States, 50% of all the bridges were built before 1940 and approximately 42% of those present structural deficiency [1]. As a result, many researchers have been investigating techniques and methods to evaluate the structural condition for objective decision-making. Successful visual inspection of these structures depends on considering all possible damage scenarios at all critical locations, not an easily accomplished task even for an experienced inspector. Despite all these limitations, visual inspection remains today the most commonly practiced damage detection method. o objectively evaluate the condition of existing structures and to better design new ones, researchers are exploring novel sensing technologies and analytical methods that can be used to rapidly identify the onset of structural damage [2; 3].

### *Definition of Structural Health Monitoring*

As there is not a standard approach for SHM, there is not a unique, collectively accepted definition as well. We adopt the following: SHM is, the measurement of the operating and loading environment and the critical responses of a structure to track and evaluate the symptoms of operational incidents, anomalies, and/or deterioration or damage indicators that may affect operation, serviceability, or safety reliability. It is possible to capture long-term structural behavior with continuous or discrete intervals of monitoring, capturing seasonal and environmental changes that are not readily apparent from intermittent tests. Health monitoring concept uses integrated local and global non-destructive experimental technologies together with advanced structural analysis and modeling techniques to complement inspections and provide continuous information regarding bridge state parameters, loading environment, and state of health [4] .

### *Performance Assessment Methods based on Structural Health Monitoring*

There have been extensive studies with different approaches towards the common aim of improving bridge network maintenance practices by increasing the accuracy of the predictions, improving the structural safety and reducing the life-cycle cost. With the exception of some recent studies by the authors, [5; 6; 7] and studies by Hong Kong researchers [8], structural reliability approach has not been fully studied by using SHM long term data to the extent of the authors' knowledge. So far, all reliability methods reviewed make use of condition ratings based on visual inspections or theoretical/numerical models. Supplementing the reliability models with sensor data or NDE results has been suggested in some references; however, a comprehensive reliability approach using a complete SHM application needs to be further investigated. There is a need for integrating SHM and reliability analysis as a framework composed of a comprehensive SHM application used for probabilistic analysis of system and component reliability for efficient bridge management and decision-making. Integrating reliability analysis in a SHM framework is a tool to achieve a comprehensive and advanced bridge management practice. The promise of integrating both approaches as complementing applications is given conceptually in Figure 1.

Load rating analysis of bridges is another condition rating method which is more practical and easier to conceptualize. In load rating analysis, the live load that structures can safely carry is determined. Bridges are mainly rated at two different stages, namely Inventory Rating and Operating Rating. Inventory rating is the capacity rating for the vehicle type used in the rating that will result in a load level which can safely utilize an existing structure for an indefinite period of time. Inventory load level approximates the design load level for normal service conditions. Operating rating will result in the absolute maximum permissible load level to which the structure may be subjected for the vehicle type used in the rating. This rating determines the capacity of the bridge for occasional use. Allowing unlimited numbers of vehicles to subject the bridge to the operating level will compromise the bridge life. This value is typically used when evaluating overweight permit vehicle moves. In calculating the both stage bridge rating, all statistical data and probabilistic approaches are embedded in load resisting factors.

Current SHM technology is capable of providing rapid or even real-time condition assessment of structures. Accurate and comprehensive data is produced, which greatly eliminates the uncertainties involved in the traditional structural appraisal methods. Better accuracy means improved operational safety, as well as instant notification of unexpected distress. These tools should be complemented with probabilistic structural analysis approaches for evaluation and estimation of uncertainties, and determination of structural system reliability based on SHM data. These advances in bridge condition assessment will lead to higher level management practices by performance projections, accurate project/network level cost-benefit evaluation and life-cycle cost analysis for maintenance optimization and decision making.

While reliability methods are quite powerful for quantifying risk and uncertainty, they require a great deal of input data to execute [9]. Also, current reliability techniques depend mostly on visual inspection results, which create additional uncertainty. As indicated in recent studies [5], reliability of a structure can change significantly if data from monitoring is used.

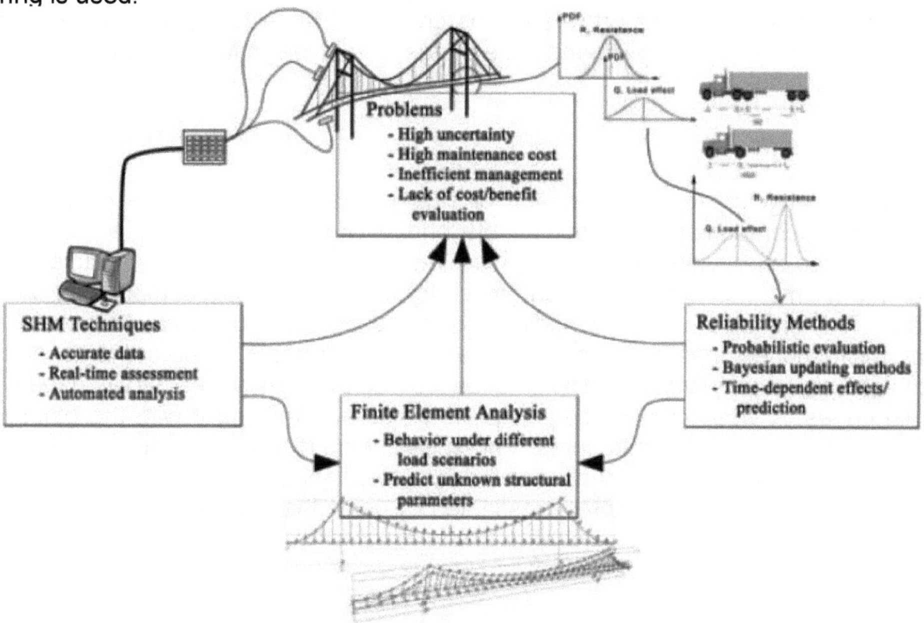


Figure 1: Approaches for Bridge Management

### *Objective and Scope*

As explained in the previous section, movable bridges experience major deterioration as compared to regular fixed bridges due to their complex structural, mechanical and electrical systems. The authors investigated the most common issues associated with movable bridges and based on the study, a comprehensive SHM plan was designed and installed on a movable bridge in Florida. As a part of this ongoing project, this paper presents preliminary analysis results of field data collected from dynamic strain gage. Traffic induced strain data is investigated for reliability estimation and bridge rating and compared with analytical results obtained from a Finite Element (FE) model. Finally, findings and future work are discussed.

## SHM SYSTEM DESIGN FOR SUNRISE BRIDGE

Florida has a large population of movable bridges due to the waterways and coastal topography. Most of these bridges are owned by the Florida Department of Transportation (FDOT). The FDOT has an inventory of 98 movable bridges including 3 lift type, 94 bascule type, and 1 swing type bridges. The research presented in this paper was conducted on a bascule bridge over Florida Inter Coastal water way, known as Sunrise Bridge (Figure 2).

Figure 2: Sunrise Bridge

### Bridge Description

Sunrise Bridge was constructed in 1989. It has double bascule leaves, each 73'10" (22.49m) long approximately, and 53'4" (26.15 m) wide, carrying three traffic lanes and opening about 15 times a day. The bascule leaves are lifted horizontally at the point of the trunnions, which are the pivot points on the main girders. The weight of the span is balanced with a counterweight that minimizes the required torque to lift the leaf. In the closed position, the girder rests on a support referred as Live Load Shoe (LLS) on the pier and traffic loads are not transferred to the mechanical system. The movable bridge also involves fixed components, such as reinforced concrete piers and approach spans.

### Design of the sensor network

As a part of the ongoing research project for FDOT, main issues for the maintenance of electrical, mechanical and structural components of the movable bridge were identified. Based on these, an extensive sensor network is designed and implemented to monitor various parts of the bridge. A total of 168 sensors are deployed to the bridge for monitoring the electrical, mechanical and structural components as well as collecting environmental data. The electrical and mechanical components are monitored with accelerometers, strain rosettes, tiltmeters, microphones, infrared temperature sensors, ampmeters, video cameras, and pressure gages. Structural components are mainly monitored with accelerometers, high speed strain gages and slow speed vibrating wire strain gages. A video camera detecting the traffic and relating it with the other measurements is also installed. Finally, a weather station measures wind speed, wind direction, humidity, temperature, barometric pressure, and rain. Figure 3 shows the installed monitoring system and the sensor locations at the main girder. In this paper, the structural parts of the bridge are investigated by using the data coming from the strain gages at the main girders.

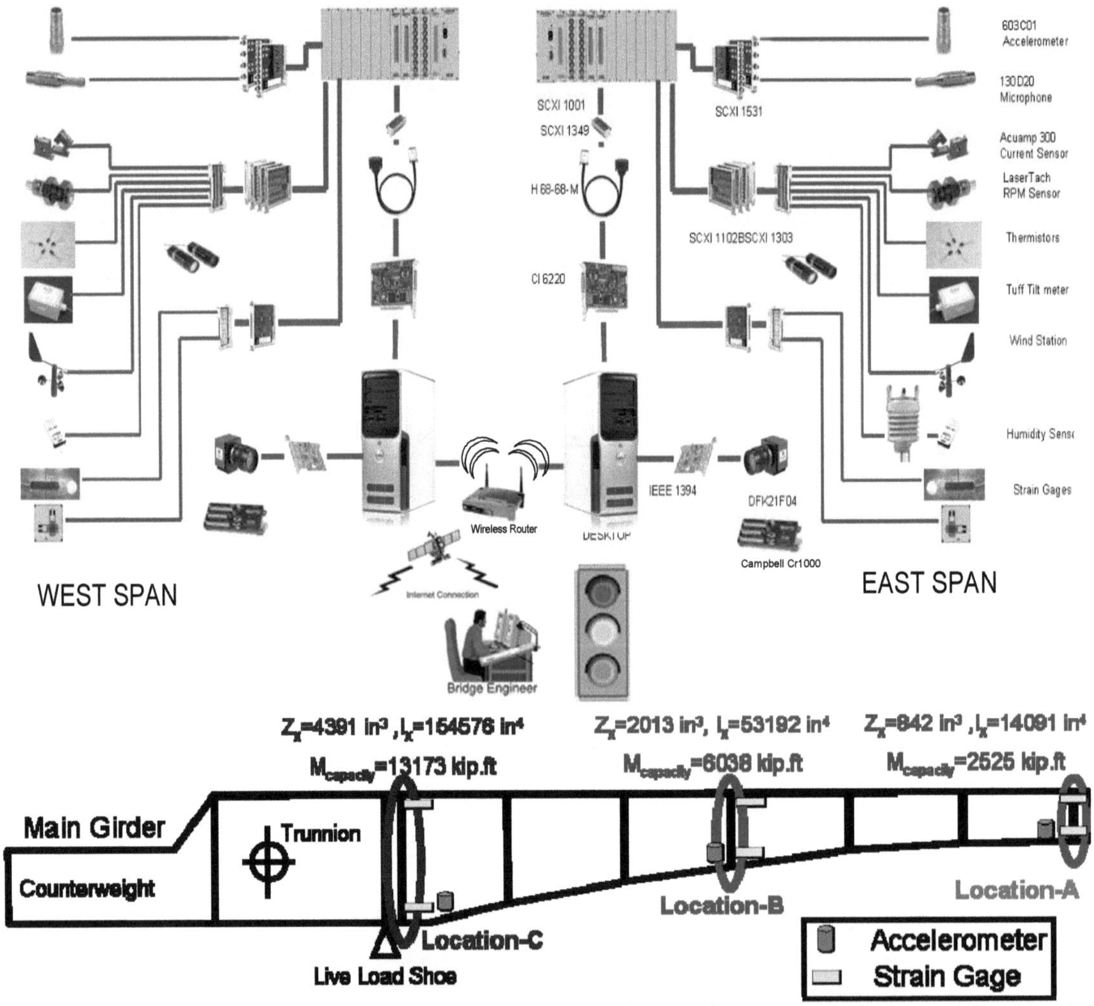


Figure 3: Sensor network used in the movable bridge project and sensor locations at the main girder

### *Finite Element Modeling of the Bridge*

The analytical model used in this study was developed using a finite element analysis program. Construction plans and details of the Sunrise Bridge were closely studied to ensure a proper modeling of the superstructure. There are a few main components of the bridge superstructure that are critical to accurately model the local behavior of the deck and secondary beams as well as the global behavior. The first main component of the bridge was the main girders where boundary conditions were imposed at the trunnion and live load shoe locations. The second main component of the system to be created was the floor beams, sidewalk brackets, and diagonal bracing which were composed of frame elements. These elements were connected to main girders and each other with rigid links at the centroids. Once all secondary beams were created, the deck of the bridge was constructed. The deck was modeled using 4-node quadrilateral shell elements and connected to the main girders and secondary beams using rigid links. Finally, solid elements were created to model the concrete counterweight. The model and its parameters can be seen in Figure 4. It was shown that all the static and dynamic comparison results of the FE model has very reasonable representation of the existing bridge even before rigorous model updating [10].

Figure 4: Final FEM of Sunrise Bridge

## DATA ANALYSIS AND LONG TERM MONITORING

### *Real-time Data Analysis*

The quality of test data is an important consideration in conducting modal identification on civil infrastructure. In addition to possible errors due to the experimental setup, the uncertainties due to incomplete knowledge of a structure's behavior and its interactions with the natural environment greatly affect the reliability of the system identification results.

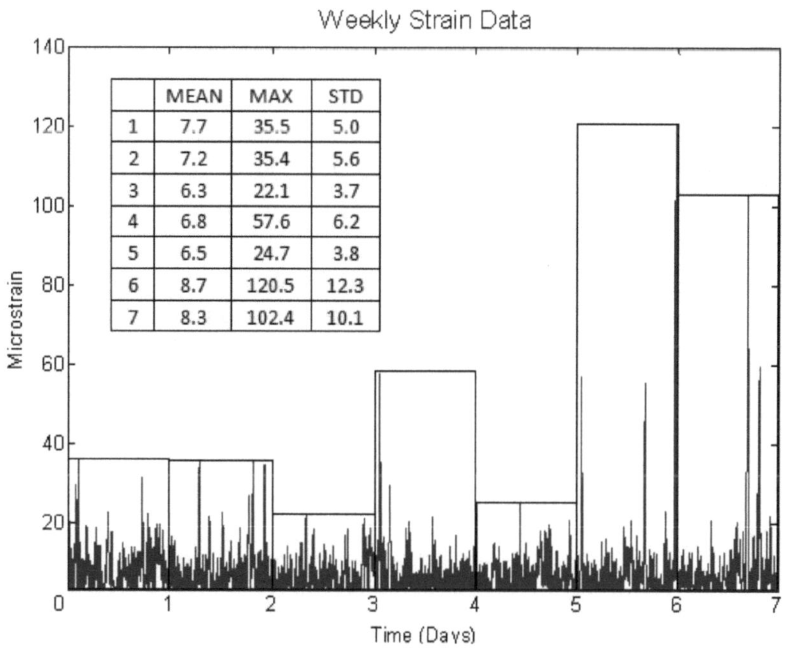


Figure 5: Weekly Strained Data Developed at Location C

In this study, the strain measurements obtained from the strain sensor at Location C which is shown in the Figure 3 was used to investigate the effect of real time data on bridge responses in terms of bridge rating and reliability. For this study, only strain responses for seven days are considered. The data is shown in Figure 5 with maximums, minimums and standard deviations determined for each day. As shown in the figure, the maximum of each day might differ six times as much as others depending on the intensity and type of vehicle crossing the bridge.

The concern for long-term strain monitoring is the peak values, since the maximum effects govern the structural capacity. Extreme values of strain are observed during heavy traffic or excessive environmental effects, that the bridge suffers through its life span. From the strain data, the maximum strain values were picked for each data block of 1-day duration as shown in Figure 5. Considering the data collected, the histogram in the Figure 6 was generated to see the strain frequencies at defined section.

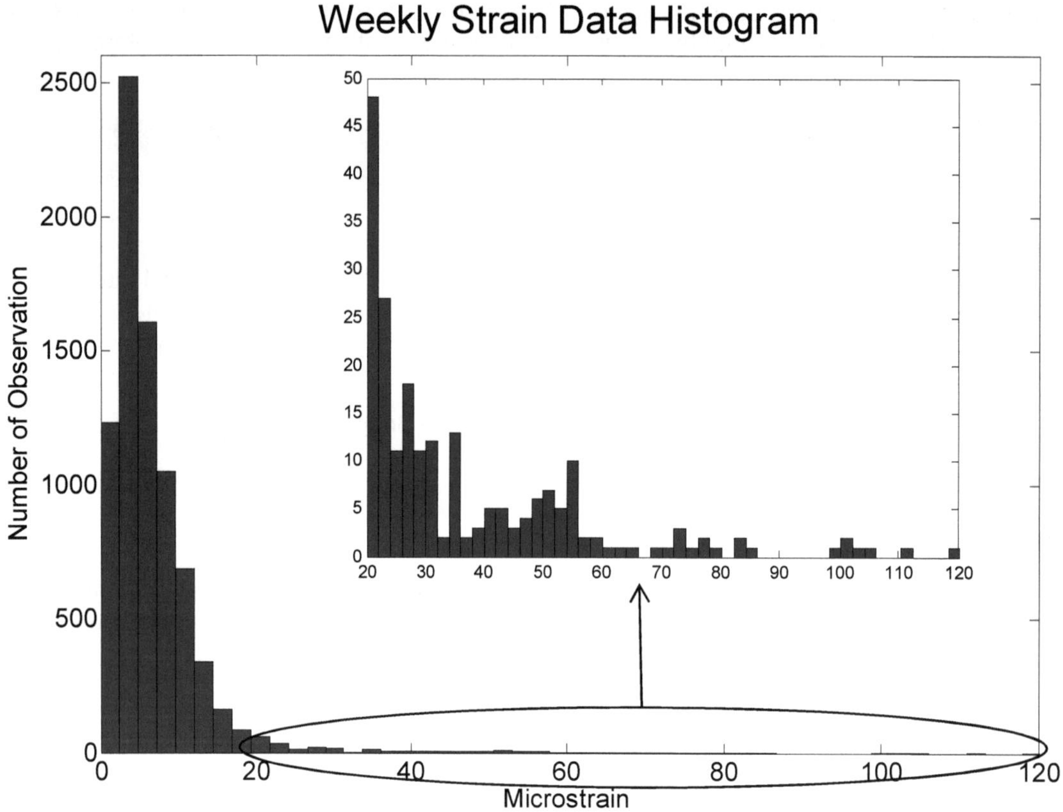


Figure 6: Frequency of Weekly Strained Data Developed at Location C

***Long Term Deterioration Effect***

All structures undergo certain deterioration effects due to normal aging and overloads. For maintenance operations and bridge safety, it is crucial to estimate the future condition of bridges. Corrosion penetration and associated section loss was considered as the main factor in capacity degradation over the lifetime of the bridge. Section loss due to corrosion is a major cause of deterioration for steel profiles, especially at locations closer to salt water and with high humidity. Numerous studies exist in the literature have setup the framework for reliability analysis under deteriorating bridge sections and multiple limit states. The following corrosion penetration model was used in the simulation to define the degradation effect over time:

$$C(t) = A \times t^{B} \tag{1}$$

where C(t) is the corrosion penetration depth in 10-6 m, t is time in years, and A and B are statistical random variables obtained from [10]. The statistical parameters of A and B are given in **Table 1**.

Table 1: Parameters of the Corrosion Model

| Random Variabale | A | B |
|---|---|---|
| Interior Girders | | |
| Mean Value | 34.0 | 0.65 |
| Coefficient Variation | 0.09 | 0.10 |
| Correlation Coefficient | 0.0 | |
| Exterior and Interior- Exterior Girders | | |
| Mean Value | 80.2 | 0.593 |
| Coefficient Variation | 0.42 | 0.40 |
| Correlation Coefficient | 0.68 | |

The corrosion penetration was assumed to be following the pattern shown in Figure 7. According to this pattern, due to pooling, the corrosion progresses along the top surface of the bottom flange and ¼ of the depth of the web. A similar example was presented in Estes and Frangopol [11].

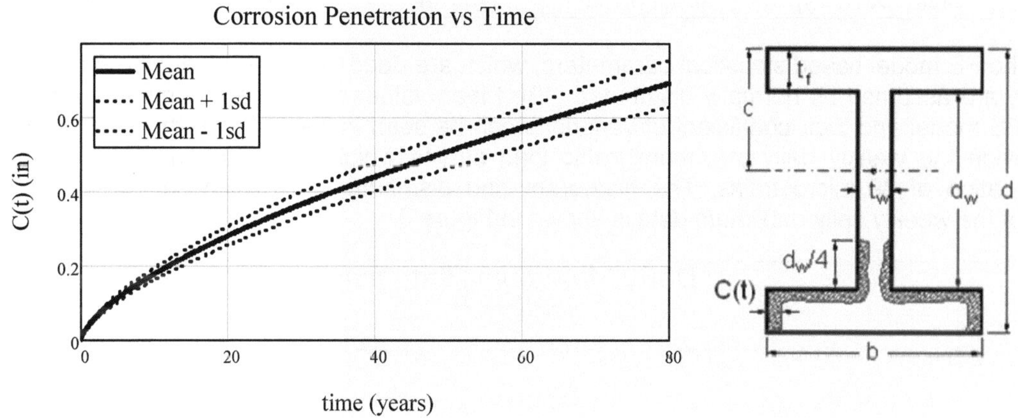


Figure 7: Corrosion penetration model and corrosion pattern

Using the corrosion penetration function and the corrosion pattern, loss of section can be calculated. Mean value and standard deviation for the cross-sectional area was evaluated using a linear approximation with Taylor Series. Similarly, mean value and standard deviation of neutral axis location and the elastic section modulus were calculated with respect to time. Degradation of the elastic section modulus is shown in Figure 8. Therefore, the long-term deterioration effects due to corrosion were modeled as statistical variables. It is noted that in the model the uncertainty increases with time, as expected.

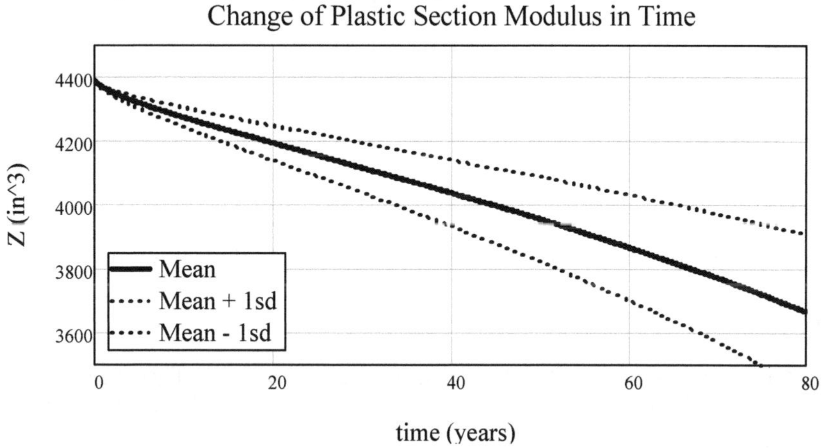


Figure 8: Deterioration of plastic section modulus over time

## ASSESSMENT OF THE BRIDGE

### *Statistical Quantities*

Probabilistic modeling of the structural capacity and load effects requires determination of the statistical parameters representing the uncertainties in the calculation of each parameter. Sources of variability are generally categorized as material factor (material properties), fabrication factor (imperfections) and analysis factor (assumptions, approximations) [11], and these have been quantified for many cases by statistical studies in the literature. At least the mean value and the variance (or standard deviation, or coefficient of variation) should be known for reliability analysis. In this study, statistical distributions are used for the random variables.

Table 2: Statistical Parameters

| *FE Model Based* | Plastic Strain Capacity | Dead Load Strain | HL-93 Truck Load Strain | Lane Load Strain |
|---|---|---|---|---|
| Mean (microstrain) | 1410 | 138 | 168 | 87 |
| c.o.v. | 0.11 | 0.10 | 0.18 | 0.18 |
| Distribution Type | Lognormal | Normal | Normal | Normal |

For the FE model based statistical parameters, which are dead load, live load (HL-93 truck and lane load) and capacity, are assumed as normally distributed [12]. Mean values for these parameters which are identified through the FE model and their coefficient of variations can be seen in Table 2. On the other hand, SHM based data will provide the weekly daily maximum traffic load strains which has a mean of 57 microstrains and a standard deviation of 39 microstrains. The histogram and distribution fit (which can be assumed lognormal distribution) of the weekly daily maximum data is shown in Figure 9.

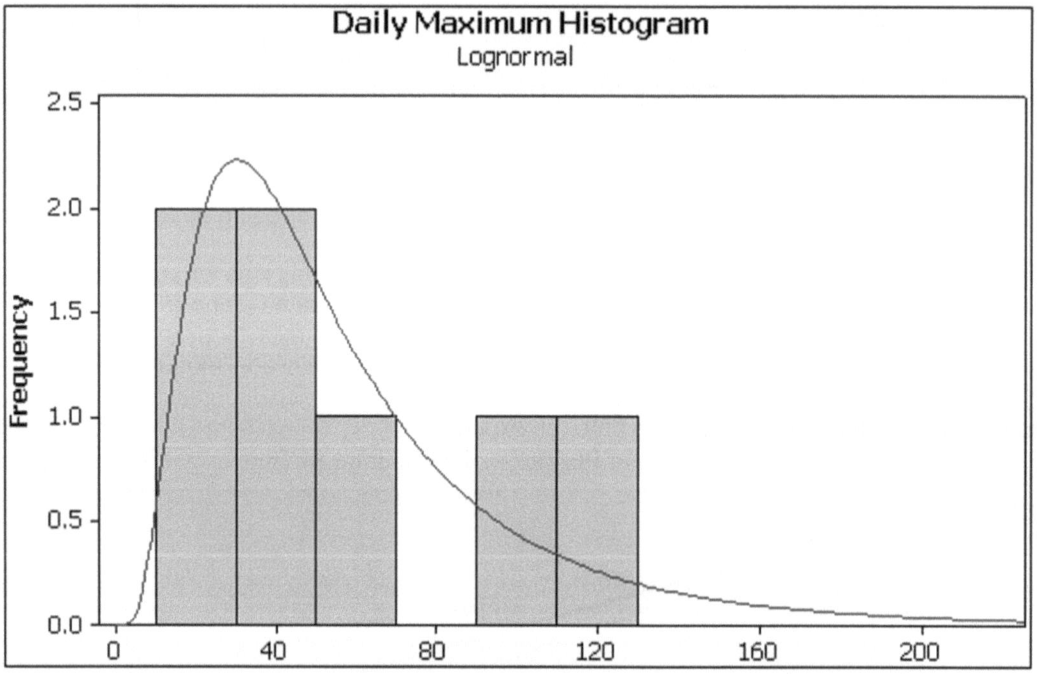


Figure 9: Histogram and distribution fit of daily maximum data

### *Component Reliability*

Component reliability indices are calculated based on strain values. The main girders constitute the main load carrying components; therefore, they are the most critical members. The performance function according to the plastic strain capacity is;

$$g(\varepsilon_C, \varepsilon_{DL}, \varepsilon_{LL}) = \varepsilon_C - (\varepsilon_{DL} + \varepsilon_{LL}) \quad (2)$$

The defined limit state functions are evaluated at each joint to obtain the probability of failure. Reliability index is defined as the distance from the origin to the failure surface in terms of standard variates. For linear limit state functions, such as;

$$g(X_1, X_2, \ldots, X_n) = a_0 + a_1X_1 + a_2X_2 + \ldots + a_nX_n = a_0 + \sum_{i=1}^{n} a_iX_i \tag{3}$$

The reliability index can be calculated simply with the following [13];

$$\beta = \frac{a_0 + \sum_{i=1}^{n} a_i \mu_{X_i}}{\sqrt{\sum_{i=1}^{n} (a_i \sigma_{X_i})^2}} \tag{4}$$

The design point is calculated to be the minimum distance from the origin to the limit state surface in reduced variates space. Therefore, iteration is required for finding the design point. According to the procedure of the Rackwitz-Fiessler method [14], equivalent normal statistical parameters of the non-normal random variables are calculated. Since the strain capacity and the traffic load strain coming from the SHM data are assumed as lognormally distributed (Figure 9).

Reliability index for each year is found for 75 years, which is the assumed lifetime of the bridge. The result shows the change of reliability index based on the strain measurements, over 75 years. The decrease in the reliability index is a result of deterioration effects and traffic increase. Deterioration effects are the decrease in the capacity and the decrease in the dead load whereas the traffic load is increasing with years. Traffic load increase is calculated from [15] which indicate that the expected 75 year maximum traffic load is as follows:

$$\Delta_{increase} = mean + 5.33 \times std.\ deviation \tag{5}$$

FE model based and SHM based reliability index variation over time is shown in Figure 10 and Figure 11. The prediction of the reliability index over the lifetime of a bridge enables maintenance decisions carried out more effectively. The bigger impact will be on the bridge safety, since the reliability curve based on strain measurements indicates the remaining life and reflects any structural problems that the bridge undergoes, therefore, reducing unexpected failures.

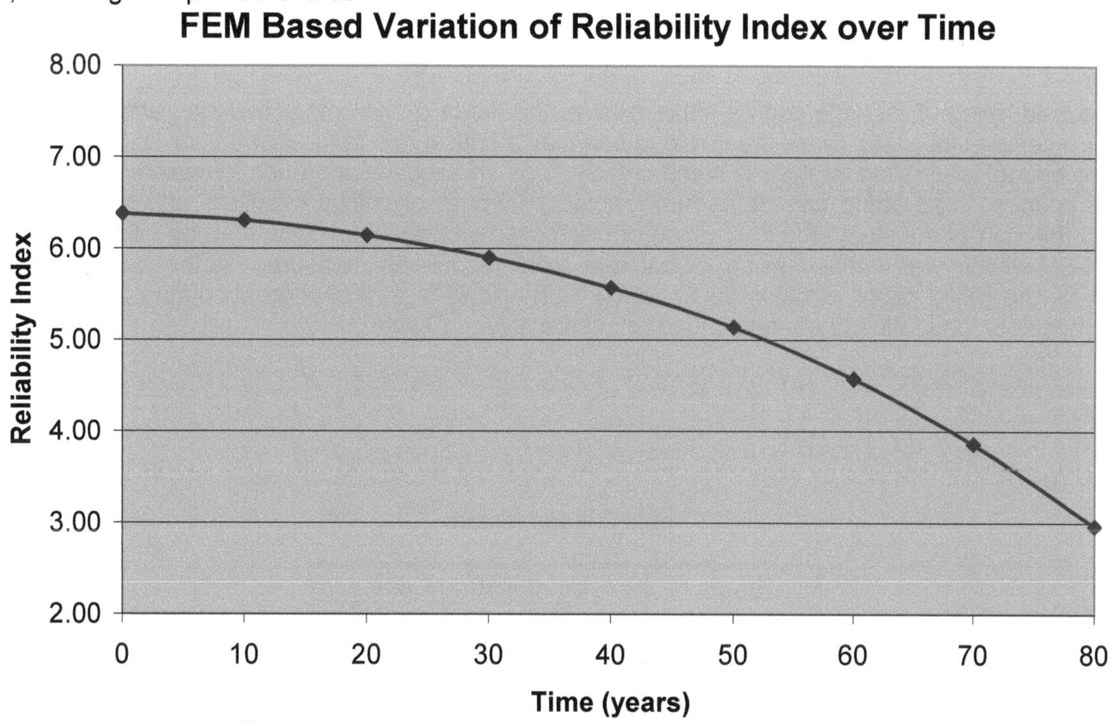


Figure 10: FE model based variation of reliability index over time

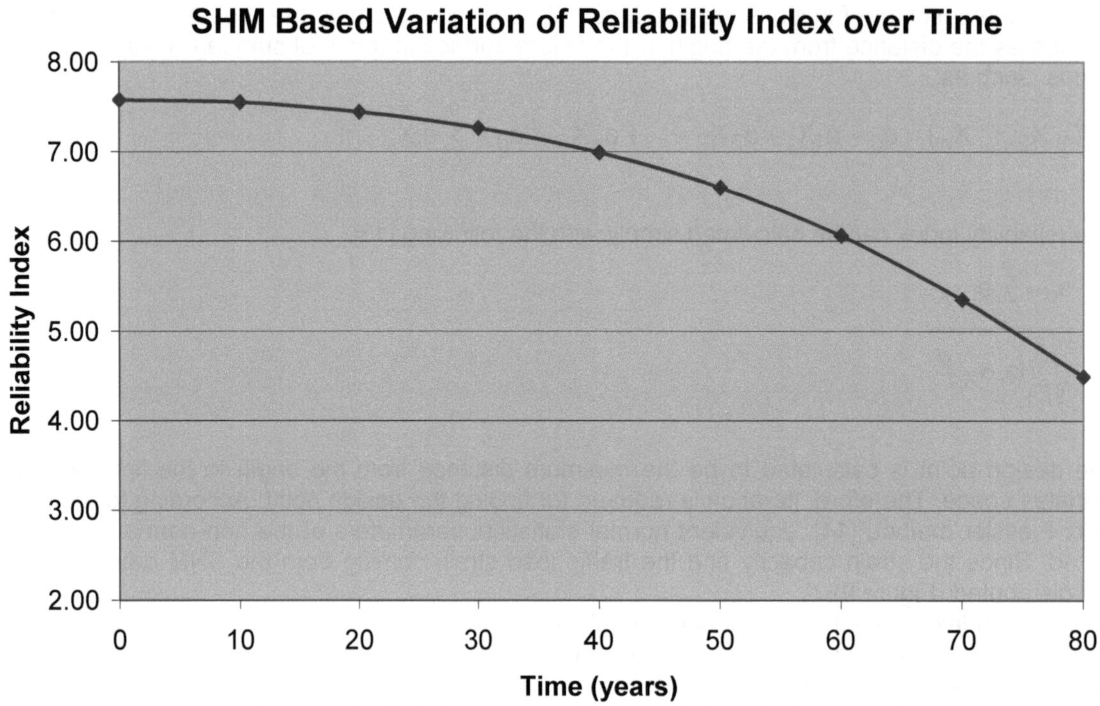


Figure 11: SHM based variation of reliability index over time

Both graphs (Figure 10 and Figure 11) look like same because the only difference is initial traffic load are different. In Figure 10, one HL-93 truck and lane load is applied to the FE model whereas in Figure 11, weekly daily maximum strains determined using SHM system are used. SHM based strain data is much less than HL-93 truck plus lane load. However, this data and the corresponding reliability illustrate the estimation based on the measurements. The longer the measurement time, the more accurate the estimation will be. For the future traffic load, an increase is also applied to the design truck (HL-93) and design lane load because reliability index is independent from the load rating calculations. In load rating, the bridge designed under design loads for its life time period therefore, no design truck load increase needed for load rating. More discussion can be found in the next section.

***Load Rating***

The load rating of a bridge can be expressed as the factor of the critical live load effect to the available capacity for a certain limit state. Since the live load can play a critical role in the distribution of loads on a short to medium size bridge, load rating analysis is commonly used as an effective approach to evaluate live load carrying capacity of bridges. Load rating can be carried out for a number of critical locations and components of the structure. In this part of the study, the load rating of the movable bridge was calculated by following the [16] and using the FEM which was verified using global and local responses measured at the bridge. The general formulation for the rating factor is shown in Equation 6. The load factors change according to the type of load rating, i.e., inventory load rating or operating load rating are given in Table 3.

$$RF = \frac{C - \gamma_{DC} DC - \gamma_{DW} DW \pm \gamma_p P}{\gamma_L LL(1 + IM)} \tag{6}$$

Table 3: Load factors

| Load Factor (γ) | Load Rating Case | |
|---|---|---|
| | Inventory | Operating |
| DC | 1.25 | 1.25 |
| DW | 1.25 | 1.25 |
| LL+IM | 1.75 | 1.35 |

For this study, moving truck load (HL-93) was simulated on the FE model and the live load moments were obtained for each truck location. Main girders were selected for load rating instead of floor beams because main girders are more critical from the structural system perspective. The transverse cross-section of the deck (Figure 12) shows the location of the lanes, design lane load and axle position for the HL-93 standard truck loading whereas Figure 13 shows the critical loading scenario for the truck simulation.

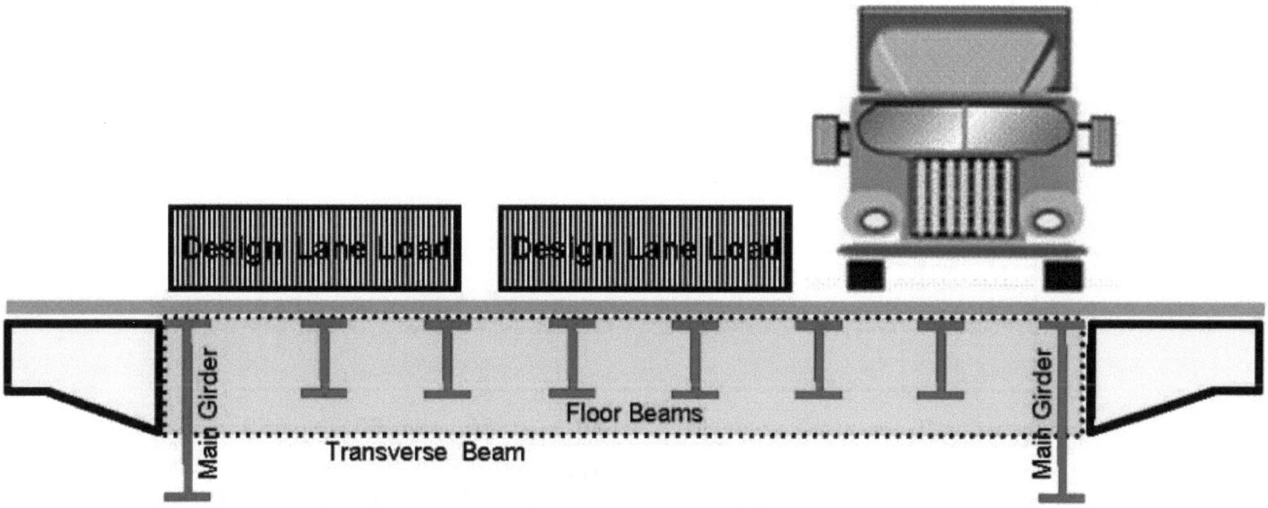


Figure 12: Vehicle loading for the Bridge

Since a 3D FE model was used, axle loads were defined as individual point loads and lane loads were defined as a distributed load equivalent to 0.64 kip/ft as mentioned in [16]. The yield strength of the steel is taken as 36.0 ksi. The dynamic impact factor is used as 33% for both inventory and operating ratings.

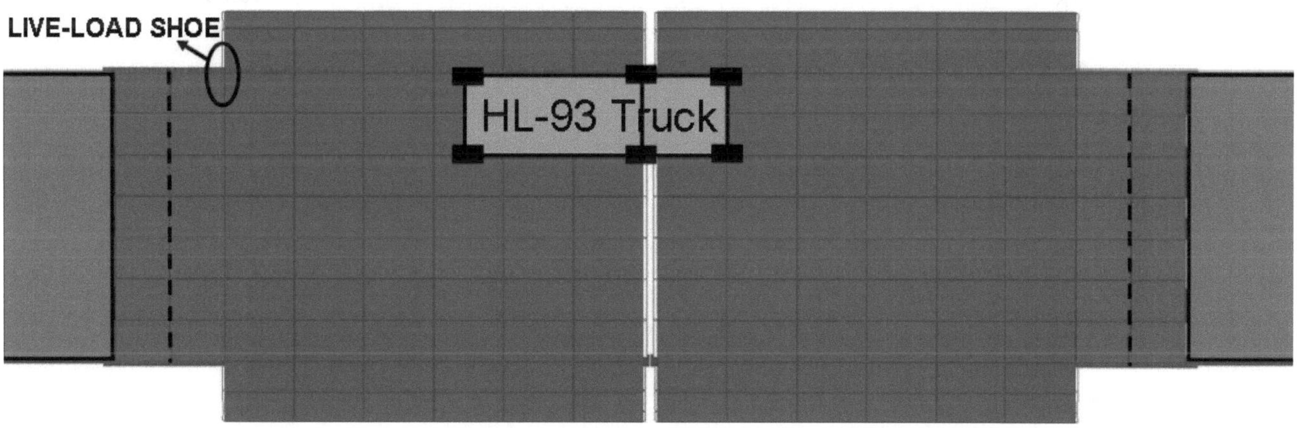


Figure 13: Critical HL-93 truck loading for the Bridge

The load ratings based on the HL-93 truck on the East North main girder were calculated at the live load shoe because it is seen that the cross-section at the live load shoe (Location-C) is more critical than the sections at Location-A and Location-B due to cantilever type configuration of the movable bridges (Figure 3).The capacities of the sections were calculated based on the ultimate moment capacity, which can be obtained by multiplication of the yield strength ($F_y$) and plastic section modulus ($Z_x$). A moving load simulation is also conducted to determine the load rating at the most critical location as function of truck location. The most critical load rating was seen at Location-C when truck position's front axle is at 84 ft., as shown in Figure 13. It should be noted that the corrosion parameters are also analyzed for the section at Location-C.

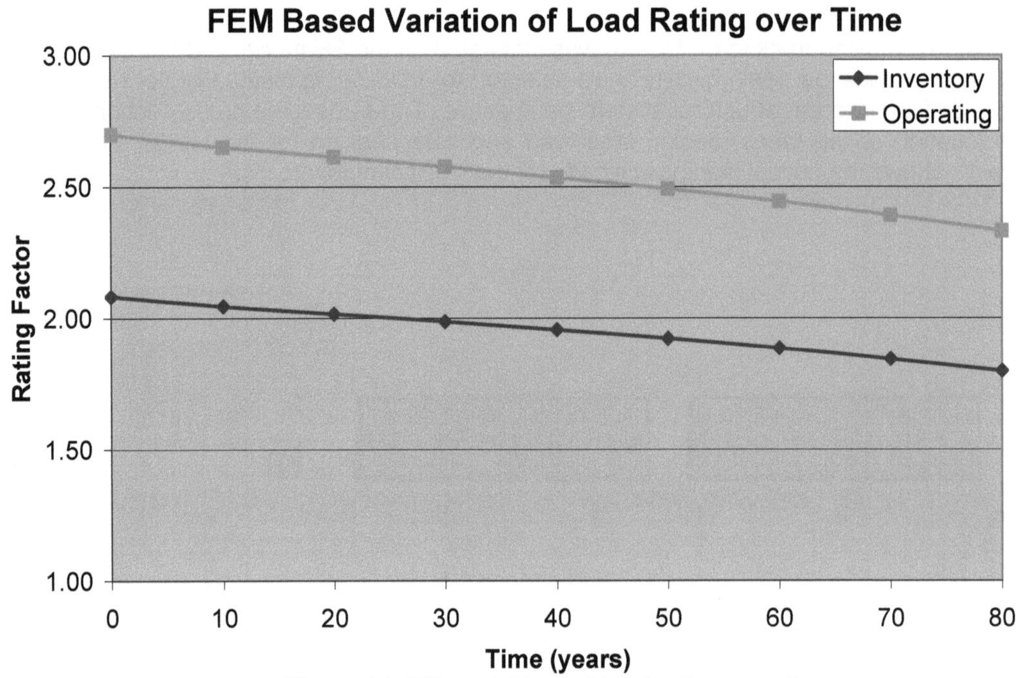


Figure 14: FE model based load rating over time

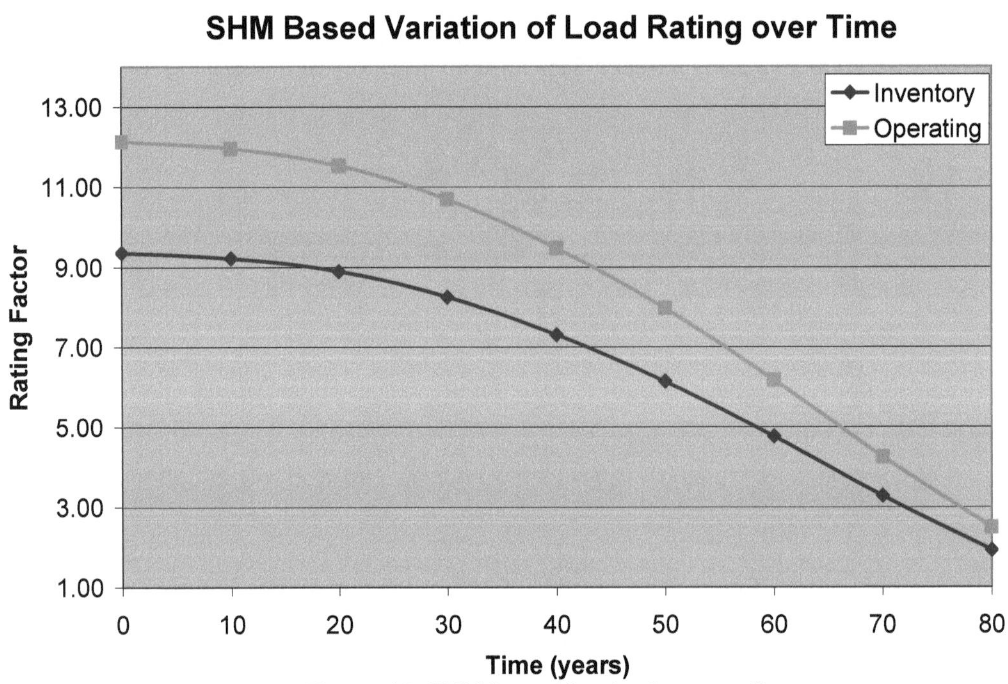


Figure 15: SHM based load rating over time

FE model based and SHM based load rating variation over time is shown in Figure 14 and Figure 15.It should be noted that for the FE model based load rating there is no traffic increase because the load rating of a bridge is designed for 3.5 and bridge is expected to survive for 75 years with this initial load rating. Therefore, Figure 14 shows only the deterioration effects on the load rating for FE model based analysis. In Figure 15, the deterioration effect and the traffic load increase can be seen. Load rating starts with a higher value because the SHM based strain data is lower than the design truck and lane load strain data. Not only deterioration effect, but also traffic load increase in lifetime period of the bridge SHM based variation of the load rating is decreasing to 2.81 for the operating load rating and 2.17 for the inventory load rating (Figure 15).

## CONCLUSIONS AND RECOMMENDATIONS

In this study, a detailed FE model of the Sunrise Boulevard Bridge is created to obtain the dead load strain, HL-93 truck load strain and lane load strain. Also the capacity strain of the analyzed section is calculated from the FE model. With the help of section properties a deterioration model is used for the section and the deterioration values are obtained for the life time (75 years) of the bridge. Then, based on thedeterioration effects on plastic section modulus, neutral axis, moment of inertia and section area are calculated again for the life time of the bridge. Due to these deteriorations and the possible traffic load increase in the future reliability and load rating of the bridge is investigated.

There are pros and cons for bridge condition assessment techniques. The main disadvantages of reliability analysis is the need for long term data to obtain a distribution for the variables along with higher level calculations. On the other hand, load rating is easy to calculate when a quick analysis is needed, .

The study shows that structural health monitoring data is very applicable to reliability analysis as well as the load rating of bridge structures. The prediction tools used for determining the time-variant reliability are based on previous statistical studies on corrosion, which has to be updated, as more data from monitoring is available.

The recommended future studies will be to collect long term monitoring data for better describing the probability distributions (the form as well as the parameters) that constitute the epistemic and aleatory uncertainties. Moreover, for this study, only component reliability index are investigated therefore, and the system reliability estimations will be obtained in future studies by incorporating different elements and different failure modes.

## ACKNOWLEDGEMENTS

The research project described in this paper is supported by the Florida Department of Transportation (FDOT) Contract # BD548/RPWO 23 and Federal Highway Administration (FHWA) Cooperative Agreement Award DTFH61-07-H-00040. The authors would like to thank Mr. Marcus Ansley, P.E., the Head of Structures Research at FDOT for his support and guidance throughout the project. The writers also greatly appreciate the valuable feedback provided by Mr. Alberto Sardinas at FDOT District Four, who has shared his experience. The authors would like to express their profound gratitude to Dr. Hamid Ghasemi of FHWA for his support of this research. The support of both agencies and their engineers is greatly recognized and appreciated. The authors would also like to acknowledge the following for their contributions of several other colleagues and students. The opinions, findings, and conclusions expressed in this publication are those of the authors and do not necessarily reflect the views of the sponsoring organizations.

## REFERENCES

[1] Catbas FN, Ciloglu SK, Celebioglu A, Popovics JS, Aktan AE. Fleet Monitoring of Large Populations: Aged Concrete T-Beam Bridges in Pennsylvania. In, Proceedings of SPIE, Health Monitoring of Civil Infrastructure Systems. 2001, pp. 500-509.

[2] Liu SC, Tomizuka M. Vision and Strategy for Sensors and Smart Structures Technology Research. In, Proceedings of the 4th International Workshop on Structural Health Monitoring. Stanford, CA; 2003, pp. 42-52.

[3] Aktan AE, Pervizpour M, Catbas FN, Barrish RA, Grimmelsman KA, Qin X, Kulcu E, Ciloglu SK, Curtis J, Haza- Radlitz GV. Integrated Field, Theoretical and Laboratory Research for Solving Large System Identification Problems. In, Advances in Structural Dynamics. Hong Kong; 2000.

[4] Aktan AE, Catbas FN, Grimmelsman KA, Tsikos CJ. Issues in Infrastructure Health Monitoring for Management. Journal of Engineering Mechanics, ASCE 2000; 126(7): 711-724.

[5] Susoy M, Catbas FN, Frangopol DM. Implementation of Structural Reliability Concepts for Structural Health Monitoring. In, Proceedings of the 4th World Conference on Structural Control and Monitoring. San Diego, CA; 2006.

[6] Catbas FN, Frangopol DM. Concepts and Issues of Structural Health Monitoring for Structural Reliability and Decision Making. In, Proceedings of the World Forum on Smart Materials and Smart Structures Technology. Chongqing & Nanjing, China; 2007.

[7] Messervey TB, Frangopol DM, Estes AC. Reliability-based Life-cycle Bridge Management using Structural Health Monitoring. In: Cruz PJS, Frangopol DM, Neves LC, eds., Bridge Maintenance, Safety, Management, Life-Cycle Performance and Cost. 2006, 545-546, and full paper on CD-ROM, 2007.

[8] Hua XG, Ni YQ, Ko JM. Structural Reliability Assessment using Long-term Monitoring Data. In, World Forum on Smart Materials and Smart Structures Technology (SMSST). Chongqing & Nanjing, China; 2007.
[9] Estes AC, Frangopol DM. Updating Bridge Reliability Based on Bridge Management Systems Visual Inspection Results. Journal of Bridge Engineering 2003; 8(6): 374-382.
[10] Catbas FN, Dumlupinar T, Gokce HB, Gul M, Zaurin R. Structural Health Monitoring Case Study on a Movable Bridge, Transportation Research Board Conference Proceedings 2010.
[11] Nowak AS, Collins KR. Reliability of Structures: McGraw-Hill 2000.
[12] Nowak AS. Calibration of LRFD Bridge Code. ASCE Journal of Structural Engineering 1995; 121(8): 1245-1251.
[13] Ang, A.H.-S. and Tang, W.H. 1984. Probability Concepts in Engineering Planning and Design Vol.II. New York: Wiley.
[14] Rackwitz R, Fiessler B. Structural Reliability under Combined Random Load Sequences. Computers and Structures 1978; 9(489-494).
[15] National Cooperative Highway Research Program (NCHRP), Report 454 (2001).
[16] American Association of State Highway and Transportation Officials (AASHTO). (1998). LRFD bridge design specifications, 2nd Ed., Washington, D.C.

Proceedings of the IMAC-XXVIII
February 1–4, 2010, Jacksonville, Florida USA


# Aerodynamic performance of bridges equipped small wind turbines

Soon-Duck Kwon
Associate Professor, Dept. of Civil Eng., Chonbuk National University
664-14 Dukjin, Chonju, Chonbuk, 561-756, South Korea

Seung-Ho Lee and Han Kyu Lee
Research Assistant, Dept. of Civil Eng., Chonbuk National University
664-14 Dukjin, Chonju, Chonbuk, 561-756, South Korea

## ABSTRACT

This study addresses a methodology to use small wind turbines for dual purposes, improving aerodynamic performance of flexible bridges and wind energy harvesting. A way to proper placement of small wind turbines on flexible bridges were proposed according on the analogy of conventional aerodynamic appendages. From the wind tunnel tests of the Rio-Niteroi bridge in Brazil, it was found that the wind turbine attached like fairing was effective to reduce the vortex-induced vibration of bridge and the optimal spanwise interval of the wind turbine was 3~4.5 time of turbine diameter. Moreover the aerodynamic coefficients of the bridge were improved after installation of the wind turbines. Present results showed the general availability of wind turbine for improvement of aerodynamic performance and energy supply of flexible bridges although the capacity of wind power generation was strongly dependent on wind characteristics of the bridge site.

## INTRODUCTION

Wind-induced vibrations are often critical in the safety and serviceability of long-span bridges. Wind tunnel testing technique has been commonly used to predict such vibrations. If undesirable vibrations have observed during the wind tunnel tests, the outer shape of the bridge section has been generally changed to improve the aerodynamic performance of bridges.

Although wind tunnel testing was performed prior to design, some bridges have been suffered after completion because of vortex-induced vibration or poor aerodynamic stability. The vortex-induced vibration has been actually observed in the Erskine cable-stayed bridge in UK [1], The Kessock cable-stayed bridge in UK [2], Bronx-Whitestone suspension bridge in USA [3], the Great Belt suspension bridge in Denmark [4], the Rio-Niteroi steel box girder bridge in Brazil [5], Trans-Tokyo Bay Crossing steel box girder bridge in Japan [6], and so on.

In order to mitigate the wind-induced vibrations, the aerodynamic means have been commonly used in design of new bridges, yet either aerodynamic countermeasures or mechanical countermeasures has been adopted for retrofit of existing structures. For aerodynamic countermeasures, aerodynamic appendages such as faring, guide vane, splitter, etc have successfully been used for this purpose by streamlining the sections.

The elevated bridges are good place for energy harvest because of the accelerated wind velocity at boundary layer flow. However improper placement of wind turbines, a bluff appendage, may deteriorate the aerodynamic performance of flexible bridges. Even though there have been little attempts to install wind turbines on bridge structures [7], those were limited to short span bridges which could not be oscillated due to winds.

This paper presents a solution to use small wind turbines for dual purposes, improving aerodynamic performance of flexible bridges and wind energy harvesting. A way to proper placement of small wind turbines on flexible bridges were proposed according on the analogy of conventional aerodynamic appendages. To verify the effectiveness of the proposed method, the wind tunnel tests were performed for a continuous steel box girder similar to the Rio-Niteroi Bridge. Moreover the annual energy production (AEP) of the wind turbines was also presented from the probability model of wind speed.

## PLACEMENT OF WIND TURBINES ON BRIDGES

As mentioned previously, the aerodynamic performance of flexible bridges may be deteriorated after installation of wind turbines. Therefore a feasibility study is required before applying the wind turbines into the extremely flexible cable-supported bridges. The Rio-Niteroi bridge in Brazil [5] shown in Figure 1 is used in present study. This three continuous spans (200m+300m+200m) bridge built in 1974 consists of steel twin box girders, and stands 65m above sea level. The girder height is 7.42m at the middle of center span.

For sustained cross winds around 15~16.5m/s, the Rio-Niteroi bridge experienced for many years the vortex-induced vertical vibrations with amplitude reaching ±25cm in the middle of center span [8]. Battista et al. [8] reported design and installation of the tuned mass dampers to prevent the vibrations. As an alternative for the mechanical vibration control system, the wind turbines have been studied in present study for both aerodynamic countermeasure and energy supply.

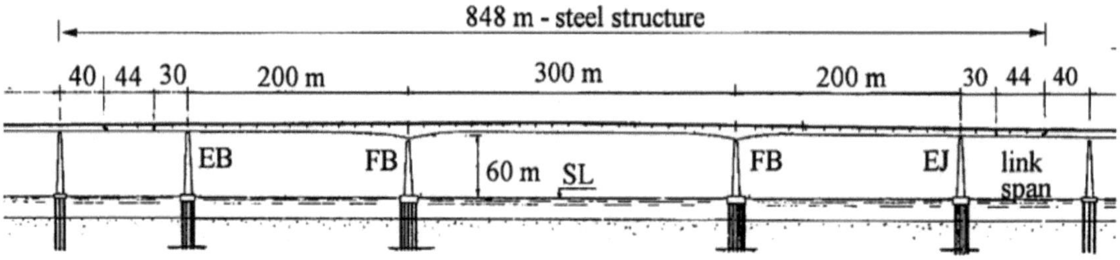


Figure 1: The Rio-Niteroi bridge in Brazil [5]

Controlling the aerodynamic vibrations, passive flow control strategies are commonly selected because of their reliability, simplicity, cost effectiveness and easy maintenance. The extensive examples for improvement of aerodynamic performance for actual bridges were provided by Wardlaw [9], JSCE [10], and Simiu and Miyata [11]. Figure 2 shows typical aerodynamic control means used in deep girder bridges. Simple aerodynamic appendages to the leading and trailing edges such as fairings, flap, and horizontal plate or skirts have been used to stabilize galloping and vortex-induced vibrations.

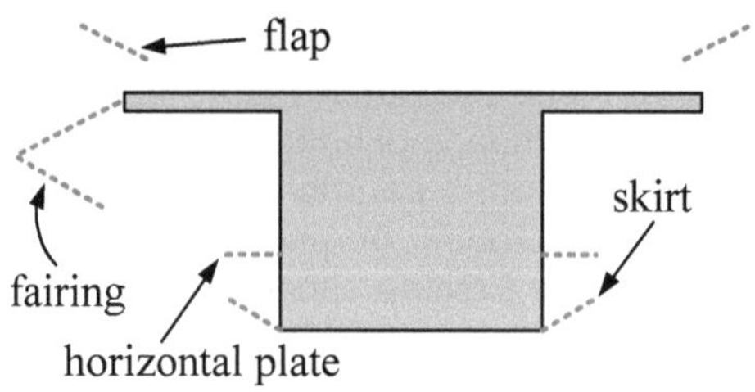


Figure 2: Aerodynamic improvements for deep girder bridges [9,10,11].

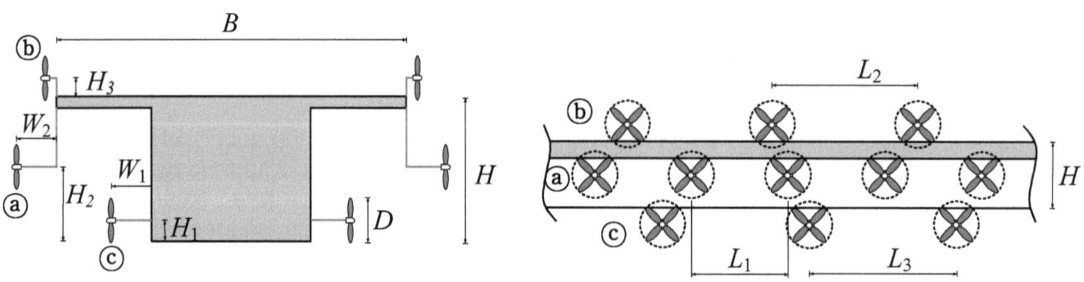


(a) Cross sectional arrangement (b) Spanwise arrangement

Figure 3: Placement of the small wind turbine on bridge girder

For making the wind turbine similar roles with the aerodynamic appendages, the following guidelines to attach the wind turbine to bridge girder can be proposed in present study. Figure 3(a) shows the general ways for placement of the small wind turbine on the analogy of the aerodynamic appendages.

ⓐ Fairing analogy : A small wind turbines attached at the leading and trailing edges may work like a fairing which suppresses flow separation and streamlines the girder.

ⓑ Flap analogy : A small wind turbines attached at the both edges of the deck floor may work like a flap which promotes reattachment to the upper surface of the girder.

ⓒ Horizontal plate analogy : A small wind turbines attached at the web of deep box girder may work like a horizontal plate (or skirt) which promotes reattachment to the lower surface of the girder.

Flexible bridge may not need to be fully equipped the aerodynamic appendages along the bridge axis. Sometimes partially installed the aerodynamic appendages might show better performance to suppress the vortex-induced vibrations. El-Gammal et al. [12] performed an experimental study to investigate the effectiveness of the spanwise sinusoidal perturbation method in controlling vortex-induced vibrations in a plate girder bridge section model. El-Gammal et al. [12] showed that the vortex-induced vibration was decreased proportional to the wave steepness. As shown in Figure 3(b), this concept can be applied to determine the spanwise interval of the wind turbines along bridge axis.

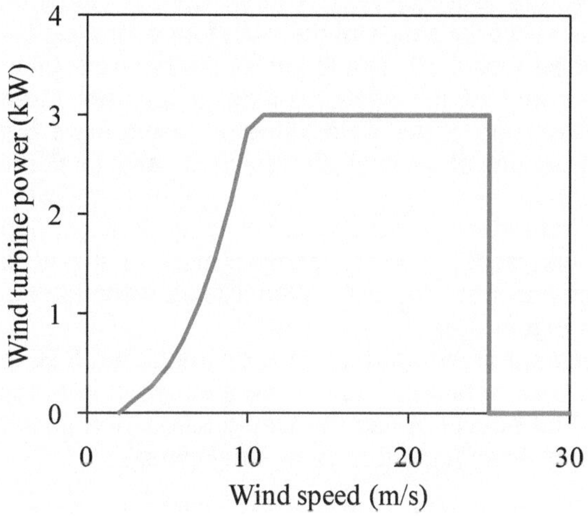

Figure 4: Power performance curve of the small wind turbine, TA36-3000

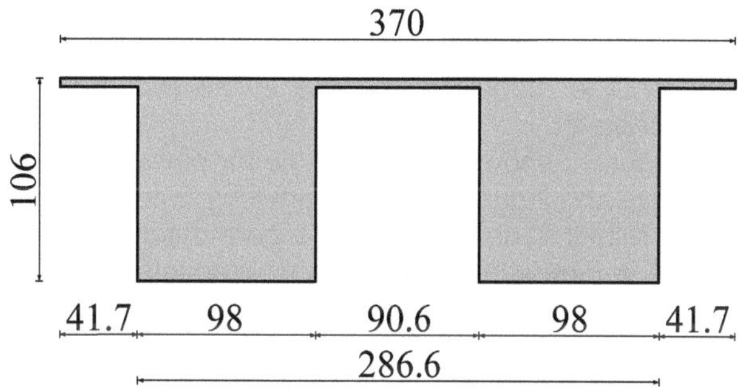

Figure 5: Wind tunnel model of deep girder for a continuous steel box bridge

## WIND ENERGY ASSESSMENT

The power in the wind is converted into the mechanical-rotational energy of a wind turbine rotor which would reduce the speed of the air mass. Therefore both wind data and wind turbine data are demanded to assess the wind energy potential. In wind energy assessment, the Weibull probability distribution is commonly used and widely adopted [13]. The probability density function, $f_V$, is given as follows.

$$f_V(V) = \frac{k}{c}\left(\frac{V}{c}\right)^{k-1} e^{-\left(\frac{V}{c}\right)^k} \tag{1}$$

where $V$ is wind speed, $c$ is the scale parameter, and $k$ is the shape parameter. The scale and shape parameter can be estimated by using the maximum likelihood method. To determine the probability of there being no wind that cannot be considered in the definition of the Weibull distribution, a slight modification is given to the number of hours for one year [14].

$$T = 365 \times 24 \times \{1 - f_V(V = 0)\} \tag{2}$$

where $f_V(V=0)$ is the probability of observation of null wind speed. Wind blowing at some directions may not be suitable for wind energy harvesting. In this case, the wind direction can be binned into several segments and corresponding probabilities of the wind speed can be evaluated [15]. The annual energy production (AEP) considering the wind direction is obtained in the following integration which is the output power at each wind speed, $P_{wt}(V)$, multiplied by the probability of the wind speed, $f_V$ and wind direction, $f_\theta$ [16].

$$E_{AEP} = T\sum_{j=1}^{n}\int_{0}^{\infty} P_{wt}(V) f_{V_j}(V) f_{\theta_j}(\theta) dV \tag{3}$$

In addition to the annual energy production, the availability factor and the capacity factor are open used to assess performance of wind turbine. The availability factor is defined as an operation percentage of a wind turbine. The capacity factor is ratio of actual energy output to the rated power of a wind turbine.

In present study, it was assumed that a small horizontal-axis wind turbine was attached in bridges. The technical data for the wind turbine was adopted from the TA36-3000 [17] which has the rated power of 3kW. Rotor diameter, cut-in speed, and rated wind speed of the wind turbine were respectively 3.6m, 2.7m/s, and 11m/s. Figure 4 shows the power performance curve for the wind turbine used in present study.

## WIND TUNNEL TESTS

The experiments were performed in a wind tunnel of the KOCED Wind Tunnel Center, Chonbuk National University, Korea. As shown in Figure 5, the cross sectional shape of the wind tunnel model was simplified from that of the Rio-Niteroi bridge. The model scale was 1:70. The length of the bridge section model was 0.9m, and end plates were attached at both end of the section model to prevent three dimensional air flow at the tips. The model was mounted outside of the wind tunnel by using eight coil springs. The model could move in the two degrees of freedom of vertical displacement and torsional displacement.

Figure 3 shows a general rule for placing the small wind turbine at deep girder. Fairing analogy, flap analogy, and horizontal plate analogy were applied to this model. For proper placement of the wind turbines in this deep girder, the parameters to be determined from the wind tunnel tests were $H_1$~$H_3$, $W_1$~$W_2$ in Figure 3(a), and the spanwise intervals, $L_1$~$L_3$ in Figure 3(b).

Table 1 show test cases and results based on arrangement of the wind turbines on the girder. SD0 in the table represents the girder without wind turbines. The parameters in Table 1 are expressed in terms of the wind turbine diameter ($D$) or the girder height ($H$). The effects of flap, horizontal plate, and fairing analogy were focused at first, and then extended to spanwise arrangement of those analogies.

**Table 1: Test cases and results for the wind turbines attached at deep girder (H: girder height, D: wind turbine diameter, WW: windward, LW: leeward).**

| Case | Wind turbines | | | | | | | | | Maximum amplitude(m) |
|---|---|---|---|---|---|---|---|---|---|---|
| | $H_1$ | $H_2$ | $H_3$ | $W_1$ | $W_2$ | $L_1$ | $L_2$ | $L_3$ | location | |
| SD0 | × | × | × | × | × | × | × | × | × | 0.34 |
| SD1 | × | × | 0.1$H$ | × | × | × | 3.5$D$ | × | WW+LW | 0.42 |
| SD2 | 0.5$D$ | × | × | 0.8$D$ | × | × | × | 3.5$D$ | WW+LW | 0.37 |
| SD3 | -0.5$D$ | × | × | 0.8$D$ | × | × | × | 3.5$D$ | WW+LW | 0.39 |
| SD4 | × | 0.5$H$ | × | × | 2$D$ | 1.7$D$ | × | × | WW | 0.29 |
| SD5 | × | 0.5$H$ | × | × | 2$D$ | 1.7$D$ | × | × | WW+LW | 0.19 |
| SD6 | × | 0.5$H$ | × | × | 2$D$ | 2.3$D$ | × | × | WW+LW | 0.15 |
| SD7 | × | 0.5$H$ | × | × | 2$D$ | 3.5$D$ | × | × | WW+LW | 0.10 |
| SD8 | × | 0.5$H$ | × | × | 2$D$ | 4.6$D$ | × | × | WW+LW | 0.08 |
| SD9 | × | 0.5$H$ | × | × | 2$D$ | 6.9$D$ | × | × | WW+LW | 0.26 |
| SD10 | × | 0.5$H$ | × | × | 1.5$D$ | 2.3$D$ | × | × | WW+LW | 0.15 |

| | | | | | | | | | | |
|---|---|---|---|---|---|---|---|---|---|---|
| SD11 | × | 0.5*H* | × | × | 1.5*D* | 3.5*D* | × | × | WW+LW | 0.15 |
| SC12 | × | 0.5*H* | × | × | 1.5*D* | 4.6*D* | × | × | WW+LW | 0.25 |
| SC13 | × | 0.5*H* | × | × | *D* | 3.5*D* | × | × | WW+LW | 0.29 |

Figure 6 show the test results for the wind turbine attached like upper flap, horizontal plate, and lower flap which are corresponding to ⓑ and ⓒ in Figure 3(a). As can be seen in the figure, these positioning of wind turbine did not decrease the vortex-induced vibration. When the wind turbine was attached at bridges like a fairing which is ⓐ in Figure 3(b), the test results are given in Figure 7. The reduction of the vortex-induced vibration is expected in case of fairing turbines installed at both windward and leeward rather than turbines at windward only.

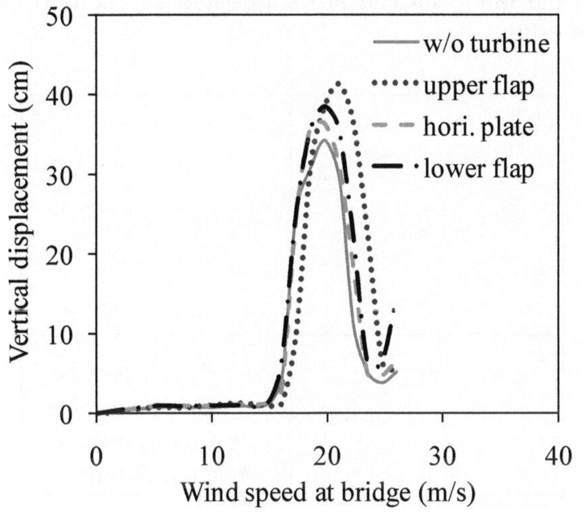


Figure 6: The results of upper flap, horizontal plate and lower flap analogy

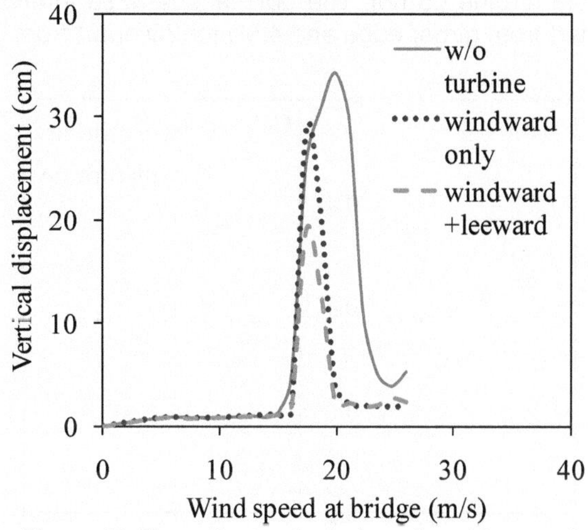


Figure 7: The effects of windward and leeward turbines for faring analogy

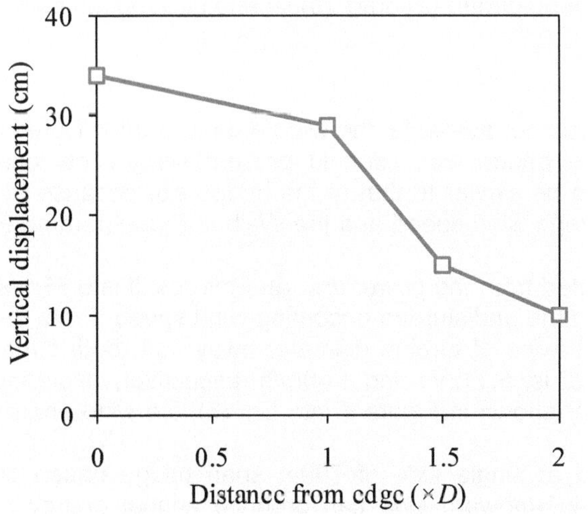


Figure 8: The effects of distance from girder edge for faring analogy

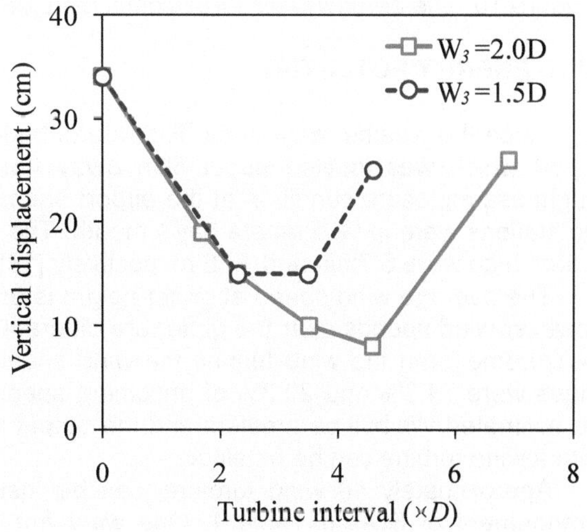


Figure 9: The effects of spanwise intervals for faring analogy

From the above results, it was found that fairing turbine was effective to reduce the vortex-induced vibration of the deep girder. A comprehensive study was performed for the fairing turbine. Figure 8 shows the control efficiency of faring turbine according to distance from girder edge. The peak displacements of the vortex-induced vibration were inversely proportion to the distance of faring turbine to girder edge. The test was limited at twice times of fan diameter because long support to install the fairing turbine is not desirable.

The effects of spanwise fairing interval are shown in Figure 9. The sparse placement of the wind turbines along bridge axis shows slightly better performance for suppressing the vibration compared with dense placement. The optimal spanwise intervals were respectively 3D and 4.5D when faring turbines were attached at 1.5D and 2D apart from girder edge.

The aerodynamic force coefficients for the deep girder with or without wind turbines are given in Figure 10. The wind turbine arrangement is SD8 in Table 1. The coefficients were normalized by girder width. It is generally known that negative slope of lift coefficient may not be desirable because of galloping instability [11]. The original section has negative slope of lift and moment coefficients, yet the slope has improved to positive at the bridge section with wind turbine.

The new finding from the experimental results can be summarized as follows. In deep girder, the fairing turbine contributes to decrease the vortex-induced vibration, yet the flap turbine and horizontal plate turbine do not. The optimal spanwise intervals of the faring turbine are suggested as 3D for 1.5D apart from girder edge and 4.5D for 2D apart from girder edge.

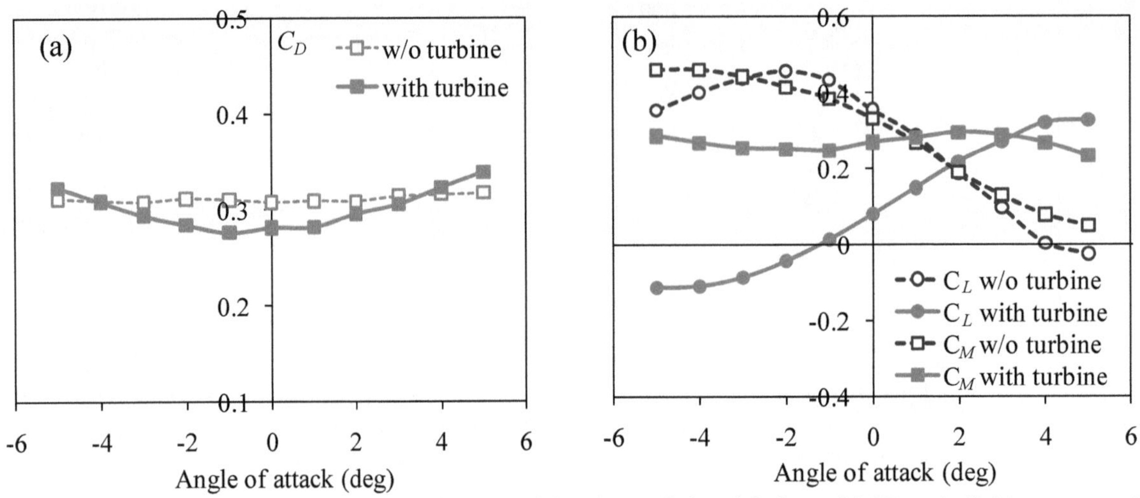


Figure 10: The aerodynamic force coefficients of the deep girder; (a) drag, (b) lift and pitching moment

**WIND ENERGY POTENTIAL**

Since the weather data at the Rio-Niteroi bridge was not available, the wind data at Santos Dumont airport which was located about 5km away from the bridge was used in present study. The local roughness exposure condition at the airport seemed to be similar to that of the bridge site because the two stations were in Guanabara bay's mouth. The average wind speed and the Weibull shape parameter at 50m high were 6.7m/s and 1.86 respectively [18].

The average wind speed at girder height is estimated from the power law, and the result is 6.88m/s. However wind speeds near the girder are different from the undisturbed oncoming wind speed. From the measurements in the wind tunnel, the wind speeds at twice of turbine diameter away from both girder edges were 74.3% and 20.3% of oncoming speed, that is, 5.11m/s and 1.40m/s respectively. Inserting the estimated Weibull parameters and the output power shown in Figure 4 into Eq. (3), the wind energy from a wind turbine can be obtained.

Approximately 40 wind turbines can be installed at single side of 700m span bridge based on arrangement of SD8 in Table 1. One wind turbine at windward side can produce annual energy of 6.44MWh. However a leeward wind turbine can only produce annual energy of 0.08MWh because of very low wind speed behind the girder. Total annual energy produced by 80 wind turbines equipped at both side of deep girder is expected as 260.9MWh. In this bridge, the availability factor and the capacity factor for entire year are respectively 0.783 and 0.245 for windward turbine. The results show the general availability of wind turbine characteristics. However those are 0.067 and 0.003 for leeward turbine respectively. It is clear from the factors that wind turbines at leeward side may not contribute energy production but aerodynamic performance.

**CONCLUSIONS**

This paper presents a solution to use small wind turbines for improving aerodynamic performance of flexible bridges as well as wind energy harvesting. A way to proper placement of small wind turbines on flexible bridges were proposed according on the analogy of conventional aerodynamic appendages. To verify the effectiveness of the proposed method, the wind tunnel tests were performed for deep bridge girder. Moreover the annual energy productions of the wind turbines were evaluated from the wind data at the bridge and its probability model.

From the wind tunnel tests of deep bridge girder, it was found that the fairing turbine contributed to decrease the vortex-induced vibration, yet the flap turbine and horizontal plate turbine did not. The optimal spanwise intervals of the faring turbine are suggested as 3D for 1.5D apart from girder edge and 4.5D for 2D apart from girder edge. The annual energy production of the deep girder bridge was evaluated as 260.9MWh. Since the wind velocities around the girder where the wind turbines were installed were lower than oncoming free stream velocity, the energy efficiency was not much high. This results show the general availability of wind turbine for improvement of aerodynamic performance and energy supply of bridges although the capacity of wind power generation was strongly dependent on wind characteristics of the bridge site.

**ACKNOWLEDGMENT**

This research was supported by the grant (09CCTI-A052531-02) from the Ministry of Land, Transport and Maritime of Korean government through the Core Research Institute at Seoul National University for Core Engineering Technology Development of Super Long Span Bridge R&D Center.

**REFERENCES**

[1] Hay, J.S., The wind induced response of the Erskine Bridge, Conference at the Institute of Civil Engineers (1981) 6, 81-90.
[2] Owen, J.S., Vann, A.M., Davies, J.P., Blakeborough, A., The prototype testing of Kessock Bridge: response to vortex shedding, J. of Wind Engineering and Industrial Aerodynamics (1996) 60, 91-108.
[3] Barelli, M., White, J., Billington, D.P., History and aesthetics of the Bronx-Whitestone Bridge, Journal of Bridge Engineering (2006) 11(2), 230-240.
[4] Larsen, A., Esdahl, S., Andersen, J.E., Vejrum, T., Storebaelt suspension bridge - vortex shedding excitation and mitigation by guide vanes, J. of Wind Engineering and Industrial Aerodynamics (2000) 88, 283-296.
[5] Battista, R.C., Pfeil, M.S., Reduction of vortex-induced oscillations of Rio-Niteroi bridge by dynamic control devices. J. of Wind Engineering and Industrial Aerodynamics (2000) 84, 273–288.
[6] Fujino, Y., Yoshida, Y., Wind-induced vibration and control of Trans-Tokyo Bay Crossing Bridge, J. of Structural Engineering (2002) 128(8), 1012-1025.
[7] BBC, http://news.bbc.co.uk/2/hi/science/nature/7340528.stm, (2008).
[8] Battista, R., Pfeil, M., Velihovetchi, N., Maciel, A., Multiple controllers of wind-induced oscillations of a long span bridge, 17th Congress of IABSE, Chicago, (2008).
[9] Wardlaw, R.L., The improvement of aerodynamic performance, Aerodynamic of Large Bridges (1992) Balkema, 59-70.
[10] JSCE, Wind Resistant Design of Bridges – Code Practice and Recent Developments, Structural Engineering Series 12, Japanese Society of Civil Engineers (2003) (in Japanese).
[11] Simiu, E., Miyata, T., Design of Buildings and Bridges for Wind, (2006) Wiley.
[12] El-Gammal, M., Hangan, H., King, P., Control of vortex shedding-induced effects in a sectional bridge model by spanwise perturbation method, J. of Wind Engineering and Industrial Aerodynamics (2007) 95, 663–678.
[13] IEC, Wind turbines. Part 12-1: Power performance measurements of electricity producing wind turbines, IEC 61400-12-1, International Electrotechnical Commission (2005).

[14] Kwon, S.D., Uncertainty analysis of wind energy potential assessment, Applied Energy (2009) (in press).
[15] Manwell, J.F., McGowan, J.G., Rogers, A.L., Wind Energy Explained: Theory, Design and Application, (2002) Wiley.
[16] Chang, T.J. Wu, Y.T., Hsu, H.Y., Chu, C.R., Liao, C.M., Assessment of wind characteristics and wind turbine characteristics in Taiwan, Renewable Energy (2003) 28, 851-871.
[17] Travere Industries, 2009. http://www.travere-industries.com/.
[18] Pereira, E.B., Lima, J.H.G., Solar and Wind Energy Resource Assessment in BRAZIL, National Institute for Space Research – INPE (2008).

Proceedings of the IMAC-XXVIII
February 1–4, 2010, Jacksonville, Florida USA


# Method for Vehicle Identification and Classification for Bridge Response Monitoring

Kanwardeep Singh Bhachu[1], J. David Baldwin[1], Kyran D. Mish[2]

[1]School of Aerospace & Mechanical Engineering, [2]School of Civil Engineering & Environmental Science, University of Oklahoma, Norman, OK 73019, USA.

## NOMENCLATURE

$D_{HW}$ = Distance Headway (Feet)
$D_{HW\,Min}$ = Minimum Distance Headway (Feet)
$Gmax$ = Maximum value of acceleration (g)
$H_g$ = Height of the vibration signature (g)
$H_{g\,minTruck}$ = Minimum Height of Vibration Signature for the Truck (g)
$H_m$ = Gross weight of heavy wt. truck class
$L_d$ = Limit of Discrimination
$L_m$ = Gross weight of light wt. truck class
$L$ = Length of Vehicle (Feet)
$M_v$ = Vehicle Gross Weight (lbs.)
$M_m$ = Gross weight of medium truck wt. class
N = Vehicle count/ Number of vehicles detected
$N_{AD}$ = Average Vehicle Detections (Accelerometer Data)
$N_{VD}$ = Average Vehicle Detections (Video Data)
$N_{TD}$ = Number of total detections
$THD$ = Threshold of Detection (g)
$T_{SE}$ = Time of Solitary Existence (sec)
$T_{HW}$ = Time Headway (sec)
$T_{HW\,Min}$ = Minimum Time Headway (sec)
$T_{VP}$ = Vehicle Passage Time (sec)
$V$ = Speed of Vehicle (mph, ft/sec)
$V_{min}$ = Minimum Speed of Vehicle (mph, ft/sec)
$\Delta = N^{AD} - N_{VD}$ (Difference in detection results)
$\mu$ = Mean of the acceleration values in complete time history (g)

## ABSTRACT

Loading-induced deterioration of highway bridges, e.g. fatigue, is a well known phenomenon that results in reduced residual strength and service life. To estimate the residual strength and remaining life, it is essential to determine the real vehicle loading, i.e. count, spatial location, weight and velocity, on the bridge. We have developed an algorithm that takes acceleration response data (time histories) of the bridge under traffic load and gives the automatic vehicle count (N). Vehicles of interest are single unit trucks and tractor semi-trailer units as they are considered to cause maximum damage to bridge structures. The Automatic Vehicle Detection algorithm (AVD) models vibration signature of trucks on the basis of their physical parameters, i.e. gross weight, velocity and length.

Two tuning parameters of the AVD [Threshold of detection (THD) & Time of solitary existence ($T_{SE}$)] are varied to match detection results from the video traffic analysis. It is shown that with THD = 0.20 g and $T_{SE}$ = 1.86 sec, the average detections by algorithm and video traffic are off by very small value of $\Delta$= 0.8. Thus, these values can be used to detect trucks amongst other vehicles in the acceleration time history. The resulting vehicle count/weight statistics associated with these vehicles will be available to subsequent stress and fatigue analyses on the structure.

## INTRODUCTION

According to the bureau of transportation statistics (BTS) in 2008, there were total of 601,396 bridges in US, 12% (71,461) out of these were structurally deficient and 13% (79,933) were functionally obsolete [1]. The costs associated with reducing the number of such SD/FO bridges is in the billions of dollars.

Highway bridges accumulate damage over time due to excitation caused by the transient vehicle loads. These transient loads are due to moving vehicles and vary in both time and space. Continuous response monitoring of these bridges will lead to development of analysis techniques that will provide good estimates of real traffic loading of the bridges.

Because there are numerous types of vehicles that traverse highway bridges daily, vehicles should be differentiated on the basis of their size, shape, mass and axle configuration. Of all the vehicles, trucks carrying large payloads are known to induce the most significant dynamic responses in the bridge structure thus causing the most damage. Therefore, identifying the passage of vehicles with high GVW (gross vehicle weight) is considered to be of utmost importance to the response monitoring efforts.

We have developed a method that uses cost effective accelerometer data to classify the highway traffic and extract relevant vehicle information. A video data capture is used to calibrate the accelerometer data findings. Once enough correspondence between these data streams is achieved an accelerometer data series alone can be used for the vehicle classification purposes. Our vehicle classification uses estimated GVW gross vehicle weight as the primary classification parameter i.e. “light”, “medium”, “heavy” weight vehicle, instead of “car”, “bus”, “truck”, etc.

### Vehicle detection

The first step in the process of traffic characterization is identification of total number of “Vehicles of Interest” (VOI) in a given time-history. An Automatic Vehicle Detection (AVD) algorithm has been developed that employs a vector destruction method (VDM) to identify events that correspond to the passage of a VOI during the acceleration time history. AVD gives us the first desired information about the traffic i.e. total number of vehicles (vehicle count) that have passed during an interval of time. The results from this algorithm are discussed below.

Several physical and control parameters are required to model the vibration signature of a VOI in the AVD algorithm. These are tuning parameters used in the AVD algorithm which it possible to discern vehicles of interest (VOI) amongst other vehicles in the acceleration time history.

The required physical, control and tuning parameters are:

**1. Control Parameters**

- 1.1 Classification of Vehicle of Interest (VOI).
- 1.2 Traffic Flow
    - 1.2.1 Minimum Headway

**2. Physical Parameters**

- 2.1 Vehicle length limits.
- 2.2 Vehicle gross weight limits

**3. AVD algorithm tuning parameters**

- 3.1 Threshold of Detection (THD)
- 3.2 Time of Solitary existence ($T_{SE}$)

### Vehicle of interest (VOI)

Deterioration of the highway bridges is known to be caused by daily vehicular traffic consisting of a variety of vehicles like cars, SUV’s, buses, light trucks, construction machinery, and heavy transport trucks, etc. It has been well established that out of all these vehicles, the heaviest trucks cause the most damage to the bridge structure by inducing substantial dynamic responses in the structure. Thus we are most interested in this study in finding heavy transport trucks. VOI is described as a control parameter because its selection defines the physical parameters which in turn control values of the tuning parameters used in the AVD algorithm. When performing a vehicle classification analysis using raw acceleration data from the bridge structures, we choose to define the vehicle classes on the basis of GVW, rather than by identifying them as "big truck", "bus", "car", etc. Our initial working classifications are given in Table 1.

Table 1: Vehicle classification based on GVW [($H_{m1}$-$H_{m2}$ lbs) > ($M_{m1}$ – $M_{m2}$ lbs) > ($L_{m1}$-$L_{m2}$ lbs)]

| **Vehicle Class** | **Vehicle Gross Weight Range (lbs.)** |
|---|---|
| Heavy weight | ($H_{m1}$-$H_{m2}$) |
| Medium weight | ($M_{m1}$ – $M_{m2}$ ) |
| Light weight | ($L_{m1}$-$L_{m2}$) |

The actual values of GVW in each class will be chosen to segregate the heaviest (and thus most damaging) vehicles from the others. This classification is necessary because raw accelerometer data alone may not be able to distinguish between a two-axle truck and a five-axle truck-semitrailer running empty with the same GVW at the same speed. But if the classification is based on the gross vehicle weight rather than on configuration then both of the above trucks would come under the same weight vehicle class. Further evaluation of highway traffic characteristics will be necessary to establish values for these categories.

**Traffic flow**

Traffic flow, i.e., vehicle count, serves as a second control parameter in the AVD algorithm. Traffic flow is different during different times of the day and that directly controls other parameters like minimum time headway ($T_{HW}$), speed of vehicles (V), and time of solitary existence ($T_{SE}$). The different traffic conditions will lead to different traffic flows that in turn will have direct impact on the vehicle speed and minimum headway. For example, congested traffic will lead to closer following distances and lower-than-posted vehicle speeds, whereas sparse traffic could result in longer following distances and higher-than-posted vehicle speeds. In this paper, we are concentrating on the freely flowing traffic for simplicity.

**Headway**

Headway is the difference in time, or distance, between the leading and following vehicle in a highway traffic situation. Time headway ($T_{HW}$) and distance headway ($D_{HW}$) are interrelated through vehicle speed (V) by

$$T_{HW} = \frac{D_{HW}}{V} \qquad (1)$$

Time headway ($T_{HW}$) is an important parameter because it allows us to distinguish between two closely following vehicles in an acceleration time history. $T_{HW}$ can vary from (0 - ∞) based on the actual physical distance between the two vehicles and relative speed. AVD can distinguish between the two vehicles if they maintain $T_{HW}$ equal to or greater than minimum time headway $T_{HW\,Min}$ i.e. $T_{HW} \geq T_{HW\,Min}$. Minimum time headway ($T_{HW\,Min}$) is predefined in AVD and its value is set to the value with which most trucks are going to follow each other. Any pair of trucks traveling so that $T_{HW} < T_{HW\,Min}$ will be considered as single vehicle. Based on observations of highway traffic and video traffic analysis, it was decided to set a fixed limit for minimum distance headway i.e. $D_{HW\,Min}$ between the two trucks to 40 ft. The headway parameters are illustrated in Figure 1.

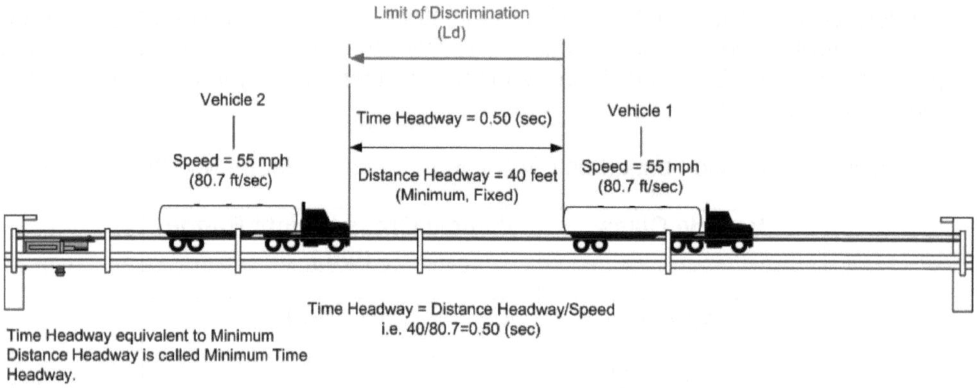


**Figure 1:** Definition of headway in time and distance domains

The definition of minimum time headway is constrained by the minimum distance headway of 40 feet. Higher values of time headway, however, are of no concern as AVD can differentiate between two vehicles which are separated by more than the minimum time headway. The lower bound, i.e. minimum time headway, would occur when trucks follow each other at the minimum possible speed in regular free flow traffic, i.e. not less than 55 mph and maintaining 40 foot distance headway. Specifically, with $D_{HW\,Min}$ = 40 ft, and $V_{Min}$ = 55 mph = 80.7 ft/s, we find that (from Equation 1) $T_{HW\,Min} \approx 0.5$ sec.

These headway measures are illustrated in Figures 1 and 2.

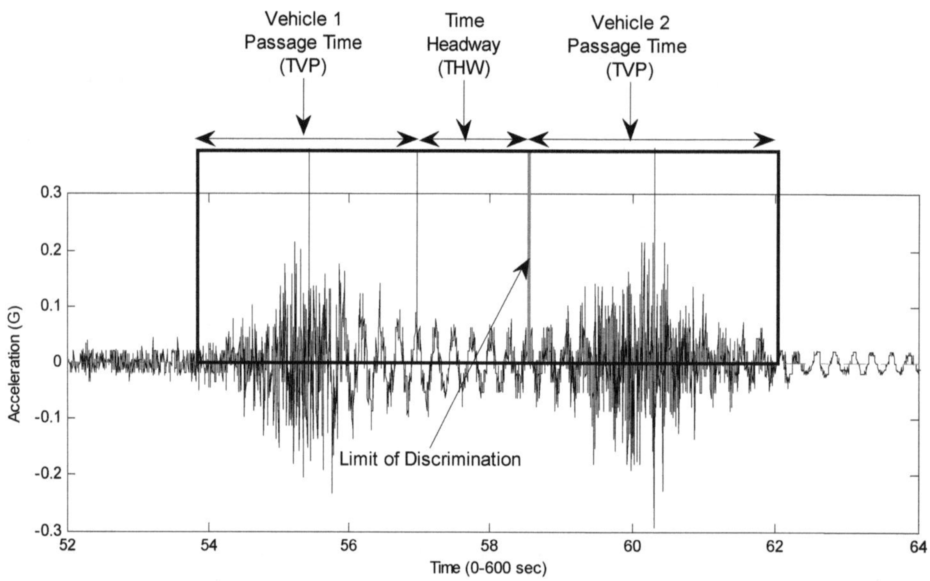


**Figure 2:** Interpretation of time headway in acceleration time history

### Allowable vehicle limits

The FHWA-allowable gross weight for heavy trucks in Oklahoma is restricted to 80,000 lbs. whereas longer combination vehicles are allowed to have up to 90,000 lbs. of gross vehicle weight [2]. FHWA also enforces limits on the length of truck semi-trailer/trailer vehicles to be between 48 and 53 feet. Typical total lengths for truck-semitrailer combination are in the range of 65-70 feet.

### AVD algorithm tuning parameters

There are two tuning parameters in the AVD algorithm, the threshold of detection (THD), and the time of solitary existence (TSE). These parameters are adjusted to make AVD algorithm capable of finding vehicles of interest in the acceleration time history.

Threshold of detection (THD) is the value of acceleration response above which AVD algorithm can find the maximum responses associated with vehicles. If the maximum response associated with vehicle's vibration signature lies above or at the value of THD, detection will be made. The THD value has to be set higher than the value of truck-induced acceleration $H_g$, so that only trucks are detected among other vehicles in the acceleration time history. The time of solitary existence (TSE) is the length of an acceleration history corresponding to the presence of a single vehicle of interest. As will be shown below, TSE is related to vehicle speed and time headway. These parameters are illustrated in Figure 3.

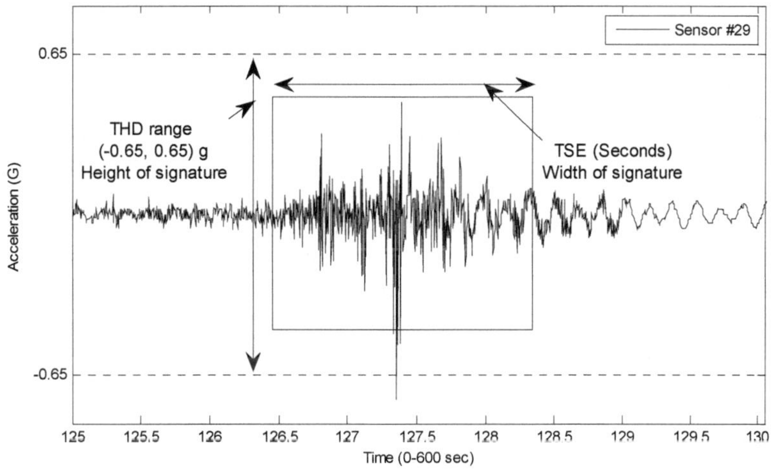


**Figure 3:** $H_g$ lying outside THD range.

### Response/vibration signature modeling

In order to automatically identify the truck semi-trailer units from the acceleration time history, it is important to define resulting vibration signature in terms of mass, length and velocity of the vehicle. A typical vibration signature caused by a single vehicle will have particular characteristics, e.g., height ($H_g$) and width ($T_{SE}$) of the acceleration response. It is considered that physical parameters of the truck such as gross weight/mass, length and velocity can be directly related to the dimensions of the vibration signature. The height of the vibration signature is function of vehicle mass ($M_v$) and speed (V), while width of the signature ($T_{SE}$) is function of vehicle length (L) and speed (V). The uniqueness in response corresponding to the vehicle having particular physical parameters can be used to detect and discriminate between different vehicles.

The AVD algorithm will also distinguish trucks from other types of vehicles like cars, SUV's, etc. A very lightly loaded truck (possibly empty) will still weigh more than a car or pickup truck and there will be particular value of acceleration response ($H_{gMin\ truck}$) that will differentiate trucks from other vehicles in the acceleration time history. AVD will always look for a value of $H_g \geq H_{gMin\ truck}$ in order to find and count trucks among other vibration signatures present in acceleration time history.

In the Figure 4, the height of the vibration signature ($H_g$) is equal to the difference between the mean value of acceleration ($G_{mean}$) and the maximum value of acceleration ($G_{max}$) that occurs within the width of the vibration signature. The maximum value of acceleration can be $\pm$ $G_{max}$ and depending on the algebraic sign of $G_{max}$ and $H_g$

will be oriented up or down from the mean. In Figure 4, the mean of acceleration is on the order of 0 g and $G_{max}$ is 0.4594 g within $T_{SE}$.

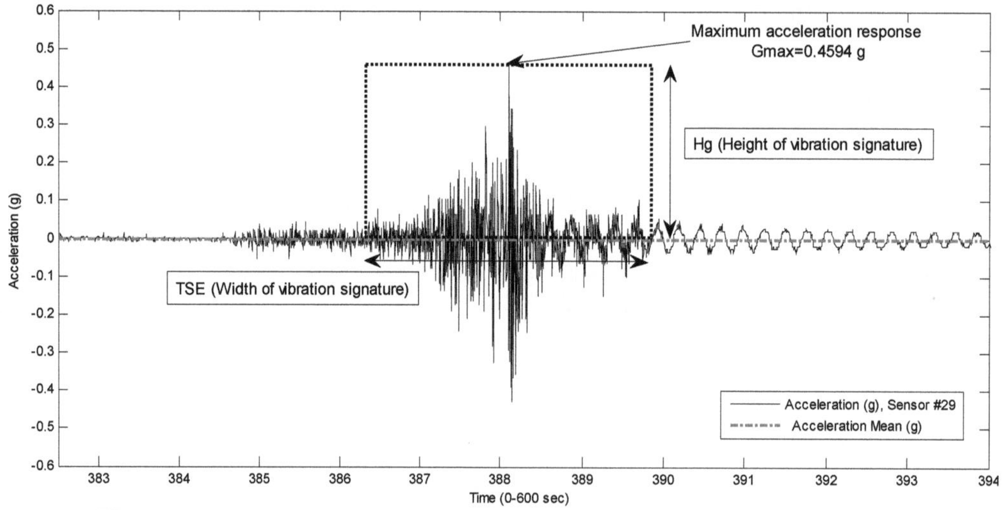


**Figure 4:** Vibration signature modeled in vehicle physical parameters.

Vehicles with different combinations of the physical parameters $M_v$, V, $L_v$ will generate vibration signatures having different heights and widths. Thus a vehicle of interest (heavy truck) will generate a vibration signature having some particular height and width, which can be checked against standard response parameters used to classify vehicles into predefined classes. Hence, the experimentally-determined value of $H_{gMin\ truck}$ will define the value of THD (a tuning parameter) that enables AVD to find trucks from acceleration data and can be found experimentally.

## RESULTS AND DISCUSSION

The primary result of this study is information about the vehicle count, i.e. the total number of vehicles identified. Vehicle count results were found to vary with the change in tuning parameters and the most important algorithm tuning parameter was found to be the Threshold of Detection (THD).

### Time of solitary existence ($T_{SE}$)

The time of solitary existence ($T_{SE}$) is a period of time during which only one vehicle is presumed to be present in an acceleration time history. Therefore, the time of solitary existence ($T_{SE}$) can be formulated as 2×time headway + vehicle passage time, i.e. $T_{SE}$ = 2× (0.5) +0.86 =1.86 seconds.

$$T_{SE}= 2\times T_{HW\,Min}+ T_{VP} \tag{2}$$

Where, $T_{HW\,Min}$ is minimum time headway and $T_{VP}$ is time of vehicle passage.

Hence, the final $T_{SE}$ equation is an addition of (front) $T_{HW}$, (back) $T_{HW}$ and $T_{VP}$ as shown in Figure 5, i.e. only single vehicle can exist for 1.86/2= 0.93 seconds.

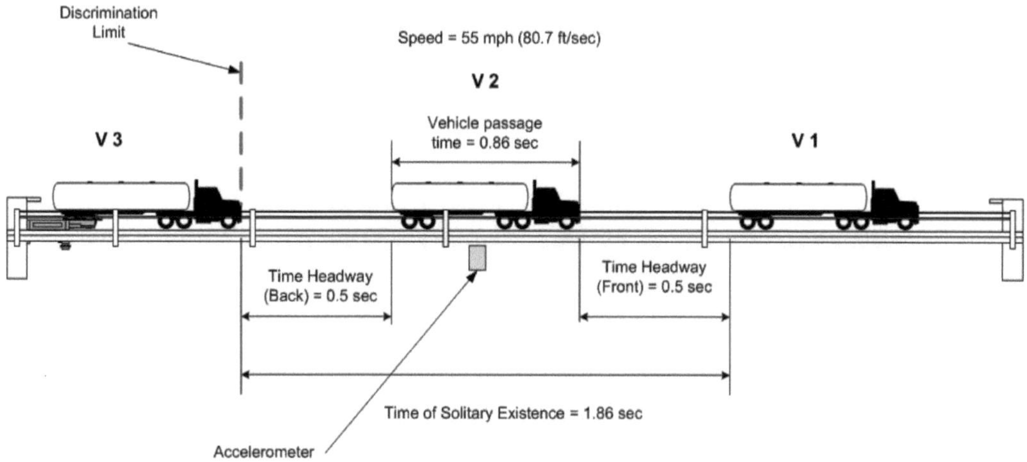


**Figure 5:** Three trucks following situation and computed $T_{SE}$.

**Tuning Parameter 1 (Threshold of Detection, THD)**

It was hypothesized that if THD is set to higher value of acceleration response, there will be very less or zero detections. Where as if THD is set about the some middle value of acceleration response there will be increase in the number of detections and if THD is set to lower value of acceleration then there will be very large detections. Analysis results corresponding to the above hypothesis are presented below by taking one acceleration file and changing the THD value i.e. (0.40, 0.25, and 0.12) g. The acceleration data used in this analysis came from the bridge monitoring system installed on the I-35 Canadian River Bridge located at Norman, Oklahoma.

It can be seen in Figure 6 that if higher value of THD is used then only three vehicles are detected from the acceleration time history. It can be visibly seen that there are only three peaks that lie outside of THD window range i.e. ± 0.40 g. These peaks correspond to the maximum value of acceleration response caused by unknown vehicles at that particular time. It can be noticed in Figure 6 that all the peaks corresponding lie within ± 0.7 g, and if THD > 0.7 there will be no detections as acceleration response lie out of ± 0.7 g range.

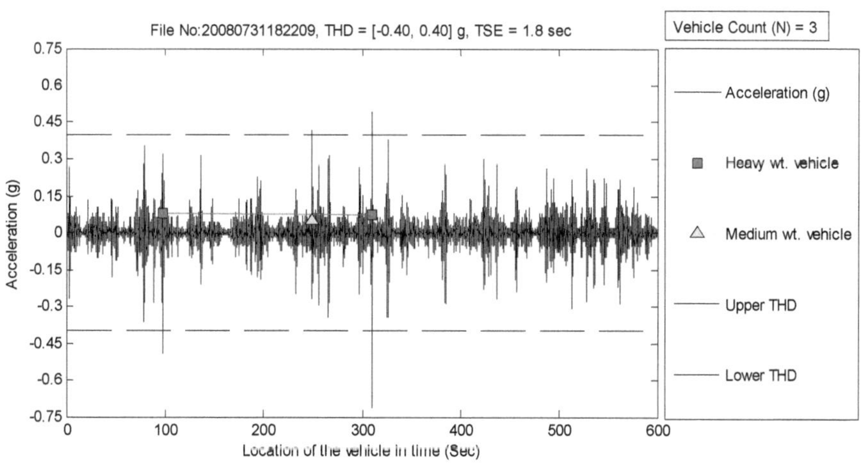


**Figure 6:** Vehicle count results by setting value of THD to 0.40g (High value).

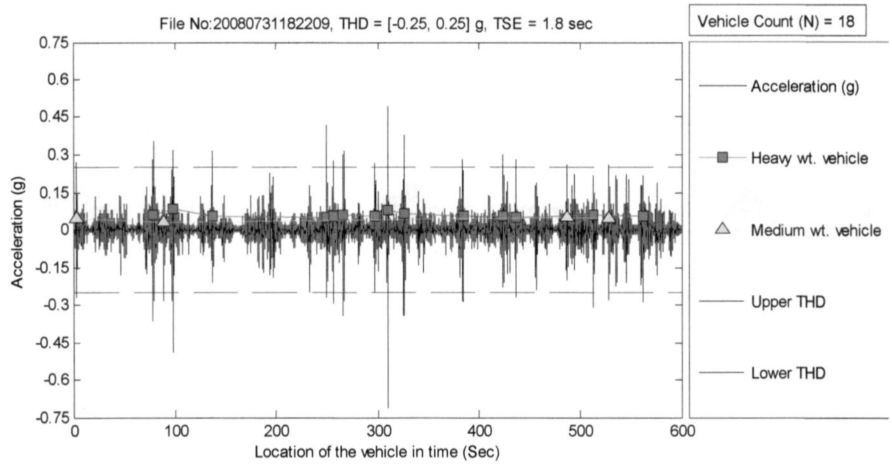


**Figure 7:** Vehicle count results by setting value of THD to 0.25g.

It is evident from the Figure 7 that if the THD window range is squeezed from ± 0.4 g to ± 0.25 g, there is increase in the vehicle count from N=3 to N=18 that confirms with the above presented hypothesis. Similarly in the Figure 8, as the value of THD window was changed from ±0.25 g to ±0.12 g, there was further increase in the vehicle count results i.e. from N=18 to N=79. Such a large increase in detections suggests that the value of THD is in the range of acceleration response likely to be caused by the smaller light weight vehicles. Such an assertion can be made because traffic always consists of large number of smaller vehicles than large vehicles like truck-semitrailer combinations.

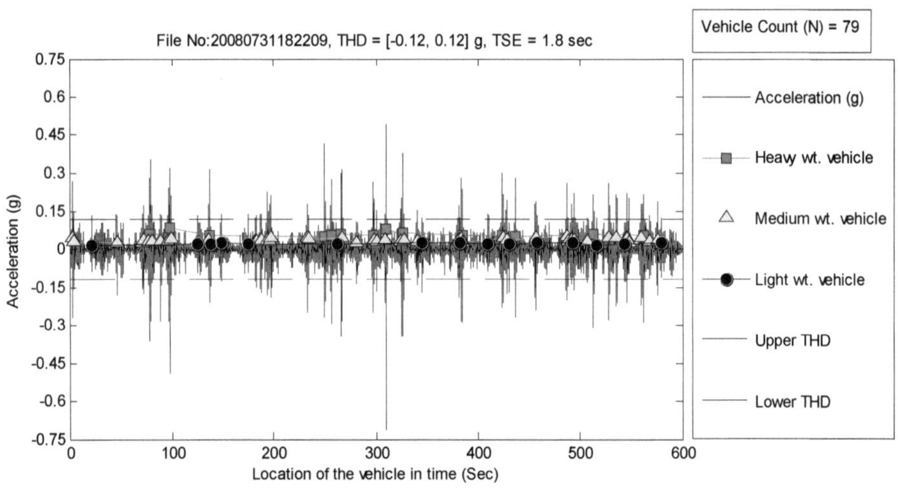


**Figure 8:** Vehicle detections made by setting THD = ± 0.12 g.

Hence, the above results show that the number of vehicle detection increased with decrease in the value of THD.

### Video Analysis

An attempt was made to capture a video record of the traffic that could be synchronized with the acceleration data file. This video record was supposed to serve as an initial source of verification on the vehicle count results but due to problems with the data acquisition hardware it could not be done. However this video data was compared against the acceleration data file taken on a different day but essentially the same time of a day.

In order to have an idea about the value of THD that would correspond to large truck semitrailer units, we have compared the detection results from the acceleration data and video data.

**Average detections from the video data**

50 minutes of video data was analyzed manually for counting the number of truck semitrailer. Video data was divided into five 10 minutes of data files such that comparison would be made with acceleration data files that were 10 minutes long.

The average number of trucks identified from the video record was 32.6 ≈ 33, while the average numbers of cars detected were 201. The comparison between the both types of vehicles indicated that number of cars is always greater than trucks. The ratio between the average number of trucks and car is 0.16 that is very low. The comparison can be seen in Figure 9.

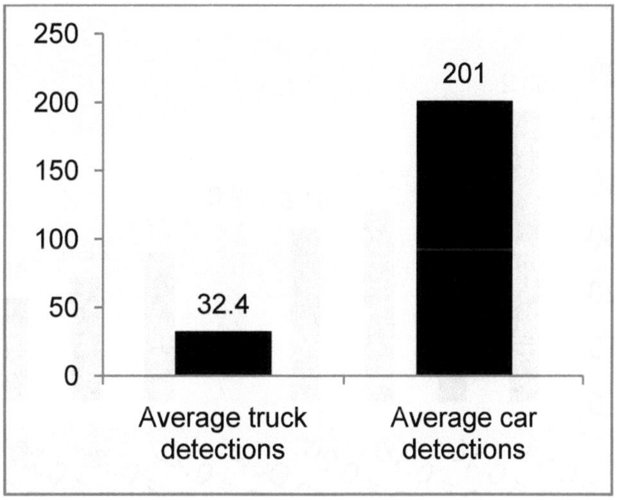


**Figure 9:** Number of detections comparison between car and trucks based on video data.

**Average detections from the acceleration data**

The value of $T_{SE}$ was kept constant while varying the value of THD from (0.14 to 0.28). Results given by AVD are presented below in Figure 10.

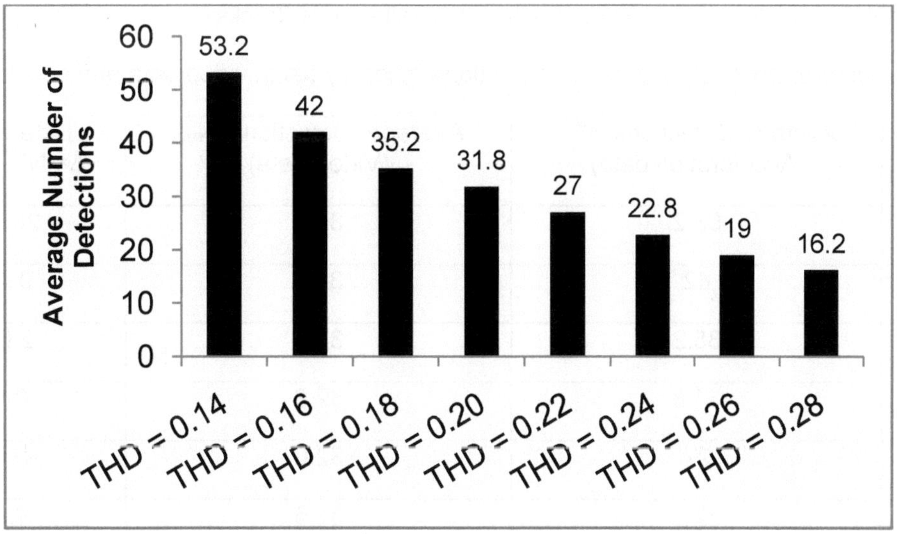


**Figure 10:** Average number of detections made while varying THD values and keeping $T_{SE}$ = 1.86 constant (using accelerometer data).

Data used to generate the above graph came by investigating five different acceleration data files that were recorded from 8:10 AM in the morning till 2:29 PM. It is seen that the number of vehicles detected by AVD are continuously decreasing as the THD level is raised from 0.14 g to 0.28 g. The average number of detections made with THD = 0.14 is about 69.5 % more than with THD = 0.28. Sharp increase in average number of detections by decreasing the value of THD to 0.14 indicates that THD is now in the range of acceleration response that is being generated by smaller vehicles. This is supported by low truck/car ratio of 0.16 (Figure 9).

**Comparison between the detection results**

Finally, the vehicle count results from both types of data were compared against each other to find the value of THD that would give vehicle count results from the acceleration data files closer to vehicle count results given by video data. I can be seen in Figure 11 that for THD = 0.20 g, the average vehicle counts ($N_{AD}$) is very close to average vehicle counts ($N_{VD}$) from video data and the difference $\Delta$ comes out to be -0.8. The difference between the two detection results can be seen in Table 2.

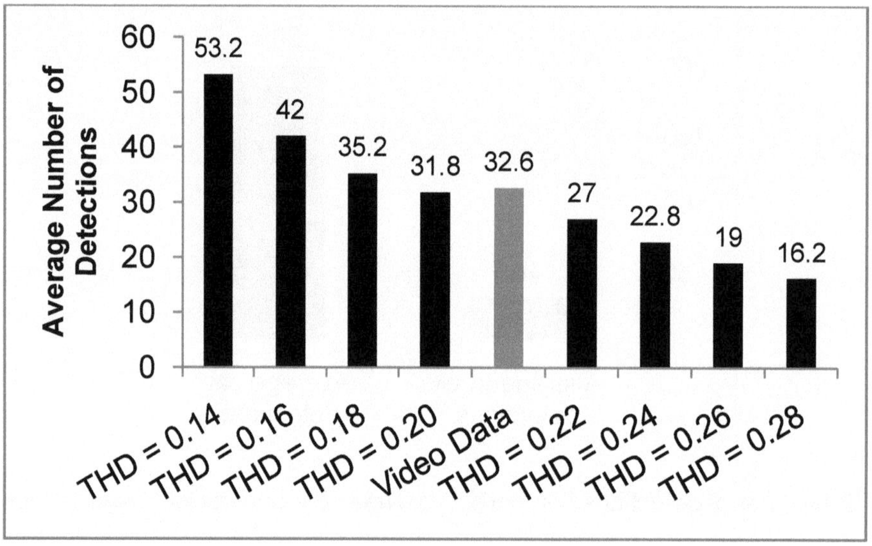


**Figure 11**: Comparison between detection made by video data and accelerometer data.

The Figure 11 shows that the average number of detections made by using video data falls very close to that obtained by using THD = 0.18, 0.20 and 0.22 (g's) . This suggests that the value of THD from 0.18 g to 0.22 g could be used to detect vehicles of interest i.e. tractor semitrailer units (trucks).

**Table 2:** Difference (Δ) between avg. no. of detections made by using video data and accelerometer data.

| THD (g) | Average Detections, $N^{AD}$ (Acceleration data) | Average Detections, $N_{VD}$ (Video data) | Difference $\Delta = N^{AD} - N_{VD}$ |
|---|---|---|---|
| 0.14 | 53.2 | 32.6 | 20.6 |
| 0.16 | 42 | 32.6 | 9.4 |
| 0.18 | 35.2 | 32.6 | 2.6 |
| 0.20 | 31.8 | 32.6 | -0.8 |
| 0.22 | 27 | 32.6 | -5.6 |
| 0.24 | 22.8 | 32.6 | -9.8 |
| 0.26 | 19 | 32.6 | -13.6 |
| 0.28 | 16.2 | 32.6 | -16.4 |

Thus, 0.20 g may be used as the value for THD that is capable of detecting trucks.

**CONCLUDING REMARKS**

An algorithm has been developed that can take acceleration data from a vibrating bridge structure and predict the total number of trucks that have passed. The initial values of the tuning parameters of the algorithm (THD and $T_{SE}$) were established in order to be able to identify the truck-semitrailer units in an acceleration time history. Our analysis indicated that $T_{SE}$ = 1.86 seconds was appropriate for detecting truck-semitrailer combinations. The initial value of (THD) was predicted by comparing average video detection and algorithm detection results and it was found that a THD of 0.20 g gave an excellent comparison between the two detection methods. We propose that THD = 0.20 g and $T_{SE}$ = 1.86 sec AVD can filter truck-semitrailer combinations amongst other vehicles in the acceleration time history. In the future, the accuracy of AVD to detect trucks will be validated by conducting experimental work involving truck test runs along with synchronized video and accelerometer data comparison. Greater accuracy on detection results can be achieved by fine tuning the parameters of AVD algorithm.

**REFERENCES**

[1] Bureau of Transportation Statistics, Condition of U.S. Highway Bridges: (1990-2008) http://www.bts.gov/current_topics/2009_03_18_bridge_data/html/bridges_us.html

[2] U.S. Department of Transportation, *Comprehensive Truck Size and Weight (CTS&W) Study*, http://www.fhwa.dot.gov/reports/tswstudy/Vol1-Summary.pdf (pp- 15, Table 5).

Proceedings of the IMAC-XXVIII
February 1–4, 2010, Jacksonville, Florida USA


# In-Situ Fatigue Sensors for Structural Health Monitoring of Highway Bridges

B. H. M. Priyantha Wijesinghe[1], Scott Zacharie[2], J. David Baldwin[2], Kyran D. Mish[1] and Thordur Runolfsson[3]
[1]School of Civil Engineering and Environmental Science, [2]School of Aerospace and Mechanical Engineering, [3]School of Electrical and Computer Engineering,
The University of Oklahoma, Norman, OK 73019, USA

## NOMENCLATURE

| | |
|---|---|
| $E$ | Elastic modulus |
| $K'$ | Cyclic strength coefficient |
| $K_t$ | Elastic stress concentration factor |
| $2N_f$ | Reversals to failure |
| $b$ | Fatigue strength exponent |
| $c$ | Fatigue ductility exponent |
| $\Delta e$ | Nominal strain range |
| $n'$ | Cyclic strain-hardening exponent |
| $\Delta\varepsilon$ | Local strain range |
| $\varepsilon_f'$ | Fatigue ductility coefficient |
| $\Delta\sigma$ | Local stress range |
| $\sigma_o$ | Mean stress |
| $\sigma_{max}$ | Maximum stress |
| $\sigma_f'$ | Fatigue strength coefficient |

## ABSTRACT

Fatigue damage in bridges has been identified as a threat to bridge safety. Fatigue cracks in steel structures are well-known to occur at stresses lower than typical design stresses. Short fatigue cracks can be difficult to detect by visual inspection and therefore can be easily overlooked. Since a larger portion of fatigue life in metals is spent on crack nucleation than on propagation, it is vital to develop methods of detecting such damage. We will discuss the initial development work of an *in-situ* fatigue sensor to detect fatigue damage in steel bridges. The concept of the fatigue sensor is built on the strain-life fatigue analysis method and Miner's damage summation rule. The fatigue sensor is made of electrically conductive material with notched sensor arms with varying stress concentration factors; each arm is designed to fail at different numbers of stress cycles to indicate the accumulation of damage in the structural member to which it is attached. The concept has been verified for constant amplitude cyclic loading in the laboratory. A finite element analysis has been carried out for the sensor and the results are compared. Future work will address the case of variable amplitude loading.

## INTRODUCTION

Fatigue damage and corrosion are among the major defects that emerge due to aging of civil structures [1]. Detecting fatigue cracks, therefore, is crucial in Structural Health Monitoring (SHM) of steel bridges. Unlike corrosion, yielding, spalling or material loss, early stage fatigue damage is frequently not visible. Because of the brittle nature of fatigue damage, it can lead to sudden structural failure. Since current practice in bridge health monitoring is largely dependent upon visual inspection [2], a question remains as to the safety and reliability of these aging bridges. Currently, various non destructive evaluation (NDE) methods are being researched, developed, and evaluated for detecting fatigue damage, including ultrasonic testing [3], infrared and thermal Imaging [4], acoustic emission [5], eddy current methods [6], dye penetrant inspection [3],[5], magnetic particle testing [7],[8], and radiographic testing. In general, these techniques can be considered as non strain-based methods. Strain-based methods include, fiber optic sensors [9], piezoelectric sensors [1] and various crack propagation gauges [10]. However, most of these techniques have certain limitations such as accessibility, automation, power supply, long term durability, environmental noise and bulkiness [1]. In practice, the majority of these techniques cannot be cost-effectively deployed in required bridge sites to obtain reliable information that can be easily interpreted by a bridge inspector. Moreover, the majority of these methods are developed based on the fracture mechanics approach thus do not address the crack nucleation stage of fatigue damage evolution. Knowing that the larger portion of life of a steel component/structure frequently is spent on crack nucleation than on crack propagation to failure, it is vital to develop methods to detect the nucleation of a fatigue crack. This paper presents ongoing research and development of certain aspects of a fatigue sensor which should be capable of detecting fatigue crack nucleation in steel structures while being simple in concept, low in cost, easy to apply, and easy to interpret.

## IN-SITU FATIGUE SENSOR

The sensor, which is made of an electrically conductive material, consists of several sensing arms; each arm is designed with different notch geometries to create different stress concentration factors (SCFs). A prototype fatigue sensing element is shown in Figure 1. When the sensor is properly attached to a structure, it will experience the same nominal strain, but different maximum strains at the notch roots due to differing notch geometries. As a result, the arms will fail progressively at different number of loading cycles thus indicating the accumulation of fatigue damage in the attached structure. By using a strain-based fatigue analysis method we can calibrate the progressive failure of the sensing arms with the accumulation of stress in the attached structure. Failure of each sensing arm can be detected by measuring its associated electrical resistance.

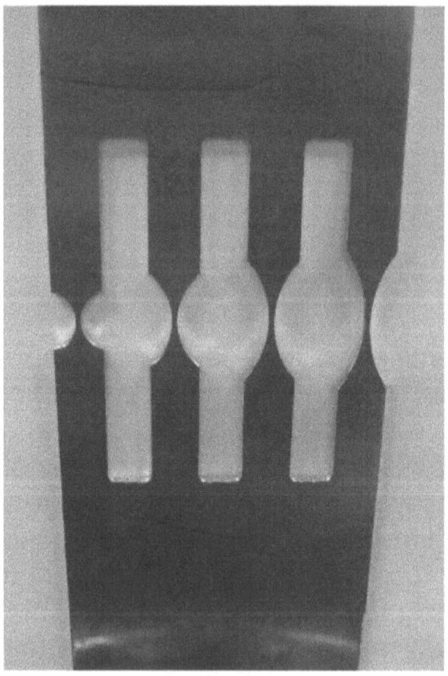

**Figure 1:** A prototype fatigue sensing element used for Proof-of-Concept

## STRAIN-LIFE FATIGUE ANALYSIS

The anticipated behavior of the in-situ fatigue sensor under development in this research is based on the strain life fatigue analysis method. Strain-based fatigue methods assume that a laboratory smooth specimen tested under strain control can simulate fatigue damage at the notch root of an engineering component. Equivalent fatigue damage is assumed to occur in the material at the notch root and in the smooth specimen when both are subjected to identical stress-strain histories [11]. Despite the fact that most engineering structures are designed for elastic nominal stresses, local stress concentrations can cause plastic strains to develop in their vicinity, thus causing more severe fatigue damage. The advantage of using the strain-life method is found in its ability to incorporate the elastic and plastic responses of the structural material. This method also provides a means to account for the cyclic plasticity at the root of a notch.

The basis of the strain-life analysis is the expression relating the total strain amplitude $\left(\Delta\varepsilon/2\right)$ in a structure to the number of strain reversals to failure $\left(2N_f\right)$, given by

$$\frac{\Delta\varepsilon}{2}=\frac{\sigma_f^{'}}{E}\left(2N_f\right)^b+\varepsilon_f^{'}\left(2N_f\right)^c \tag{1}$$

where $E$ is the elastic modulus and $\sigma_f'$, $b$, $\varepsilon_f'$ and $c$ are material properties. This expression takes no account of mean stress effects; we use the Morrow model

$$\frac{\Delta\varepsilon}{2}=\frac{\sigma_f^{'}-\sigma_o}{E}\left(2N_f\right)^b+\varepsilon_f^{'}\left(2N_f\right)^c \tag{2}$$

where $\sigma_0$ is the mean stress, and the Smith-Watson-Topper model

$$\sigma_{\max}\frac{\Delta\varepsilon}{2}=\frac{\left(\sigma_f^{'}\right)^2}{E}\left(2N_f\right)^{2b}+\sigma_f^{'}\varepsilon_f^{'}\left(2N_f\right)^{b+c} \tag{3}$$

to estimate the mean stress effects on life. The strain amplitudes $\left(\Delta\varepsilon/2\right)$ used in these equations were the notch strains, estimated using the cyclic stress-strain curve

$$\Delta\varepsilon=\frac{\Delta\sigma}{E}+2\left(\frac{\Delta\sigma}{2K^{'}}\right)^{1/n^{'}} \tag{4}$$

and Neuber's rule

$$\Delta\sigma\Delta\varepsilon=\left(K_t\Delta e\right)^2 E \tag{5}$$

where $\Delta e$ and $\Delta\varepsilon$ are the nominal and true (notch) strain ranges, respectively; $\Delta\sigma$ is the stress range; and $K_t$ is the elastic stress concentration factor. Parameters $K'$ and $n'$ are the material cyclic strength coefficient and exponent, respectively.

In the following discussions, the relevant material properties for the Al 7075-T6 sensor prototypes are given in Table 1.

**Table 1: Material Properties for Al 7075-T6 [12]**

| | |
|---|---|
| Elastic Modulus, psi | 10,298,000 |
| Poisson Ratio | 0.33 |
| Tensile Yield Strength, psi | 68,000 |
| Tensile Ultimate Strength, psi | 84,000 |
| Strength Coefficient ($\sigma_f'$ ) psi | 127,100 |
| Strength Exponent ($b$) psi | -0.0751 |
| Ductility Coefficient ($\varepsilon_f'$ ) | 0.4664 |
| Ductility Exponent ($c$) | -0.7779 |
| Cyclic Strength coefficient ($K'$) psi | 136,800 |
| Cyclic Strain Hardening Exponent ($n'$) | 0.0966 |

## EXPERIMENTAL DETAILS

The goals of the experimental procedures were twofold; first, to identify the appropriate testing load range, configuration, and material for the sensor; and second, to explore the most effective notch geometry within the limitations of machining and handling.

### Sensor Proof-of-Concept

As a proof of concept for the sensor design, specimens were made from 12 inch x 3 inch x 0.063 inch 7075-T6 aluminum sheet as shown in Figure 1. The four sensor arms were machined with circular notch radii varying from 0.25 - 1.0 inch, resulting in static SCFs as shown in Table 2. Eight prototype sensors were tested under displacement controlled tensile cyclic loading (max = 0.02 inch; min = 0.01 inch) at a 10 Hz. sinusoidal excitation frequency. Results of these tests confirmed that by varying the SCFs at the notches, a progressive failure in notched sensor arms can be obtained. A finite element analysis of the sensor prototype was used to estimate the normal stresses in each arm at the limits of displacement-controlled cycling, and from those results estimated fatigue lives were computed using the Morrow mean stress correction, the Smith-Watson-Topper (SWT) mean stress model, and no mean stress correction. Table 2 summarizes the experimental and predicted results.

**Table 2:** Results of Proof of Concept Tests

| | **Notch 1** | **Notch 2** | **Notch 3** | **Notch 4** |
|---|---|---|---|---|
| Estimated Stress Concentration Factor [13] | 9.135 | 9.180 | 9.270 | 9.562 |
| FEA Normal Stress for 0.02 in displacement, psi | 18,660 | 20,260 | 21,930 | 22,770 |
| FEA Normal Stress for 0.01 in displacement, psi | 9,550 | 10,360 | 11,230 | 11,680 |
| Average Life (*N* = 8), cycles | **7,807** | **4,400** | **2,391** | **2,336** |
| Calculated Life: SWT Mean Stress Correction, cycles | **8,194** | **4,524** | **2,627** | **1,785** |
| Calculated Life: Morrow Mean Stress Correction, cycles | 4,538 | 2,923 | 2,003 | 1,541 |
| Calculated Life: No Mean Stress Correction, cycles | 1,454,200 | 450,900 | 143,550 | 60,768 |

We draw several conclusions from the data in Table 2. Most obvious is that increasing the SCF values tended to reduce the life, as expected. However, the SCF's were not sufficiently different in this prototype to clearly distinguish the lives of each sensing element, with the experimental lives showing enough scatter to overlap the data from the nearest (SCF-wise) element. Regarding the predicted lives, the SWT model was in very good agreement with the experimental values, where neglecting the effect of mean stress on the life resulted in unacceptably non-conservative life estimates.

### In-Situ Fatigue Sensor Experiments

In-situ fatigue sensor prototypes discussed in this paper are made from 7075-T6 aluminum and are attached to steel beams as the underlying carrier material. Details of the specimens, procedure and test configuration are discussed next.

### Specimens

The fatigue sensors were cut from 0.032 inch thick 7075-T6 aluminum. This prototype sensor material was selected mainly because it is widely available, with well-characterized material behavior. Moreover, the availability of strain-life fatigue parameters was an advantage in selecting 7075-T6 over other metals for the development stages of the fatigue sensor.

Two types of sensors, each shown in Figure 2, were designed to obtain SCFs varying from 26 to 37.5 at the notches. These SCF's were obtained from both [13] and FE simulation results. However, for the final fatigue life calculations FE SCF's were used. The specimen geometry was first modeled using 3D CAD software and then machined using a CNC mill.

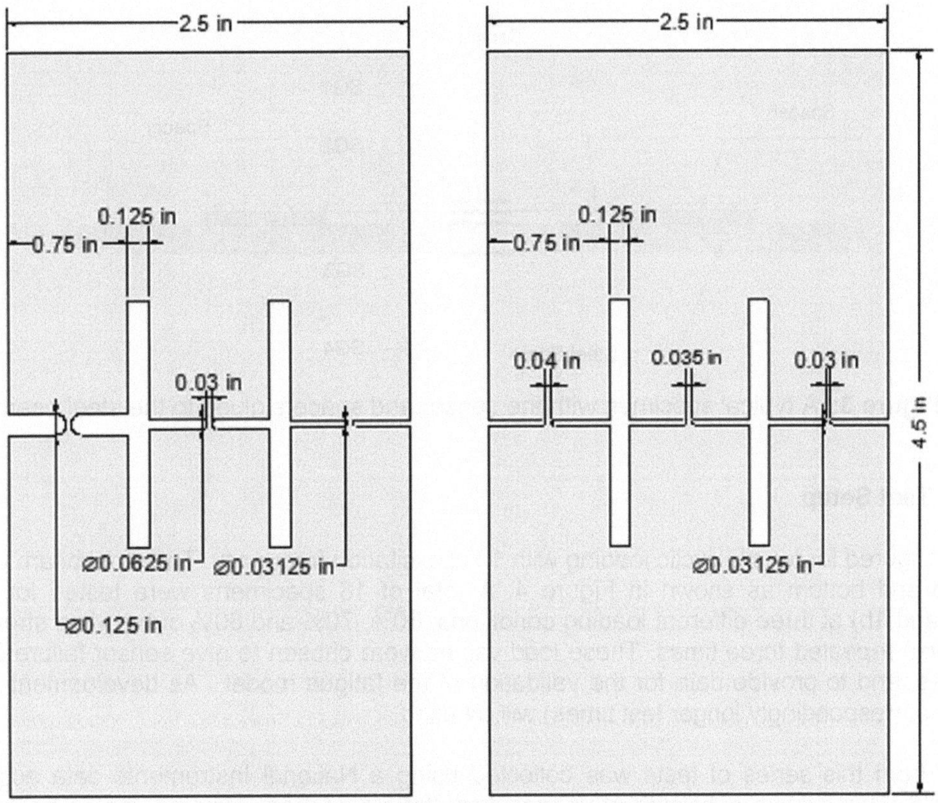


**Figure 2:** Prototype fatigue sensors Geometries
Geometry 2a (left) and Geometry 1b (right)

Sensor Geometry 2a consists of three machined notch pairs, each with a ligament between the notch roots of 0.03 inches. The notch diameters vary from 0.03125 – 0.125 inches. The 1b geometry also has three notch pairs, but these have a constant 0.03125 inch notch diameter and ligaments varying from 0.03 – 0.04 inches

Test specimens were made by attaching the aluminum sensors to 3 inch x 15 inch x 0.125 inch A1018 steel beams via aluminum spacers. The spacers (1 inch x 2.5 inch) were cut from 0.063 inch 7075-T6 aluminum and were glued to the steel beam at pre-marked locations to transfer load into the sensor. The sensors were then glued to the spacers at each end (see Figure 3). Acrylic type adhesive was used to bond spacers and the sensors to the beam. Each gluing face was abraded using sand paper and cleaned with degreaser prior to applying glue to facilitate a strong bond. A minimum 36 hour curing time was allowed before testing, per manufacturers recommendations, to gain the maximum strength of the adhesive. The adhesive layer also acts as an electrical insulator that allows measurement of electrical resistance in each sensing arm when connected to an appropriate monitoring circuit. For the experiments presented in this paper, however, this monitoring circuit was not needed, as visual observations were utilized to detect sensor arm failure.

We observed the input-output behavior of the sensors using strain gages. To measure the strain on the fatigue sensor and on the beam during testing, four foil strain gauges (SG1 – SG4) were attached as shown in Figure 3.

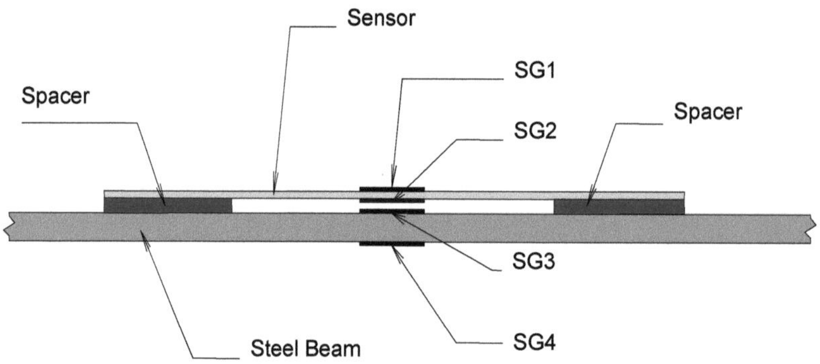


**Figure 3:** A typical specimen with the sensor and spacers glued to the steel beam

## Apparatus and Test Setup

The test was configured for tensile cyclic loading with 1 Hz excitation frequency. The steel beam was clamped 1½ inches from top and bottom as shown in Figure 4. A total of 18 specimens were tested for the two sensor geometries (2a and 1b) at three different loading conditions; 60%, 70% and 80% of the yield strength of steel (32 ksi); each test was repeated three times. These load values were chosen to give sensor failure data in relatively short time frames, and to provide data for the validation of the fatigue model. As development proceeds, lower load levels (with correspondingly longer test times) will be used.

The strain data from this series of tests was collected using a National Instruments data acquisition system, shown in Figure 5. This system consisted of voltage and strain signal conditioning, laptop-based PCMCIA data acquisition, and LabVIEW control software. The sensor failure monitoring circuit shown in Figure 5 was not required for these tests.

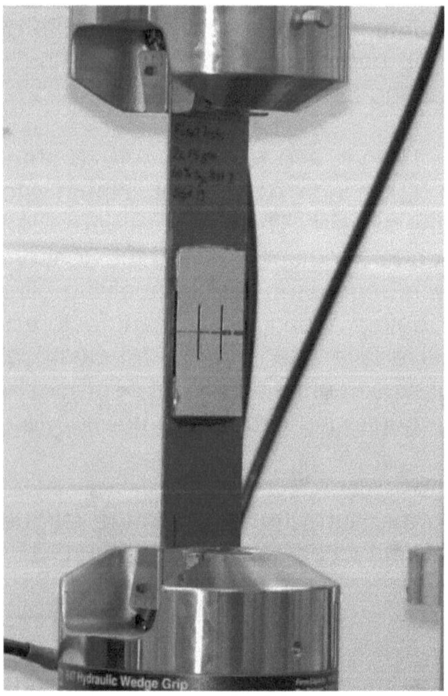

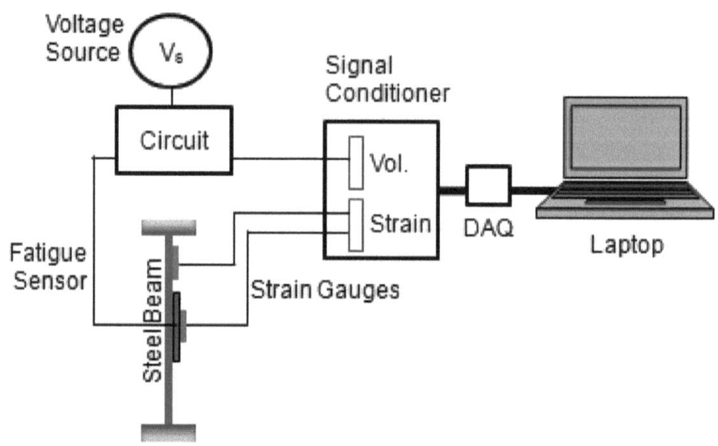


**Figure 5:** Typical fatigue sensor test instrumentation

**Figure 4:** Fatigue sensor attached to the steel beam

## FINITE ELEMENT ANALYSIS AND FATIGUE LIFE CALCULATIONS

Finite element analyses were carried out at different stages of the sensor design to (a) improve the performance in laboratory testing, (b) develop the sensor design and the geometry, (c) predict the material response, and (d) mitigate the experimental limitations pertaining to the time spent on accumulating large number of cycles. ANSYS was used for the static stress analysis of the sensor. The sensor geometry was created in Pro-E an imported to ANSYS where it was meshed with hexahedral SOLID45 elements. The mesh at the notches was refined to provide improved local solution accuracy (see Figure 6).

First, the strain output due to the applied cyclic loading was recorded using a dummy specimen for the three different loading conditions as mentioned before. An un-notched dummy sensor was instrumented as shown in Figure 3 and this sensor was tested under the same loading conditions as the actual test. Because of some slight curvature present in the sensor raw material, the strains obtained from SG1 and SG2 were then used as upper and lower limits, respectively, for the FE analysis and fatigue life calculations.

The displacements calculated from the output strain values for each loading condition were used as the boundary conditions for the sensor FE models. Element stress results obtained from the static analysis at the notch root and at the gross area of each sensing arm were then transferred to a MATLAB code to perform the fatigue life calculations. The MATLAB code calculates the fatigue life with mean stress corrections according to both the Smith-Watson-Topper and Morrow mean stress theories; based on the results of Table 2, we did not estimate the no-mean-stress lives. Material properties of 7075 T6 aluminum used in the FE and fatigue analysis were given in Table 1.

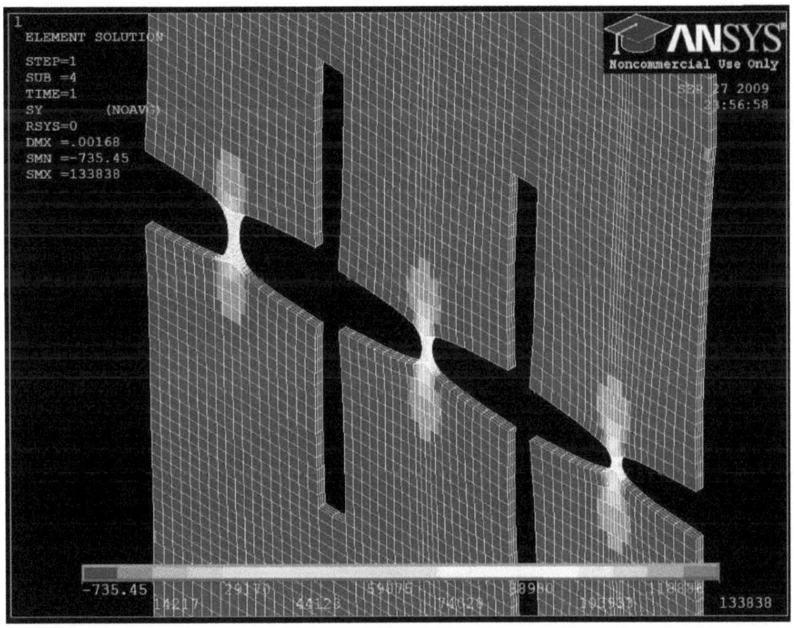


**Figure 6:** FE analysis results (axial normal stress) for geometry 2a

## RESULTS AND DISCUSSION

Experimental results for both prototype sensor geometries are plotted with results of the fatigue life simulations. Figures 7a, 7b and 7c show the data for sensor geometry 2a, and Figures 8a, 8b and 8c give the sensor geometry 1b results. Overall, the results for geometry 1b showed a better agreement with the simulated results than did geometry 2a.

Of the three notches in the Geometry 2a specimens, the largest (0.125 inch diameter) notch showed lower experimental fatigue lives compared to the simulated values at all the stress ratios tested, i.e., theory over-predicted the life, with the difference becoming larger at lower stresses (Figure 7a). Similar behavior was observed in the 0.0625 inch diameter notch geometry, but with better agreement at the highest stress level (Figure 7b). Experimental results from the smallest notch geometry (0.03125 inch diameter) agreed well overall with the simulated results (Figure 7c). However, unlike the other two notches, the experimental values for this notch were higher than theory predicted at the highest load case.

In sensor Geometry 1b, the experimental results agreed well with the theory-based predictions (Figures 8a, 8b and 8c). We noted that at lower stresses, especially at the 60% yield case, the experimental values were closer to the higher strain boundary (SG1), whereas at higher stresses they approached the lower strain boundary (SG2). In addition, the smallest ligament behavior (Figure 8c) is essentially the same as the smallest notch behavior of Geometry 2a (Figure 7c). As the geometry of both notches is the same, this demonstrates appropriate repeatability of the fatigue results obtained.

We note that the Smith-Watson-Topper and Morrow mean stress theories are much closer to each other, especially at high stress levels, than the SG1-SG2 limiting life predictions. The life estimates for sensors that are mode of flat material (i.e., no residual curvature) will presumably lie between the two sets of curves.

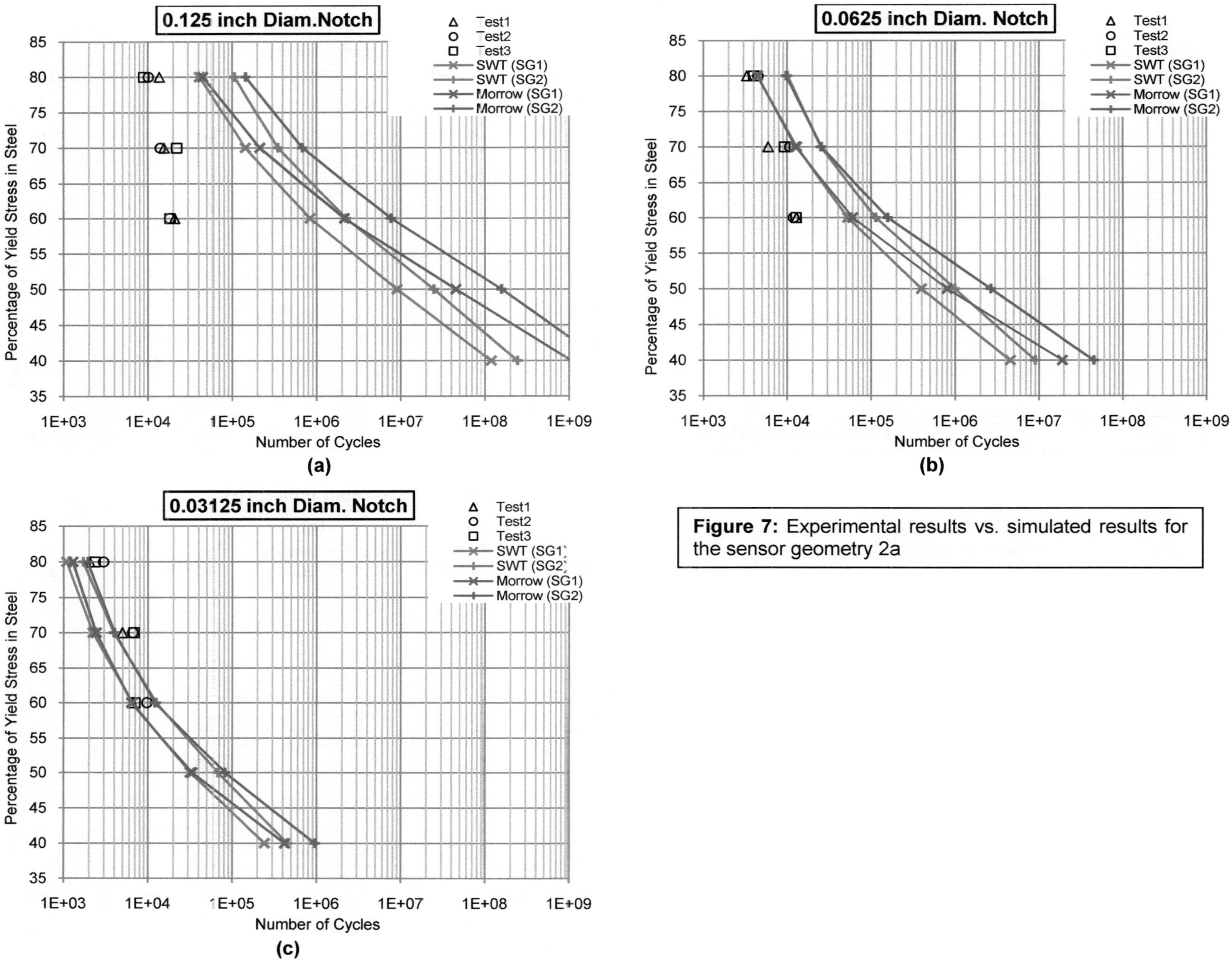


**Figure 7:** Experimental results vs. simulated results for the sensor geometry 2a

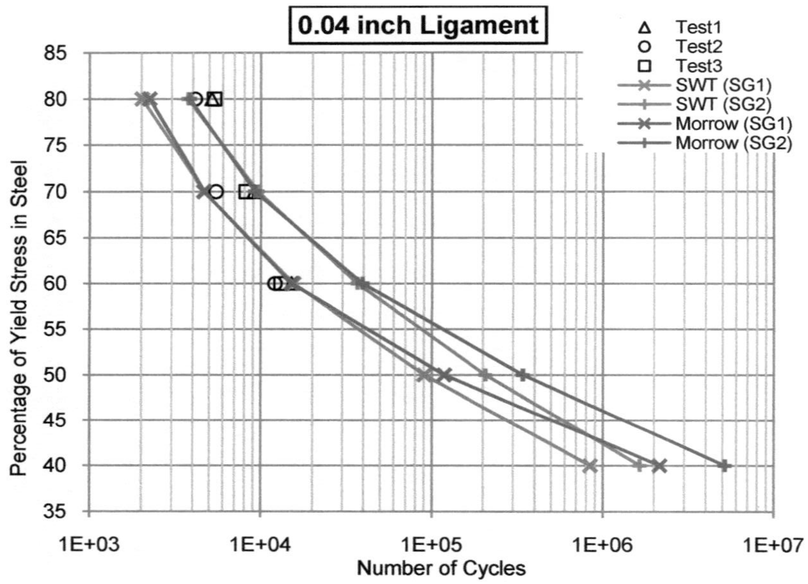


(a)

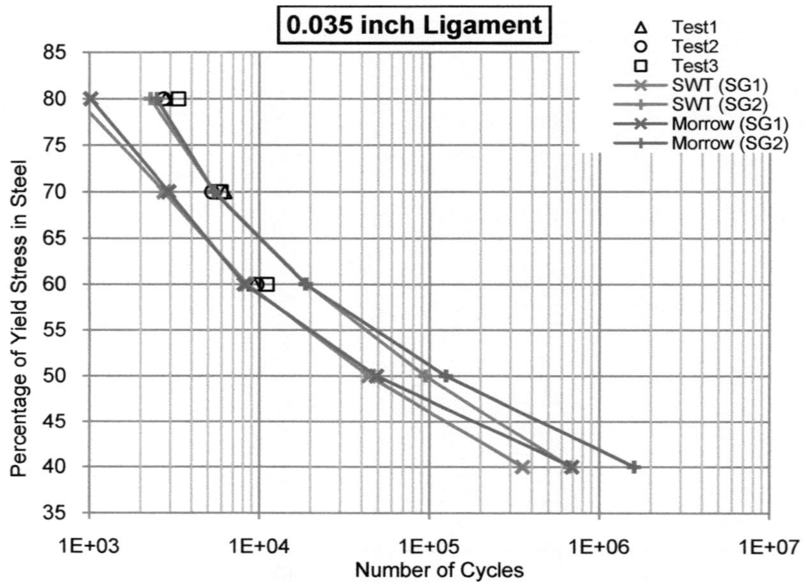


(b)

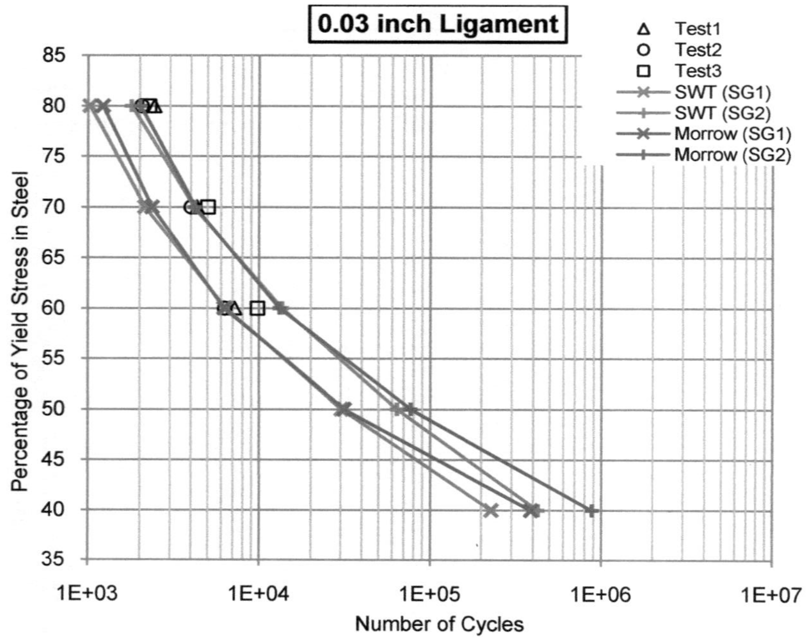


**Figure 8:** Experimental results vs. simulated results for the sensor geometry 1b

## CONCLUDING REMARKS

Results of the ongoing research and development of an in-situ fatigue sensor for detecting fatigue damage in steel bridges have been presented. Two sensor geometries with five different SCF's were tested for different ratios of yield stress in the carrier (steel beam) specimen. Constant amplitude tensile cyclic loading was applied and strain data was collected using a data acquisition system. A FE analysis was carried out for the sensor stresses and the output stresses were used in a MATLAB code to calculate fatigue life. The Smith-Watson-Topper and Morrow mean stress theories were used to account for the means stresses, and the results were plotted against the experimental values.

Overall results of the two types of sensors utilized in the research gave promising results for the development of an in-situ fatigue sensor. Smaller notches with higher SCF's (e.g., SCF > 30) were capable of providing fatigue results that can be predicted well by the simulation. Since the actual loading in a bridge is more equivalent to variable amplitude loading rather than constant amplitude, the next phase of this research will focus on the response of these sensors for variable amplitude loading.

## ACKNOWLEDGEMENT

We acknowledge with gratitude the generous financial support provided by the Oklahoma Transportation Center under the grant OTCREOS7.1-36.

## REFERENCES

1. Zhang, Y., *In Situ Fatigue Crack Detection using Piezoelectric Paint Sensor,* Journal of Intelligent Material Systems and Structures, **17**(843), 2006.
2. ODOT, *Pontis Bridge Inspection Manual for Oklahoma Bridges*, Oklahoma Department of Transportation - Bridge Division, 2006.
3. Gongkang, F., ed. *Non Destructive Testing For Steel Highway Bridges.* NDT Methods Applied to Fatigue Reliability Assessmement of Structures, ed. J. Mohammadi, American Society of Civil Engineers, 2004.
4. Jackson, C.N., Sherlock, C. N., *Nondestructive Testing Handbook: Leak Testing*, Library of Congress Cataloging-in-Publication Data, 1998.
5. Son, J., Mohammadi, J., ed. *A Review of Non-destructive Test Methods for Bridges.* NDT Methods Applied to Fatigue Reliability Assessmement of Structures, ed. J. Mohammadi, American Society of Civil Engineers, 2004.
6. Rakow, A., Chang, F. K., *An In-Situ Sensor Design for Monitoring Fatigue Damage in Bolted Joints*, in *6th International Workshop on Structural Health Monitoring (IWSHM),* DEStech Publications, Inc: Stanford, CA, 2007.
7. Gongkang, F., *Non Destructive Testing For Steel Highway Bridges*, in *NDT Methods Applied to Fatigue Reliability Assessmement of Structures*, J. Mohammadi, Editor, American Society of Civil Engineers, 2004.
8. Fisher, J.W., Kulak, G. L., Smith, I. F. C., *Fatigue Primer for Structural Engineers,* National Steel Bridge Alliance, USA, 1998.
9. Lee, D.C., Lee, J. J., Kwon, I. B., Seo, D. C., *Monitoring of Fatigue Damage of Composite Structures by Using Embedded Intensity-based Optical Fiber Sensors.* Smart Materials and Structures, **10**: p. 285-292, 2001.
10. VISHAY. *Special Use Sensors - Crack Propagation Sensors, Document Number: 91000, Revision: 18-Jul-08.* 2008, Available from: http://www.vishay.com/docs/11521/crackpro.pdf.
11. Bannantine, J.A., Comer, J. J., Handrock, J. L., *Fundamentals of Metal Fatigue Analysis*, Prentice-Hall, 1990.
12. *F.D.+E. Experimental HTML Fatigue Database*, Fatigue Design and Evaluation Committee of the SAE Experimental Web Site at the University of Waterloo, 2009.
13. Peterson, R.E., *Stress Concentration Factors*, John Wiley & Sons, Inc.,1974

Proceedings of the IMAC-XXVIII
February 1–4, 2010, Jacksonville, Florida USA


# Laser Vibrometry for Bridge Post-Repair Investigation

M. Schmieder[1]; R. Heere[2]; S.E. Chen[3]; A. Taylor-Noonan[4]

## ABSTRACT

Bridges are important elements of our traffic infrastructure. Structural failure of such elements may have devastating consequences. Therefore, monitoring the condition of such structures is of importance. Traditional monitoring (beyond visual inspection) requires the inspector or a measuring instrument to directly access the structural component to be evaluated. For locations difficult to access, laser-based vibration monitoring may now be of assistance. The authors have gathered field experience using a laser interferometer to determine the natural frequency of structural components, using ambient excitation only. Because the natural frequency is related to the stiffness of a component, some structural damages, as well as the effectiveness of structural repairs and strengthening, may be detectable. Further, repeat routine measurements at regular intervals (vibrational fingerprinting) may be able to flag structural changes which will prompt more detailed investigations by traditional means. Further, the authors have demonstrated that localised delamination of fibre composite patches on a concrete bridge can be detected using laser interferometry in combination with remote acoustic excitation. The authors are currently investigating whether this method can detect concrete delaminations, failing steel connections and other localised damages to structural elements.

The paper also will address the potential of using laser vibrometry for post-repair structural system evaluation as a validation of the rehabilitation procedure. Results from several field tests are presented. It is observed that local frequency due to voided contacts seem to be promising for damage detection.

## NOMENCLATURE

| | |
|---|---|
| **CFRP** | Carbon Fibre Reinforced Polymer |
| **DSP** | Digital Signal Processing |
| **FFT** | Fast Fourier Transform |
| **LDV** | Laser Doppler Vibrometer |
| **MTBF** | Mean Time Before Failure |
| **RC** | Reinforced Concrete |
| **SLDV** | Scanning Laser Doppler Vibrometer |

[1] Principal, IE-Consulting. 659 Foresthill Place, Port Moody, BC, Canada, V3H 3A2. ms@ie-consulting.ca. Author to whom correspondence should be addressed.

[2] Senior Materials Engineer, Metro Testing Laboratories. 6991 Curragh Avenue, Burnaby, BC, Canada, 45J 4V6. rheere@metrotesting.ca.

[3] Assistant Professor, University of North Carolina at Charlotte. 9201 University City Blvd, Charlotte, NC 28223. schen12@uncc.edu.

[4] Researcher, Metro Testing Laboratories. 6991 Curragh Avenue, Burnaby, BC, Canada, 45J 4V6. atnoonan@metrotesting.ca.

# 1 INTRODUCTION

The current process of bridge health investigation (current practice) requires a fair amount of equipment and manpower which leads to a significant cost to the operating agencies. Traffic flow is often impeded by lane closures to accommodate snooper trucks. Environmental conditions, insufficient lighting, and the inherent subjectivity of visual inspection and hammer sounding may lead to incomplete or incorrect results. Vibration and modal analysis is an emerging field, with the aim of providing detailed information on the condition and performance of a structure. Accelerometers are used in most implementations. However, placement of contact sensors poses several logistic and economic limitations. Due to the long and complicated setup and wiring process of an array of accelerometers, most in-service monitoring systems are permanently installed on the structure, while portable vibration measurement systems are used mostly for academic research projects. Permanent systems can be installed during construction when access to the structure is easy, and then left unattended to gather vast amounts of data. These systems are not without maintenance requirements: computer systems may crash and cables and sensors exposed to the weather may corrode or crack.

Another problem is that the embedded permanent instrumentation will most likely need to be replaced during the 50-75 year lifespan [1] of the structure. Current estimates for the MTBF of seismic accelerometers range from 80,000 hrs (9 years) to 300,000 hrs (34 years) [2]. The need to re-calibrate or replace sensors means access to the structure once again becomes a problem. Since a structure built to specification will most likely not require stringent monitoring in its first ten years of operation, any accelerometers installed during the construction phase may not even prove to be useful at all. Furthermore, exchanging equipment after 25 or 35 years of service may require a complete new installation due to the technical advancements in data acquisition and information technology and a lack of compatible parts with the installed system.

Scanning Laser Doppler Vibrometers (SLDV) provide a very flexible solution of remotely measuring vibration patterns without the requirement to attach instruments to a structure. This paper explores the practicality of such instruments in the process of Structural Health Investigation.

SLDV could provide a more cost-effective and convenient way of collecting vibrational data on existing structures than accelerometer setups, either permanent or temporary. The system is portable, self-contained, and does not require access to the structure or lane closures. A single portable SLDV can monitor many bridges because the long life span of a bridge means that measurements are not needed every day. Instead, measurements taken with the SLDV at intervals (every year, two years, 5 years) would provide vibrational snapshots or fingerprints of the structure that can be compared. If these snapshots are taken before and after a repair project, the effectiveness of the repair could be quantified.

In addition, SLDV could provide an affordable and reliable solution for remote delamination detection. This paper will also discuss the advantages, disadvantages and the system limitations to detect material delaminations on repaired concrete structures using carbon fiber reinforced plastic (CFRP) wrappings.

# 2 EQUIPMENT

An Ometron Type 8330 SLDV with PC-based analyzer with a 20 kHz DSP (FFT) card has been used in this study. Table 1 summarizes the specifications of the apparatus.

Table 1: Ometron Type 8330 SLDV Specifications (Brüel&Kjær, *Product Data)*

| | |
|---|---|
| Laser | He-Ne continuous wave laser, < 1mW output power, 632.8 nm (red light) |
| Working Distance | **Short Range Lens:** 50mm to 5m<br>**Long Range Lens:** 1 to 200m |
| Spatial Resolution | 1mm at 20m working distance |
| Scan Angle | 25°x 25°field of view |
| Scanning Mirror Resolution | 10.65 μRads, (0.0006102°) |

## 3 CORRELATION BETWEEN NATURAL FREQUENCIES AND STRUCTURAL BEHAVIOUR

Damage detection using vibration methods has been investigated as early as 1940s. Using multiple axial mode frequencies, Adams et al. [3] were able to identify damage location on a straight bar. Using a beam model, Chondros and Dimarogonas [4] illustrated that cracking not only introduce higher vibration amplitudes than an uncracked beam, but also introduces higher frequency vibration. Damage evaluation using SLDV has the robustness of a scanning system and is able to provide full-field local vibration signature evaluation. For damage evaluation using SLDV on bridge composite-wrap repair, the analysis scenario is very similar to previous studies on the detection of damage in composite plate-like structures; several parameters have been used including MAC (Modal Assurance Criteria) and COMAC (Co-Ordinate Modal Assurance Criteria) [5], modal stiffness and damping ratio [6] and strain curvatures [5][7][8]. The changes in the curvature mode shape were found to be more localized in the vicinity of damage as compared to the changes in the displacement mode shapes. It is clear that different modes would give different degrees of damage indication. The higher the modes, the larger the strain energy difference values. Chen et al. [9] used a SLDV to test the strain energy approach on a composite plate.

However, since the composite wrapping is attached to a structure, the scenario is actually close to the case of void detection on Frescoes. Castellini and Paone [10] first identified distinct local vibration modes due to defects on frescoes. The vibration frequencies can be directly correlated to the size of the defects. Del Vescovo and Fregolent [11] described the use of a piping system that controls the acoustic plane waves propagating to the frescoes and demonstrated that the defects can vary the acoustic field within the pipe. When no defect is present, the fresco behaves as a rigid wall, whereas the defected area would demonstrate local vibration – damage detection actually works better at the anti-resonance of the fresco. The local vibration effect is further identified by Castellini et al. [12] to exist in composite materials and a frequency sweep technique can be helpful in identifying complicated damage scenarios.

In this study, we focus on the use of vibration frequency measurements for damage detection. The natural frequency of a structure is easily measured with laser interferometry and does not require significant information of the input excitation. There is much discussion as to whether or not a realistic damage scenario for a structure will result in a measurable change in natural frequency. Little change was found in the natural frequency of a steel bridge after a 1.8m cut was made in a plate girder on the I-40 bridge [13], while the A24 bridge study found a considerable change in natural frequency after settlement was induced in a pier [14]. It is also worth noting that the natural frequency is subject to temperature and other environmental effects [15], and care must be taken to ensure that a measurement is not unduly affected by environmental conditions. Natural frequency analysis is therefore just a single indicator, and further modal shape analysis is vital for structural health monitoring. While natural frequency analysis may be able to detect global changes in the structure, modal analysis can find and locate much smaller damages. One benefit of SLDV in modal analysis is that obtaining vibration data from multiple points does not require additional hardware, therefore finely detailed mode shapes are easily attainable.

## 4 SLDV DELAMINATION DETECTION WITH REMOTE ACOUSTIC EXCITATION

The method of retrofitting structures with CFRP strips for added shear strength is gaining popularity. For quality control, the bond between the strip and the structure generally must be examined for air voids and delaminations. The current practice is to use a hammer sounding technique and mark any spot that "rings". This procedure requires little equipment and training, but it can become expensive and time-consuming because it requires direct access to the entire structure. To develop a method to perform this quality control procedure in a contactless fashion, we investigated the ability of SLDV to remotely examine a dense mesh of points for delamination. This is an automated process that only requires the operator to define the outlines of the mesh. For points with poor reflectivity, a dithering function can search for a better point close by.

## 4.1 Laboratory Results

A mockup sample comprising a hardened concrete slab portion with a single CFRP plate (1.2 mm thick) bonded to one surface was prepared. In several arbitrarily selected locations, bond between the substrate and the CFRP was prevented. Using noise excitation from a portable radio deliberately mis-tuned to produce static, the sample was scanned for vibrations between 0 and 10 kHz, with 15 trials per location, each with 0.16 s individual measurement time. The frequencies producing the most pronounced peaks were identified, and used to animate a grid model of the specimen. Delaminated areas showed a pronounced high-frequency response (above 1000 Hz) while areas without delaminations did not show any peaks at those frequencies. The largest delamination can easily be seen in the modal animation (Figure 1, near its top). Each delamination has a unique natural frequency which appears to be related to size. A smaller delamination will have a higher natural frequency. To verify that the readings correspond with the actual location of the delamination, a hammer sounding test was manually performed to locate the areas of interest. Agreement between hammer sounding and the SLDV scan was found.

Exciting the sample mechanically with a vibrator at a side perpendicular to the face carrying the CFRP plate did not result in any meaningful readings differentiating between bonded and unbonded locations.

## 4.2 Field Test: Carbon Fibre Plate

We also were able to detect CFRP delamination under field conditions. We affixed a 10 cm square carbon fibre plate to the soffit of a bridge deck, deliberately leaving the centre of the patch devoid of adhesive. Figure 2 depicts test results for the carbon fibre plate attached to the soffit of the bridge deck.

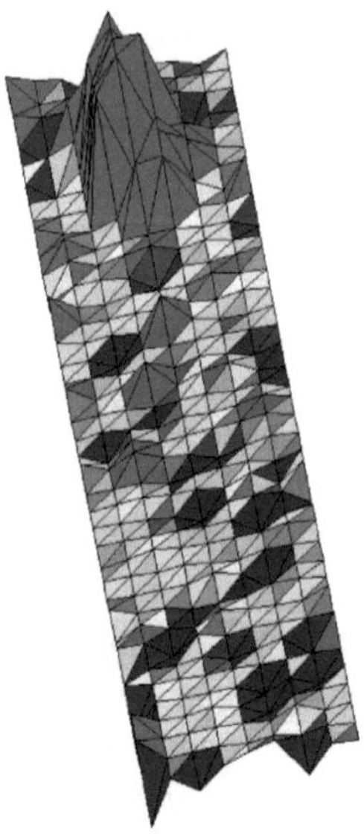

Figure 1: Mode Shape of CFRP plate at 2572 Hz

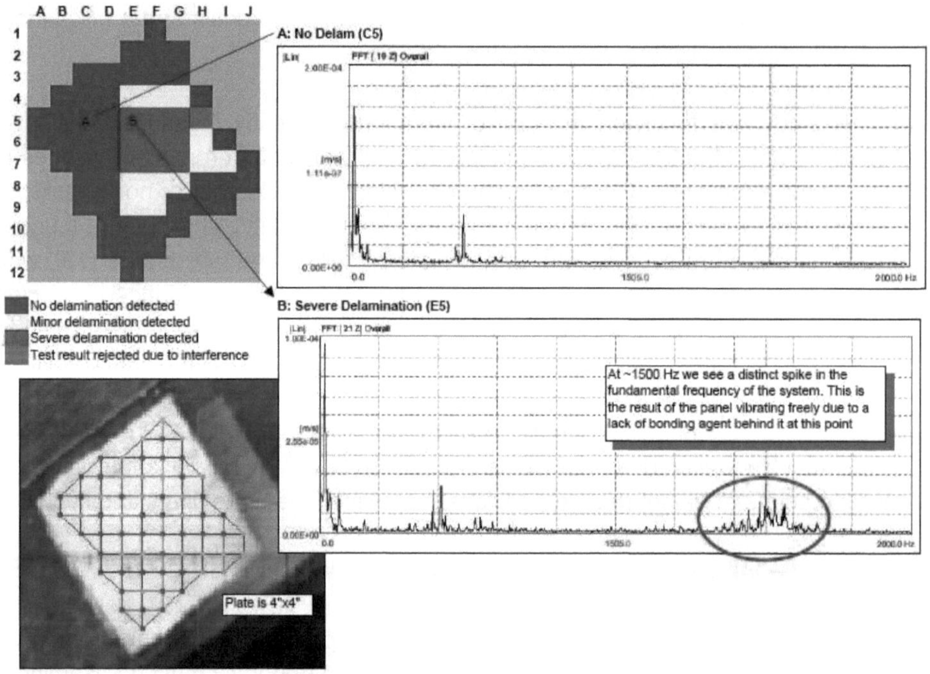


Figure 2: Delamination detection on the underside of the deck using sound impact waves

The surface of the CFRP plate was scanned in a mesh of 8 x 8 measurement points while exciting the area with the exhaust sound of a one-cylinder two-stroke gasoline engine running at estimated 50 $s^{-1}$. The distance from the excitation source to the underside of the deck was 6 m (Figure 3). The sound energy at the underside of the deck was measured at 86 dBA. As shown, the delaminated area can be clearly identified by a frequency response in the range of 1500 Hz. This is the result of the plate freely vibrating due to the lack of bonding agent between plate and concrete.

This experiment quickly demonstrated the need for a higher excitation energy. An energy of less than 84 dBA at the sample did not result in any positive readings. Energy levels above 86 dBA were suitable for detection of delaminated areas of the test specimen.

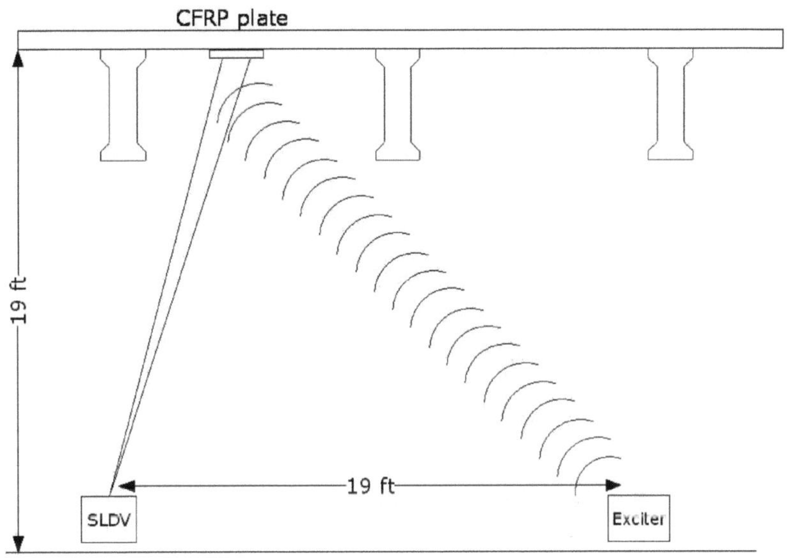


**Figure 3: Instrumentation placement**

### 4.3 Field Test: Quality Control of CFRP Reinforcement

Metro Testing Laboratories Ltd. became involved in the quality control process of a seismic reinforcement project on a RC bridge comprised of three main spans of 6.1 m each (Figure 4). The project utilized one 100 mm wide CFRP strip flanked by two strips about 50 mm wide on each of the outer girders, affixed with epoxy. Each strip is 1.2 mm thick. Flexible composite panels were wrapped over the beams (Figure 5). A hammer sounding test was carried out, and delaminations were marked on the CFRP surface using yellow chalk.

The laser test for this bridge used remote acoustic excitation, provided by a one-cylinder two-stroke gasoline engine with the muffler removed. The sound pressure was approximately 85 dBA at the surface to be scanned. With the SLDV located on the ground directly under the girder, the laser scan was conducted for frequencies between 0 and 5 kHz with a per-point measurement time of 0.32 s (repeated 5 times and then averaged). The harmonics of the bridge were seen at all measurement points. Higher frequencies were also seen at points with known delamination (from the hammer sounding test). Figure 6 shows the measurement grid upon a picture of the scanned area, and the FFT of two representative points.

It was found early on that the capability of this method depends greatly on the type and amount of excitation energy supplied to the specimen. In the transition from laboratory to field tests, this proved to be the most difficult component of the system to scale up. To achieve the acceptable excitation under field conditions, Metro Testing Laboratories Ltd. is developing a directional impact sound device which emits directional sound waves towards the measured subject. It is also obvious that the size of embedded damage can dictate the localized vibration modes, hence, a correlation between damage size and vibration frequencies should be established prior to the actual field applications.

The approach of using the frequency only method may not have the resolution required to quantify the size of damage – more advanced approaches such as using other vibration parameters – mode shapes, damping ratios and FRFs, may enhance the technique to actually quantify size of delamination.

**Figure 4: Seismic reinforcement project**

**Figure 5: Installed CFRP strips and composite wrap**

## 4.4 Conclusions of Delamination Detection Tests

In this study, the SLDV system is found well suited to delamination detection on CFRP wrapped concrete structures. The laser's scanning feature allows a dense mesh of points to be scanned with minimal setup time. A dense mesh is ideal for identifying small, localized delaminations. All analysis to find the delaminations is done in the frequency domain, so using an unknown or poorly defined acoustic excitation method appears to be suitable in principle. The results clearly indicate the identification of local vibration modes (at relatively high frequencies) can be an effective method for damage detection and quality control of concrete repairs.

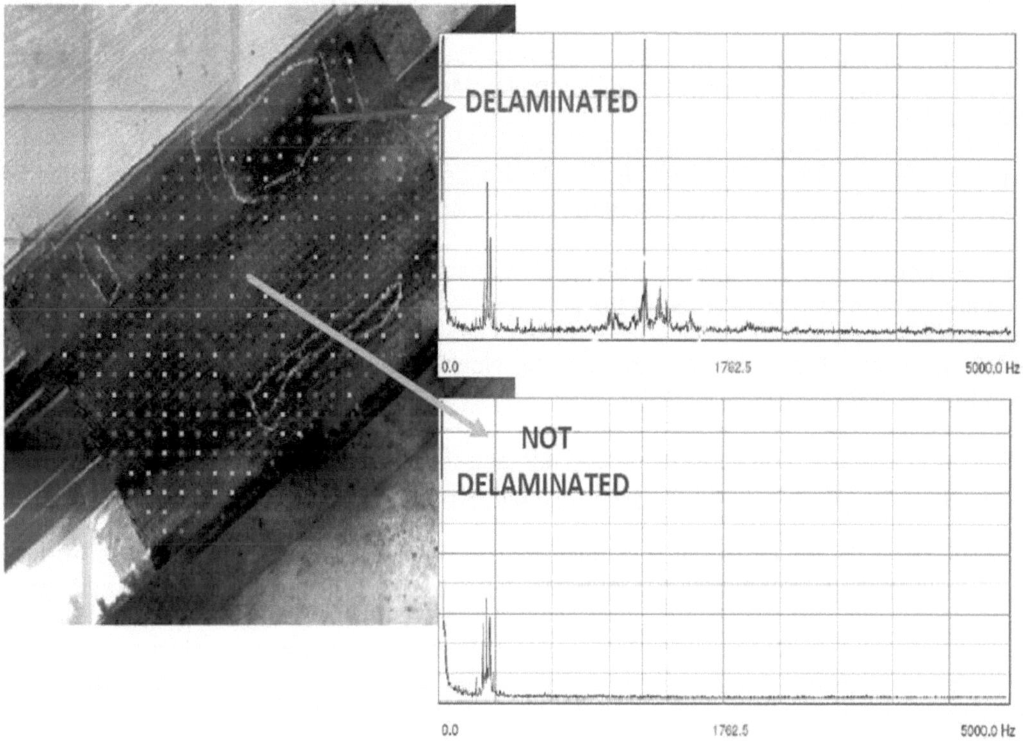


**Figure 6: Detection of delaminations between flexible composite panel and substrate**

## REFERENCES

[1] Manual on Service Life of Corrosion Damage Reinforced Concrete Bridge Superstructure Elements, NCHRP Report 558, Page 6 ff.

[2] Bâzu, M., Gălăţeanu, L., Ilian, V.E., Loicq, J., Habraken, S., Collette, J.P., Quantitative Accelerated Life Testing of MEMS Accelerometers, Sensors, Volume 2007, Issue 7, pp. 2846-2859, 20 November 2007

[3] Adams, R. D.,Cawley, P.,Pye ,C.J., and Stone, B.J., A Vibration Technique for Nondestructively Assessing the Integrity of Structures. Journal Mechanical Engineering Science, 20(2), 93-100, 1978.

[4] Condros, T.G. and Dimarogonas, A.S., A Continuous Cracked Beam Vibration Theory, J. Sound and Vibration, 215(1), 17-34, 1998.

[5] Salawu, O.S. and Williams, C., Structural Damage Detection Using Experimental Modal Analysis – A Comparison of Some Methods, IMAC, 254-260, 1993.

[6] Salane, H.J. and Baldwin Jr, J.W. Identification of Modal Properties of Bridges. ASCE *Journal of Structural Engineering*, 116 (7), 2008-2021, 1990.

[7] Pandey, A. K., and Biswas, M. (1995). Damage diagnosis of truss structures by estimation of flexibility change. *Intl. Journal. Anal. Expmtl. modal anal.,* 10 (2), 104-117.

[8] Park, S. and Stubbs, N., Reconstruction of Mode Shape Using Shannon's Sampling Theorem and Its Application to the Nondestructive Damage Localization Algorithm, Smart Systems for Bridges, Structures and Highways, SPIE, v.2446, 1995, p.281-292.

[9] S.E. Chen, S. Venkatappa, S.H. Petro and H.V.S. GangaRao, A Novel Damage Detection Technique Using Scanning Laser Vibrometry and A Strain Energy Distribution Method, *Materials Evaluation*, Vol. 58, No.12, December 2000, p.1389-1394.

[10] Castellini, P., Revel, G.M., Tomasini, E.P., Diagnostic of Composite Materials by Laser Techniques: Application on Panels with Diffierent Structures and Shapes, IMAC, 710-716, 2001.

[11] Castellini, P. and Paone, N., Development of a Measurement Procedure for Non-Invasive Detection of Frescoes' Detachments, IMAC, 1210-1213, 1995.

[12] Del Vescovo, D. and Fregolent, A., An Acoustic Non-Invasive Technique for the Diagnosis of Plaster Detachment in Fresco, IMAC, 1405-1411, 1999.

[13] Farrar, C.R. et al, Dynamic characterization and damage detection in the I-40 bridge over the Rio Grande, Los Alamos National Laboratory Report LA-12767-MS, Los Alamos National Laboratory, 01 June 1994

[14] Abdel Wahab, M.M., De Roeck, G., Damage Detection in Bridges using Modal Curvatures: Application to a Real Damage Scenario, Journal of Sound and Vibration, Volume 226, Issue 2, pp. 217-235, 16 September 1999

[15] Peeters, B., De Roeck, G., One year monitoring of the z24-bridge: environmental influences versus damage events. In Proceedings of IMAC 18, the International Modal Analysis Conference, pp. 1570–1576, San Antonio, Texas, USA, February 2000

Proceedings of the IMAC-XXVIII
February 1–4, 2010, Jacksonville, Florida USA


# Computer Vision for Structural Health Monitoring and Damage Detection of Bridges

Ricardo Zaurín, PhD Candidate
University of Central Florida
Civil, Environmental & Construction Engineering
Engr II of 116 Orlando, FL 32816
ricardozaurin@hotmail.com

F.Necati Catbas, Associate Professor
University of Central Florida
Civil, Environmental & Construction Engineering
Engr II of 211 Orlando, FL 32816
catbas@mail.ucf.edu

## ABSTRACT
Structural performance of Civil Infrastructure Systems (CIS) often decreases due to reasons such as damage, over loading, severe environmental conditions, and aging due to normal continued use. These effects will result in long-term structural damage and deterioration. As a result, novel Structural Health Monitoring (SHM) strategies are increasingly becoming more important. In this paper, integrated use of video images and sensor data in the context of SHM is demonstrated as promising technologies for safety and security of bridges. The synchronized image and sensing data are analyzed to obtain Unit Influence Line (UIL) as an index for monitoring bridge behavior under identified loading conditions. The UCF 4-span bridge model is used to explore the use of imaging devices and traditional sensing technology with UIL for damage detection. Different damage scenarios such as changes in boundary conditions and loss of connectivity between composite sections are analyzed. Experimental data is processed by means of statistical methods. Outlier detection algorithms are used to identify structural changes in large data sets obtained by monitoring and results are presented. Finally, advantages and disadvantages of the method are discussed.

## 1. INTRODUCTION
Despite limitations, visual inspection remains today the most common practice for condition assessment and damage detection for CIS. Structural Health Monitoring (SHM) is a novel approach for objectively identifying the condition and damage, and is expected to complement current visual and heuristic based assessments. SHM can be defined as the measurement of a structure's operating and loading environment to track and evaluate incidents, anomalies, damage and deterioration. SHM utilizes advanced technologies to capture the critical inputs and responses of a structural system in order to understand the root causes of problems as well as to track responses to predict future behavior. There has been extensive research on smart technologies such as wireless sensors and new algorithms for SHM. Very recently, some investigators have explored the possibility of incorporating imaging and optical devices and combining them with sensing technology. Kanda et al, applied optical motion tracking technologies to measure earthquake induced motion with surveillance cameras [1]. A test was conducted on a two-story model subjected to uni-axial seismic motion. Ten spherical 2.5 mm. markers wrapped in reflective tape were tracked by two digital video cameras. Also conventional sensors were installed for comparison. A framework was proposed by Elgamal et al (2003) combining a network sensors array, a database for storage and archival, computer vision applications for risk analysis and damage detection [2] . Achler and Trivedi (2004) discussed vehicle tracking and traffic classification [3]. Fraser (2006) presented a study where one of the main objectives was the use of video analysis of pre-recorded data, computer vision algorithms, and artificial intelligence as a mean to classify and keep records of traffic (type, number of vehicles, velocity, peak strain readings) over a fiber reinforced composite deck used as test bed [4]. This study also explored damage detection, using a one dimensional finite element model and simulating two scenarios for a damage and undamaged structure with the vehicular loads. Another framework for structural health monitoring of bridges by combining computer vision and a distributed sensor network that allows not only recording the events but identifying the change in performance possibly due to damage by using a damage index was proposed and demonstrated by Zaurin and Catbas [5, 6]. This framework consists of five integrated and closely interrelated components: the vision module, distributed sensors network array, analytical model, database and the diagnostic

module. In this framework, video stream is used in conjunction with computer vision techniques to determine the class and the location of the vehicles moving over a bridge as well as to have security surveillance on the bridge. By knowing the position and magnitude of moving loads, sensors data and video are synchronized and the structure is monitored at every instant by using operational traffic.

## 2. OBJECTIVE AND SCOPE

In the previous section, some of the different studies of integrating vision and traditional sensing technology for SHM are reviewed. The implementation of these methods presents limited results and shows there is still a need for condition assessment of structures and damage detection with conceptual damage indices. This paper describes a methodology that uses both images and sensor data in conjunction with outlier detection methods to determine the changes in structural behavior and to detect damage, especially for bridge type structures. A demonstration of the application of the method on an experimental bridge model is presented. Common damage scenarios observed in real life are simulated and studied. Data from video stream is processed to detect, classify, and track two different types of vehicles, with different load configurations, as they crawl over the bridge while sensors measure responses at various locations. Images and responses are correlated and used to create a series of Unit Influence Lines (UIL) which are used as input features to detect and localize damage in the structure (Figure 1). It is shown that this method using UIL, obtained through computer images and sensors data, can effectively identify a normalized structural response irrespective of the type and/or load of the vehicles which is used for damage detection. In addition, the new method based on statistical outlier detection from UIL's is presented along with test results. This method is shown to be successful to compile and analyze large sets of data for damage identification.

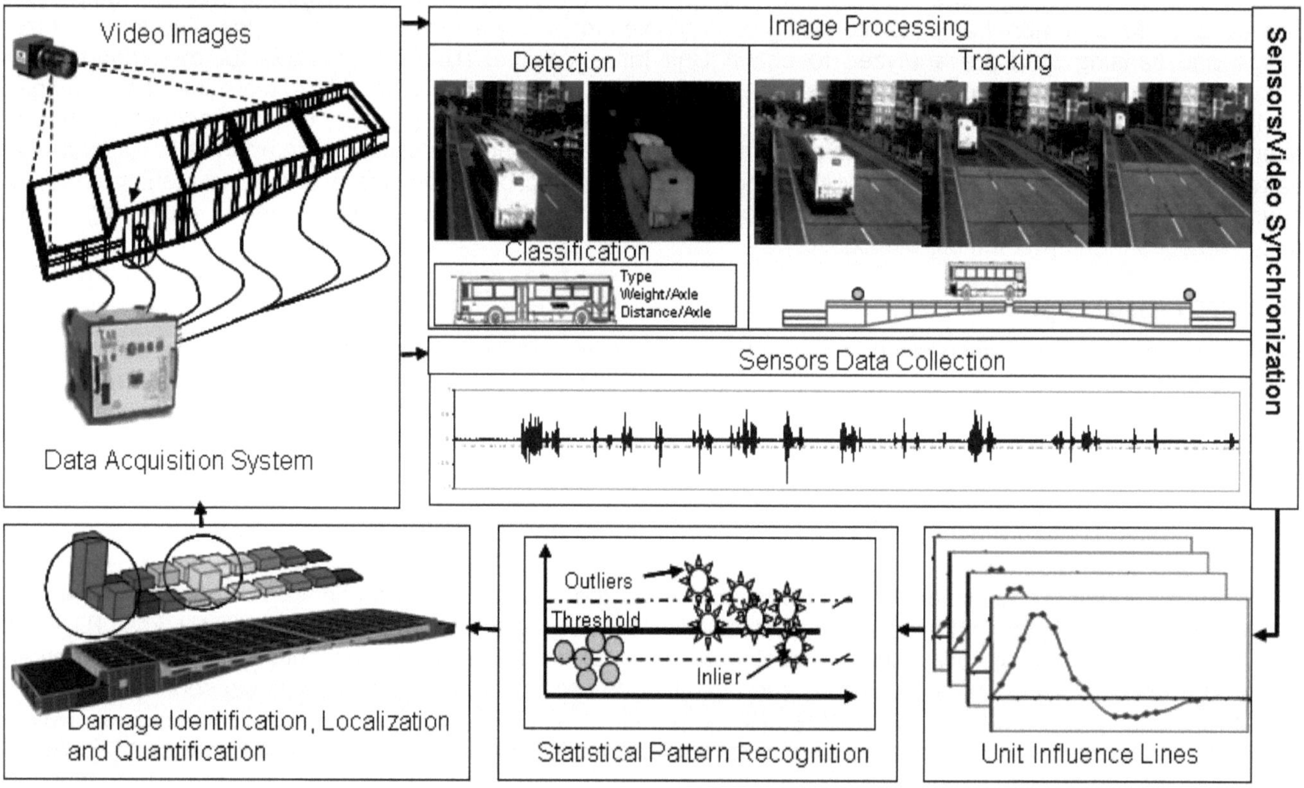


Figure 1. Overview of the Framework

## 3. THEORETICAL BACKGROUND

In this section, the theoretical basis of the proposed method is introduced. First, video images are used to detect, classify and track the vehicles (input loads) crawling over the bridge model while sensor data (responses) is captured and correlated with the loading position. By using the video image and sensor data, a set of UIL is extracted for an initial time interval ($t_0 < t < t_1$) where the structure is undamaged.

Out of this first set of UIL (represented in Figure 2 as $y_{ij}$), the mean for each UIL points ($m_{ij}$) and covariance matrix for the whole set($S$) are calculated. For each UIL, Mahalanobis distance ($Md$) is calculated to determine the variation of the set with respect to the rest of the feature sets, and a threshold is established in such a way that all UIL's within the initial set '$y$' are inliers. Then, at a new time interval ($t_1 < t < t_2$) where damage is assumed to occur, a new set of UIL's is extracted ($x_{ij}$), Mahalanobis distance is calculated and thresholded to detect possible damage on the structure. In Figure 2, a summary of the process is shown where $y_{ij}$ and $x_{ij}$ denote unit influence line coefficients obtained at two different times; "$i$" identifies the test number and "$j$" refers to the load location, for the influence line coefficient. Methodology details and examples are given in the following sections.

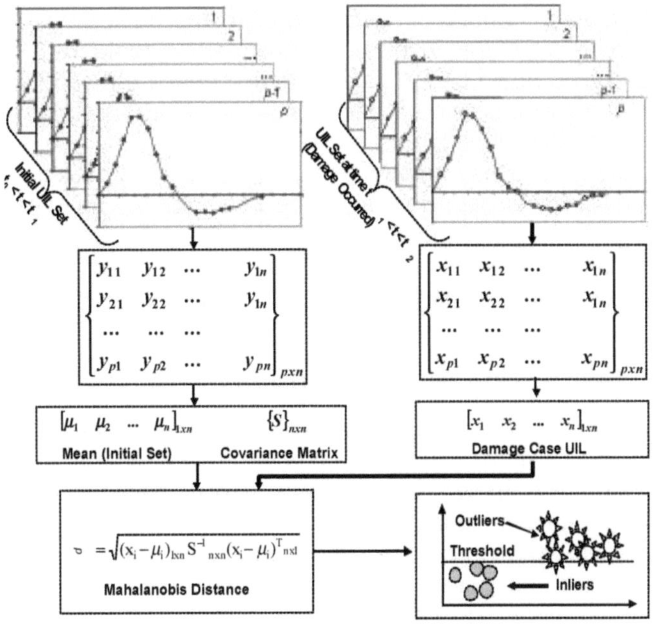


Figure 2. Overview of the Methodology

# 4. EXPERIMENTAL STUDIES

## *4.1. Test Set-Up*

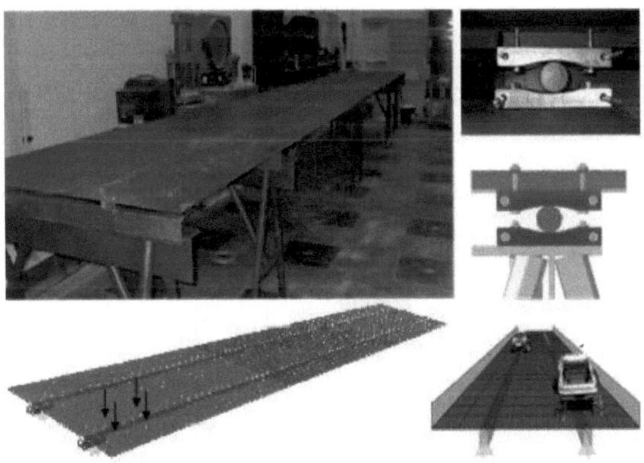

Figure 3. Experimental Set-up

An experimental setup was designed and constructed by the researchers to demonstrate the sensing and video monitoring framework along with the data analysis method. The test set up is a four span bridge-type structure consisting of two 120 cm approach (end) spans and two 304.8 cm main spans with a 3.18 mm thick, 120 cm wide steel deck supported by two HSS 25x25x3 girders separated 60.96 cm from each other. Supports were designed in such a way that they could be easily changed to roller, pin or fixed boundary conditions as shown in Figure 3. Girder and deck can be connected together by using bolts at different locations to modify the stiffness of the system and to simulate damage. Radio controlled vehicles can crawl over the deck with different loading conditions (from 4.02 kg to 15.71 kg ). Wheel axis distance and speed are also variable.

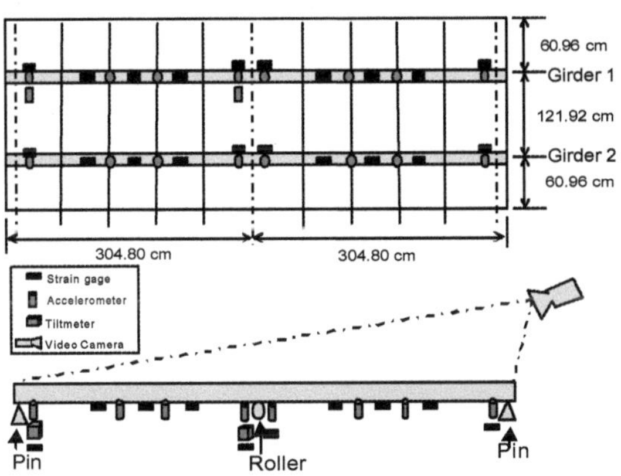


Figure 4. Sensor array

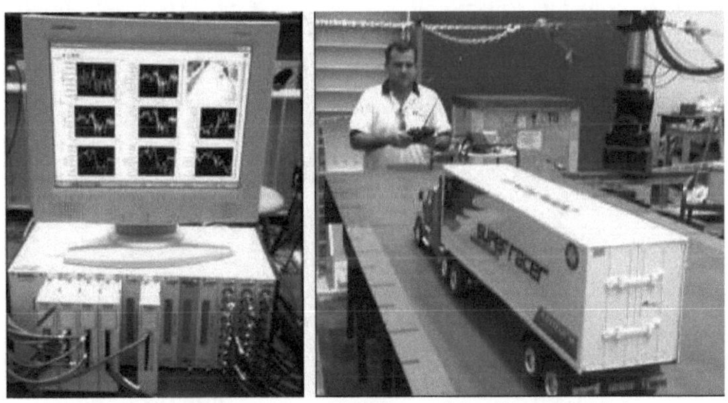

a) DAQ System b) One of the Test Vehicles

Figure 5. Data Collection

While a video camera is used to identify and track the vehicle, a set of strategically located sensors collects the data to be correlated with the video stream in real-time. It is important to mention that although the structure is not a scaled down bridge model, its responses are representative of typical values for medium span bridges. A more comprehensive description of this structure as well as the analytical model can be found in [5]. This setup is instrumented with various sensors

as shown in the instrumentation plan (Figure 4). A data acquisition system collects data while a USB type camera collects video stream data at a rate of 15 Hz. 20 foil type strain gages are sampled at a rate of 1kHz and are averaged every 100 points to minimize noise for an effective rate of 10 Hz. Figure 5 shows the DAQ and the execution of the tests. There are also sixteen accelerometers, and two dynamic tiltmeters collecting data at the same rate as the strain gages. It should be noted that only strain data is used in this study.

## *4.2. Description of the Experiment*

Two different remote controlled vehicles were used for this experiment, each one under two loading scenarios as shown in Figure 6. Every vehicle crawled over the undamaged structure for a total of 15 times for each load case. In this way, a total of 60 UIL could be extracted, ensuring a sufficient number of data points for a statistical study to determine a representative starting point per sensor. Then, the same procedure was repeated for the damaged structure.

| Weight per Axis(kg) | | | Weight per Axis (kg) | | | | | |
|---|---|---|---|---|---|---|---|---|
| Axis 1 | Axis 2 | Total | Axis 1 | Axis 2 | Axis 3 | Axis 4 | Axis 5 | Total |
| EMPTY | | | EMPTY | | | | | |
| 2.02 | 2.00 | 4.02 | 1.26 | 0.98 | 0.98 | 0.71 | 0.71 | 4.64 |
| LOADED | | | LOADED | | | | | |
| 7.18 | 7.89 | 15.07 | 1.83 | 3.22 | 3.22 | 3.72 | 3.72 | 15.71 |

Figure 6. Loading Scenarios

### 4.2.1. Damage Scenarios

Different damage scenarios were considered and studied for the UCF 4-span bridge as shown in Figure 7. These cases were chosen because they represent some of the most common issues affecting bridges, according to the Department of Transportation (DOT) engineers based on our private discussions. The first case (Figure 7a) involve changes in boundary conditions corresponding to a case usually found when rollers or pinned supports of the bridge (Figure 8a) become corroded or blocked by cinders and the structural configuration of the bridge changes. These alterations cause stress redistribution, affecting the different structural elements. In the bridge model, the case was simulated by fixing the supports as shown in Figure 8b.

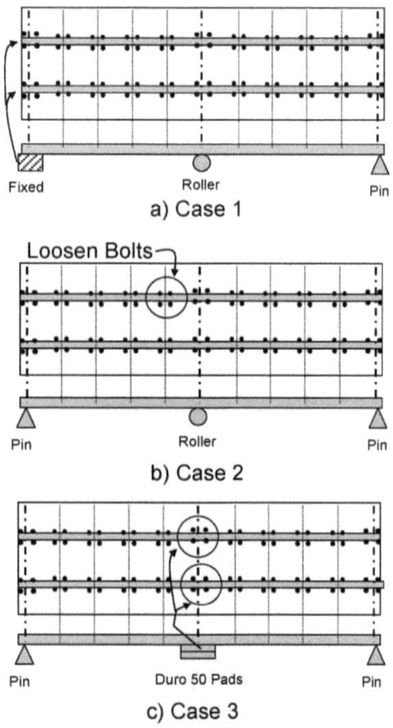


Figure 7. Studied Damage Cases

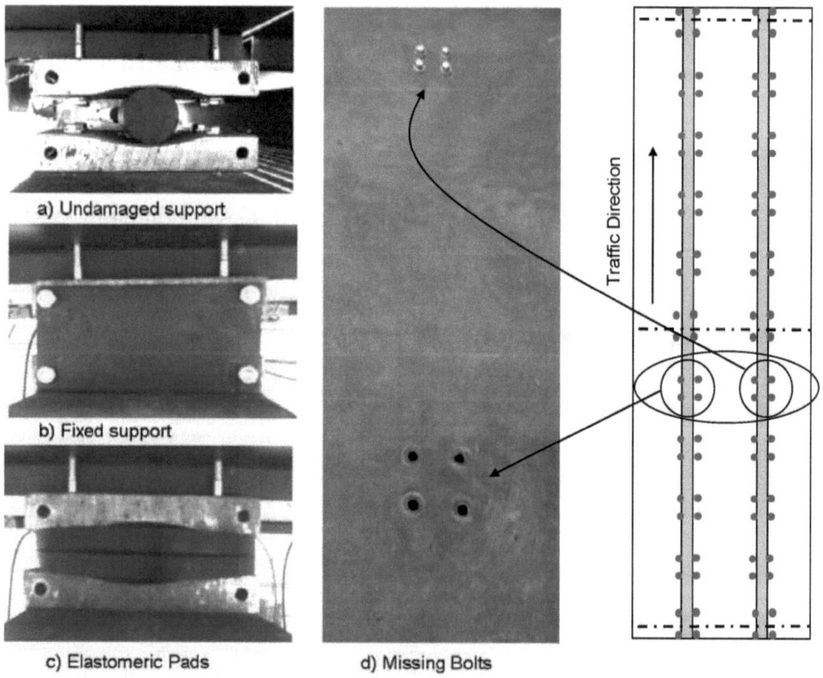


Figure 8. Simulated Damage Cases

Missing bolts and section stiffness reduction are also cases given as common by DOT engineers. Case 2 simulate the loss of connectivity between composite sections and also can generate localized stiffness reduction. For case 2, only 4 bolts were loosened (Figure 7b). Figure 8d shows the bridge model when bolts were retired for testing. Case 3 is another example of changes in boundary conditions. Here, the rollers at the central support were changed elastomeric pads (Figure 7c & Figure 8c).

# 5. DAMAGE DETECTION STATISTICAL ANALYSIS

## *5.1. Unit Influence Lines Results*

The data collected with the DAQ is typically in time domain. In order to use measured vehicle responses to extract UIL's, a direct correlation between responses and load location is needed. The raw data for the measured responses are a combination of static and dynamic responses, and noise (Figure 9a). Static responses have to be determined by filtering out the other components. The procedure used for filtering is performed by changing the time domain data into frequency domain using Fourier Transformation. Dynamic and high frequency noise components are cut out while static component is kept (Figure 9b). Resultant signal is converted back to the time domain by applying the Inverse Fourier Transformation as shown in Figure 9c. Subsequently, the computer vision algorithm is used to detect and track each one of the test vehicles by calculating the location versus time as the vehicles crawl over the bridge (Figure 9d). Finally, the structure response is correlated with the input force and location by means of synchronized computer image data and UIL are extracted as shown in Figure 9e. More detailed information can be found in [6].

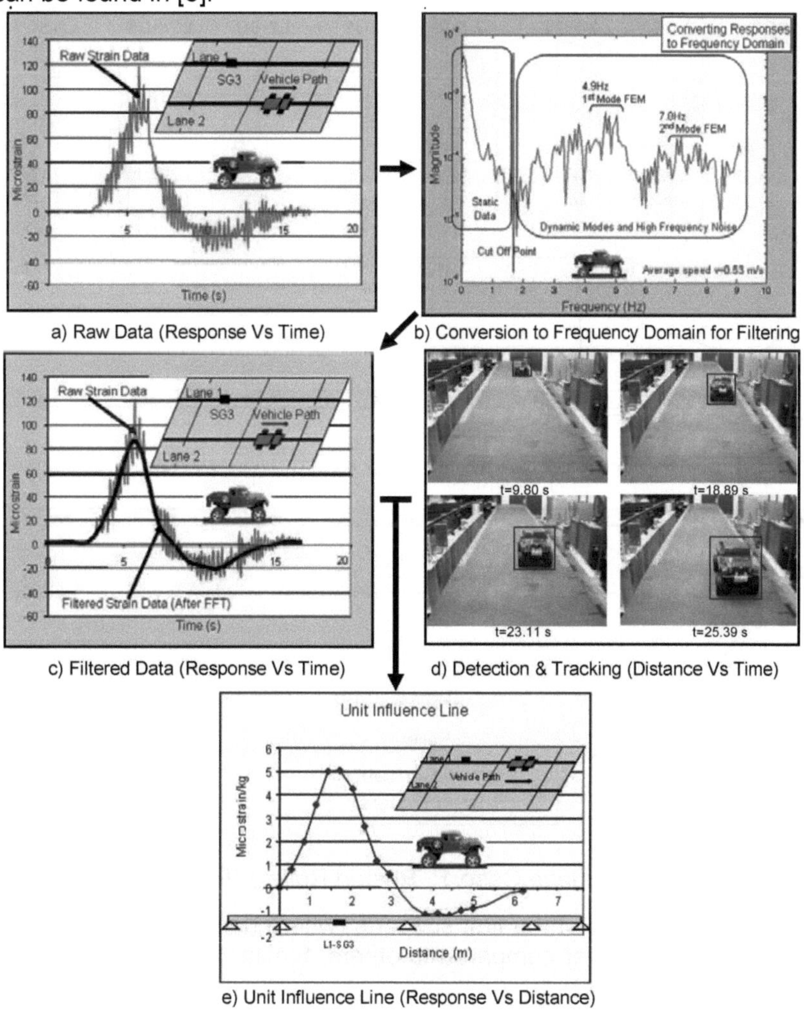


Figure 9. Procedure for Unit Influence Line Extraction

## 5.2. Outliers Detection-Clustering

After application and processing the different damage cases, influence lines were obtained as explained in the previous section. Mahalanobis distance was calculated and results were analyzed. Figure 10 shows a plot of the outliers detection for all sensors for Damage Case 1. In general, it can be seen that almost every sensor show some kind of variation with respect to the comparative case collected at $t_0 < t < t_1$. It is also noticeable that the sensors closer to the damaged area show a greater Mahalanobis distance than those farther away. Even though strain measurements provide a localized response; the method shows sufficient information for different channels to determine the damage location from the relative distances using the thresholds.

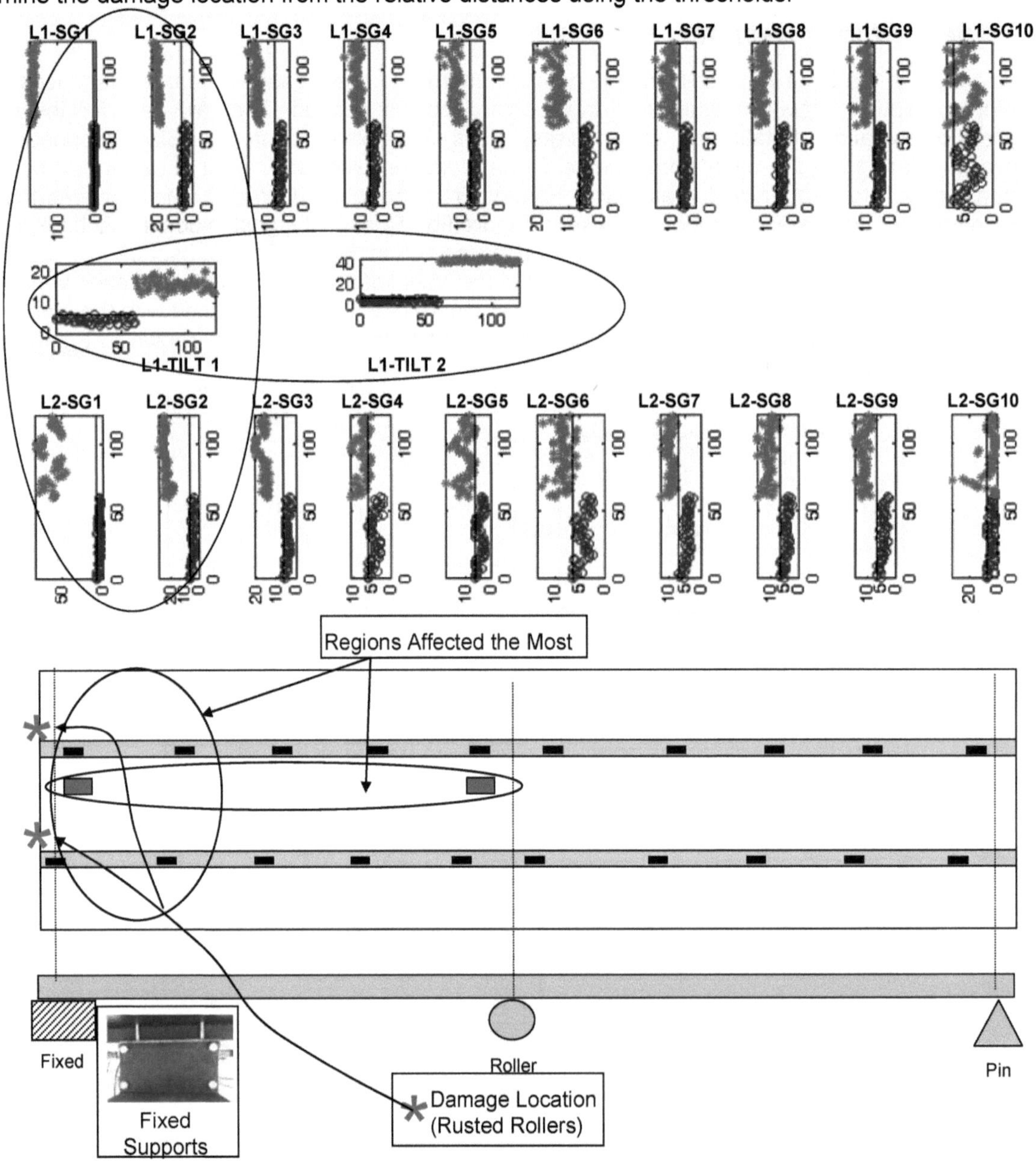


Figure 10. Damage Case 1: Rusted Rollers (First Support)

In Figure 10, the outliers near the vicinity of the first supports (where damage is induced ($t_1 < t < t_2$) indicate that the change in structural behavior is greater compared to others. Similar results were obtained for cases 2 and 3 and will be presented during the conference.

### *5.3. Identification of Damage*

In the previous section, Mahalanobis distance-based outlier detection algorithm is presented with Figure 10, showing the sensors which are more affected by a particular damage case. Although the inspection and evaluation of outliers provides valuable information, it might be tedious in some cases due to large data sets. As a result, it is more desirable to find a more effective method to not only show differences between different sets of measurements but also better pinpoint the approximate area where damage exists. Since Mahalanobis distance (*Md*) is scale invariant, it is possible to plot all sensors in a single representative graph. First, the mean of the *Md* values for the set corresponding to $t_0 < t < t_1$ and the mean of all outliers for the set $t_1 < t < t_2$ are calculated (Figure 11). If all of the calculated *Md*'s from the set ($t_1 < t < t_2$) are outliers, the distance between their mean divided by the mean of the *Md* set for $t_0 < t < t_1$ can be considered as the normalized change, denoted as $\overline{N}_d$. Therefore, $\overline{N}_d$ is a normalized indicator between *Md*'s calculated for UIL sets obtained at two different times ($t_0 < t < t_1$ and $t_1 < t < t_2$) during the monitoring of the structure.

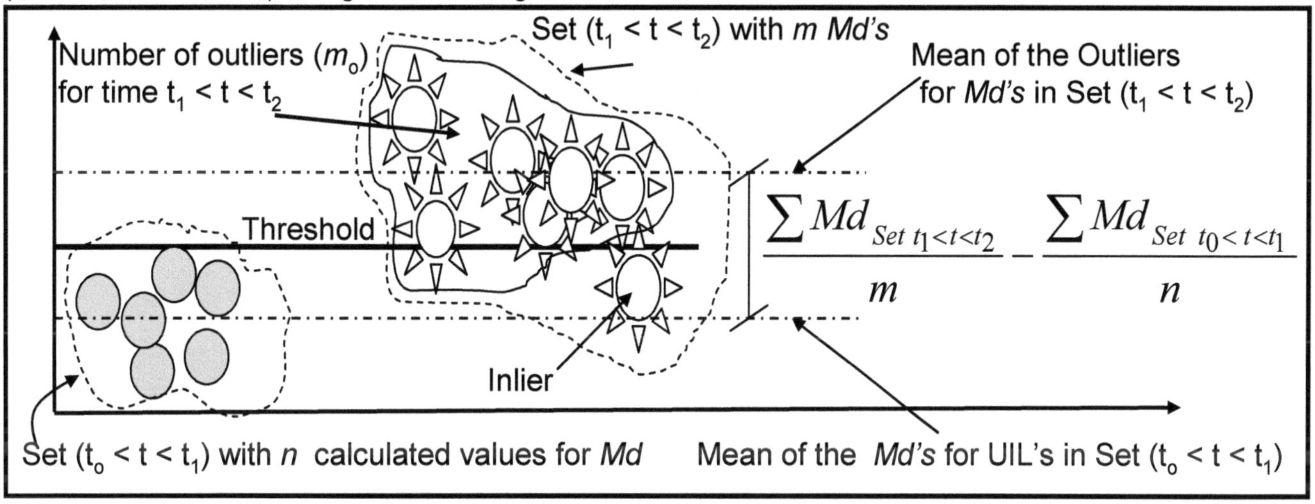


Figure 11. Distance Between two UIL Sets

If there is any inlier within the UIL set ($t_1 < t < t_2$), as shown in Figure 11, then the normalized distance $\overline{N}_d$ has to be affected by the relation between the number of outliers ($m_o$) and the total number of points within the set ($m$) as written in Equation 1.

$$\overline{N}_d = \frac{\left| \frac{\sum_{i=1}^{m_o} Md_{i(t_1<t<t_2)}}{m_o} - \frac{\sum_{i=1}^{n} Md_{i(t_0<t<t_1)}}{n} \right|}{\frac{\sum_{i=1}^{n} Md_{i(t_0<t<t_1)}}{n}} \left(\frac{m_o}{m}\right) \tag{1}$$

Figures 12, 14, and 15 show the plots of $\overline{N}_d$ for all sensors for all damage cases studied. The sensor locations are shown in the figures along with the damage location and types. The bar diagram shown on the top of the bridge and its instrumentation figure corresponds to $\overline{N}_d$ for each sensor. As discussed before, the first damage case corresponds to changes in the boundary conditions due to rust in the rollers. This damage case was simulated by fixing the supports accordingly. By simply inspecting Figure 12, it can be noticed that the plot for $\overline{N}_d$ distance shows the approximate area where the damage occurred. As can be seen from Figure 12, when the boundary conditions are fixed, the $\overline{N}_d$ corresponding at the boundary location UIL set indicates major separation from the initial case (set $t_0 < t < t_1$ corresponding to roller boundary conditions). We also notice that the two $\overline{N}_d$'s at the support location have different level of effect due to changes at the boundary conditions.

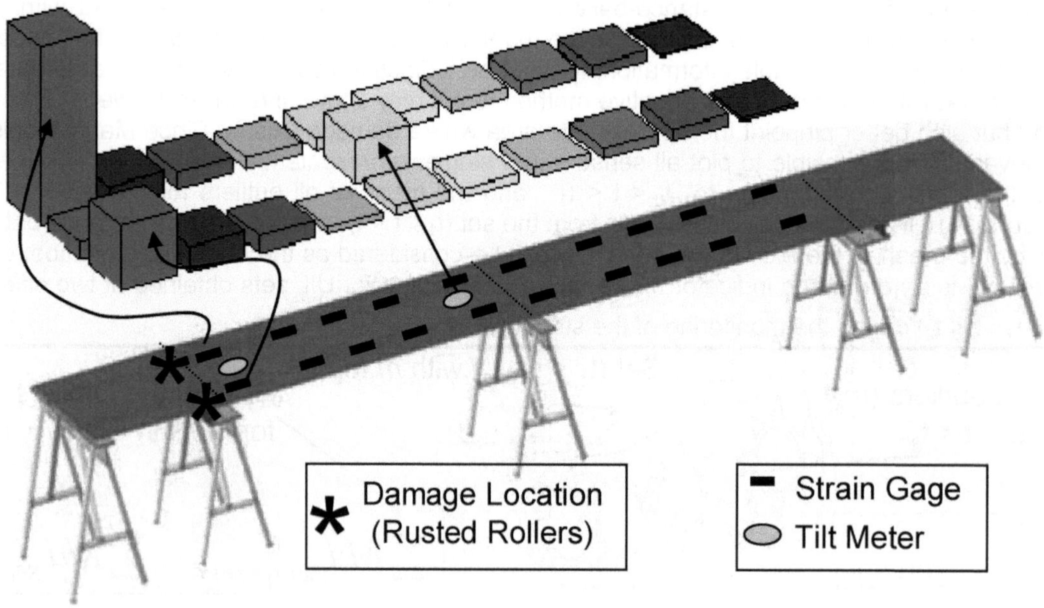


Figure 12. Damage Identification for Case 1 (Rusted Rollers on Left Support)

A closer look at Figure 12(damage case 1) shows that the tiltmeter 2 (close to the central support) presents more variation than the one located at the vicinity of the left support, where the damage was induced. At this point it is important to analyze the reason for this outcome and evaluate if this corresponds to a physical case or it is a shortcoming of the proposed method. Figure 13 shows the influence lines obtained for the undamaged and damaged case 1. The variation in the UIL's for both sensors is shown as the shaded grey area. It can be easily seen that even though the induced damaged happened at the left support, the change in UIL is greater for the tiltmeter located at the central support. The tiltmeter 1 senses a greater change while the moving load is on the first span, closer to it. However, once the load crosses the central support, the variation for the measured rotations is small. Tiltmeter 2 detects changes happening as the load moves through both spans of the continuous structure, hence $\overline{N}_d$ appears greater for this sensor.

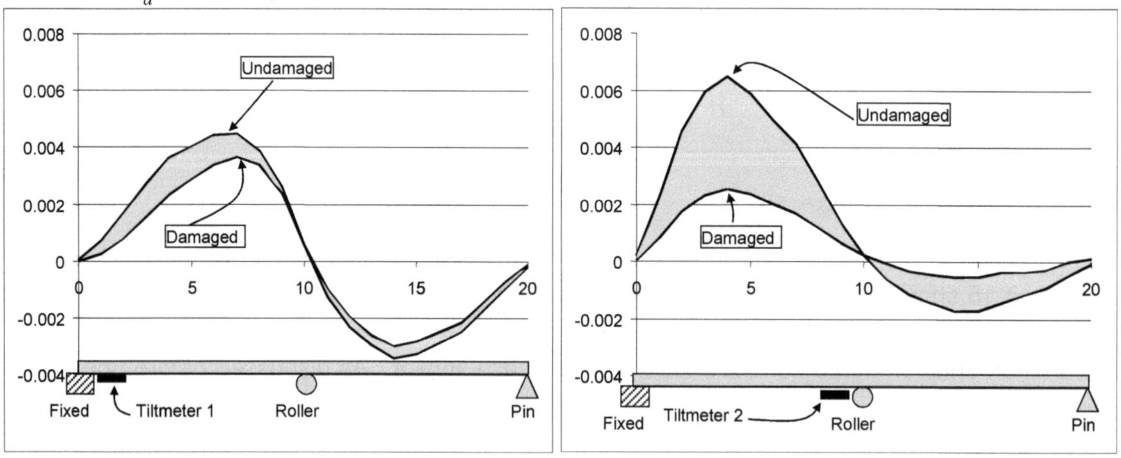


a) Tiltmeter 1 b) Tiltmeter 2

Figure 13. UIL for Tiltmeters 1 and 2 (Undamaged Case and Case 1)

Also, it is evident in Figure 12 that Girder 1 shows more relative change than Girder 2. The reason for this is that all UIL's were obtained when the vehicles were crawling on top of Girder 1. The small values for strain and rotation on Girder 2 as well as all uncertainties described before, affect Girder 2, revealing damage only at the vicinity of damage with relatively less $\overline{N}_d$ magnitude compared with the $\overline{N}_d$ magnitude for Girder 1 as shown in the bar chart in Figure 12.

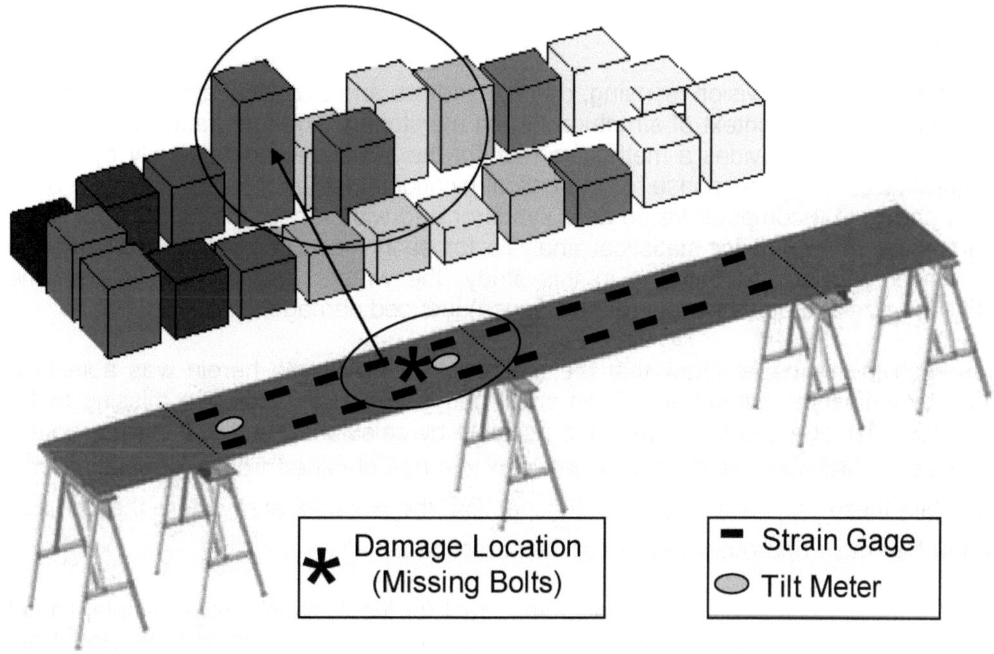


Figure 14. Damage Identification for Case 2 (4-Missing Bolts )

Figures 14 correspond to damage case 2. As previously described, these cases simulate missing bolts causing loss of connectivity between composite sections with the corresponding stiffness reduction. For case 2, only four bolts were disconnected. Once again, by simple inspection, the plot of $\overline{N}_d$ allows to detect the approximate region where the damage is present.

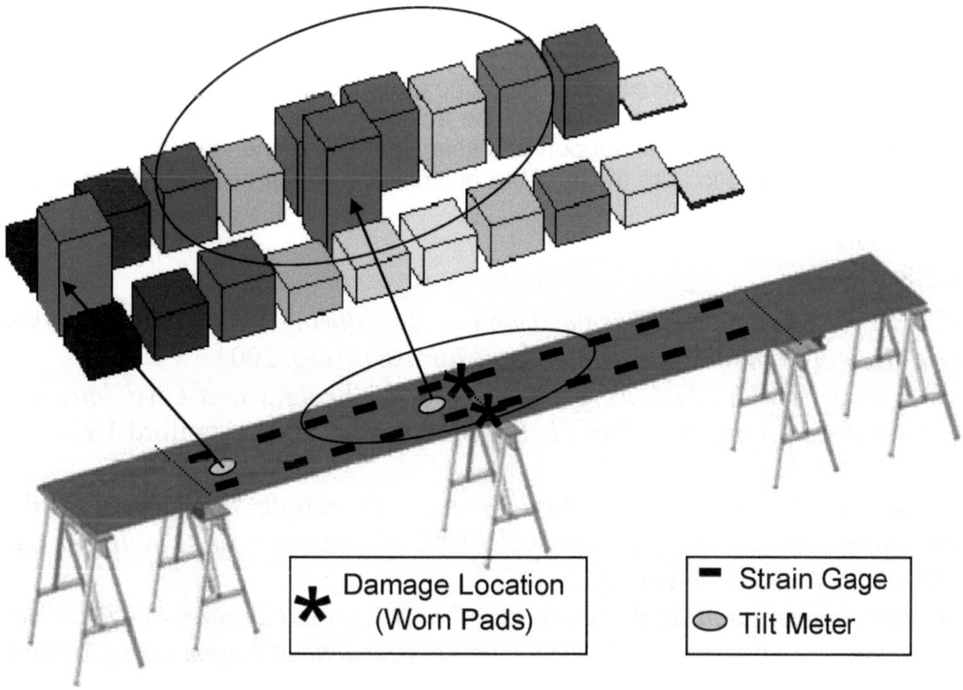


Figure 15. Damage Identification for Case 3 (Worn/Not Fully Settled Pads)

Finally, Figure 15 shows the results of the damage detection method for changed boundary conditions. Worn or not fully settled pads of the supports are simulated by replacing the central roller with elastomeric pads as shown previously. Even though this is a challenging case for detection, here is also identified the approximate region where the damage is present, leading to conducting further and more extensive examination and analysis.

## 6. CONCLUSIONS

In this paper, the use of computer vision, sensing, damage indices and statistical analysis methods are presented for damage identification in the context of structural health monitoring. The integration of computer imaging with traditional sensing technology provides a method to monitor the structures continuously by using Unit Influence Line (UIL) as a normalized bridge response for the critical locations instrumented by any type of sensor. In order to obtain UIL, it is shown that computer vision data synchronized with sensor measurements are employed. The UIL sets can further be employed for statistical analysis for each measurement location using a Mahalanobis distance based outlier detection algorithm. In this study, the writers also propose a new method to more effectively identify, localize and quantify (in a relative sense) induced damage.

The results presented in this paper show that the methodology discussed herein was able to sense damage simulations on an experimental test set-up. Even small and localized damage like missing bolts (Case 4) was successfully detected. Damage detection was also possible by calculating and plotting the normalized distance "$\overline{N}_d$" which is based on Mahalanobis distances between features obtained from a population of UIL's extracted as part of SHM. With these "$\overline{N}_d$" plots, by simple inspection, the possible area where the damage occurred can be identified and further more rigorous analysis can be prescribed.

It is also shown that the measurement type is also important for localization. For example, tiltmeters showed a clear indication of structural variations for all the studied cases. This is due to their global nature, UIL's for rotation proved to be more affected than strain when loading and damage were far from the tiltmeters. On the other hand, UIL's for strain measurement provide a more localized response making it possible to be used as a more efficient mechanism to pinpoint the damage. Although structural changes can be detected with a few sensors, the proposed method for damage localization improves if a dense spatial resolution especially in the vicinity of critical components and elements is available.

## 7. ACKNOWLEDGMENTS

The research project described in this paper is supported by Federal Highway Administration (FHWA) Cooperative Agreement Award DTFH61-07-H-00040 and the Florida Department of Transportation (FDOT) Contract # BD548/RPWO 23. The support of both agencies and their engineers is greatly recognized and appreciated. The authors would like to express their profound gratitude to Dr. Hamid Ghasemi of FHWA for his support of the advanced exploratory research program. The authors would also like to thank Mr. Marcus Ansley and Mr. Alberto Sardinas from FDOT for their support and sharing their experiences. The opinions, findings, and conclusions expressed in this publication are those of the authors and to not necessarily reflect the views of the sponsoring organizations.

## 8. REFERENCES

1. Kanda, K. and Y. Miyamoto. *"Seismic Damage Monitoring with Optical Motion Tracking"*. in *Structural Health Monitoring and Intelligent Infrastructure*. 2003.
2. Elgamal, A., et al. *Health Monitoring Framework for Bridges and Civil Infrastructure*. in *4$^{th}$ International Workshop on Structural Health Monitoring*. 2003. Stanford University, Stanford, CA.
3. Achler, O. and M. Trivedi. *Camera Based Vehicle Detection, Tracking, and Wheel Baseline Estimation Approach*. in *7th International IEEE Conference on Intelligent Transportation Systems (ITSC 2004)*. 2004. Washington, DC.
4. Fraser, M., *Development and Implementation of an Integrated Framework for Structural Health Monitoring. Doctoral Dissertation*, in *Department of Structural Engineering*. 2006, University of California at San Diego: La Jolla, CA.

5. Zaurin, R. and F.N. Catbas. *Computer Vision Oriented Framework for Structural Health Monitoring of Bridges*. in *IMAC XXV. Society for Experimental Mechanics*. 2007. Orlando, Florida.
6. Zaurin, R. and F.N. Catbas. *Demonstration of a Computer Vision and Sensor Fusion Structural Health Monitoring Framework on UCF 4-Span Bridge*. in *Internationam Modal Analysis Conference- Technologies for Civil Structures*. 2008. Orlando, Florida.

Proceedings of the IMAC-XXVIII
February 1–4, 2010, Jacksonville, Florida USA


# Experimental Investigation of Reykjavik City Footbridge

S. Živanović[1], E. T. Ingólfsson[2], A. Pavić[3], G. V. Gudmundsson[4]

[1]Civil Research Group
School of Engineering, The University of Warwick
Coventry, CV4 7AL, UK

[2]Department of Civil Engineering, Technical University of Denmark
Brovej, Building 118, 2800 Kgs. Lyngby, Denmark

[3]Vibration Engineering Section
Department of Civil and Structural Engineering, The University of Sheffield
Sir Frederick Mappin Building, Mappin Street, Sheffield, S1 3JD, UK

[4]Efla Consulting Engineers
Sudurlandsbraut 4a, 108 Reykjavik, Iceland

## ABSTRACT

This study describes experimental investigation of a 160 m long footbridge in Reykjavik. The bridge is a continuous post tensioned concrete beam spanning eight spans, the longest being 27.1 m. In plan, the structure has eye-catching spiral shape. Modal testing of the structure was conducted to identify its dynamic properties. As many as seven modes of vibration were identified in the low-frequency region up to 5 Hz. After this a series of controlled tests involving up to 38 test subjects were performed. These were designed to test vibration performance of the footbridge under various loading scenarios such as: single person either walking or jumping, group of people walking, jogging or jumping and stream of pedestrians. The severity of vibration responses of the bridge for different scenarios was then evaluated against vibration serviceability criteria defined in a guideline. It was found that the vibration performance of the bridge is quite satisfactory for wide range of loading conditions.

## 1 INTRODUCTION

To comply with vibration serviceability requirements, a footbridge has to be designed with care of realistic loading scenarios to which it could be exposed in its lifetime. Ideally these scenarios need to account for relevant load events of various probability of occurrence. Naturally, the events that occur frequently would require more stringent vibration limit with regard to users' comfort, while the limits could be relaxed for the events with low probability of occurrence [1].

The expected human activities on footbridges could normally be categorised into three groups: walking, jogging/running and jumping. The efforts of the research community have been directed towards mathematical modelling of these human activities. Most attention was paid to walking, which has been addressed in extensive literature reviews published over the last five years [2-4]. Modelling jumping force is also in a quite advanced stage [5-7] and more research is directed towards jogging/running as well [8-10]. Of these, walking is the most frequently encountered activity on footbridges, followed by occasional jogging/running and quite rare jumping loading normally associated with actions of vandals.

On the other hand extensive experimental studies that include a range of relevant loading scenarios are rare, primarily because of demanding time and human resources required for successful completion of this type of task. In addition, reliable data acquisition system and quick processing time are also required.

This paper presents a comprehensive experimental investigation of the Reykjavik City Footbridge (RCF) situated in the Icelandic capital city, with the aim of estimating modal properties of the structure and its proneness to vibration under human-induced loading.

First the description of the structure and the modal testing as well as estimated modal properties are described. Then, the tests which included up to 38 volunteers taking part in simulating various traffic scenarios are presented. The results of these tests are commented with respect to the vibration serviceability state of the structure.

## 2 DESCRIPTION OF THE STRUCTURE

The Reykjavik City Footbridge (RCF) is located in the Icelandic capital. It creates a pedestrian link, across an urban highway, between the University of Iceland and the city centre. In plan, the bridge has an eye-catching spiral shape with a total length of 160 m, divided into eight spans, where the main span is 27.1 m (Figure 1a). It is built as a continuous posttensioned concrete beam without any expansion joints, supported on guided elastomeric bearings at the end abutments and on cast-in, circular steel-concrete composite columns, with diameter of 500 mm, at the intermediate supports. At one end of the main span, spiralling steps diverge from the bridge deck to the ground to shorten the walking distance for the pedestrians. The cross section of the bridge deck is 3.2 m wide and has a variable thickness, from 170 mm at the edges, increasing to 700 mm at the centreline (Figure 1b). Four of the columns are supported on piles and the rest on a compacted fill [11].

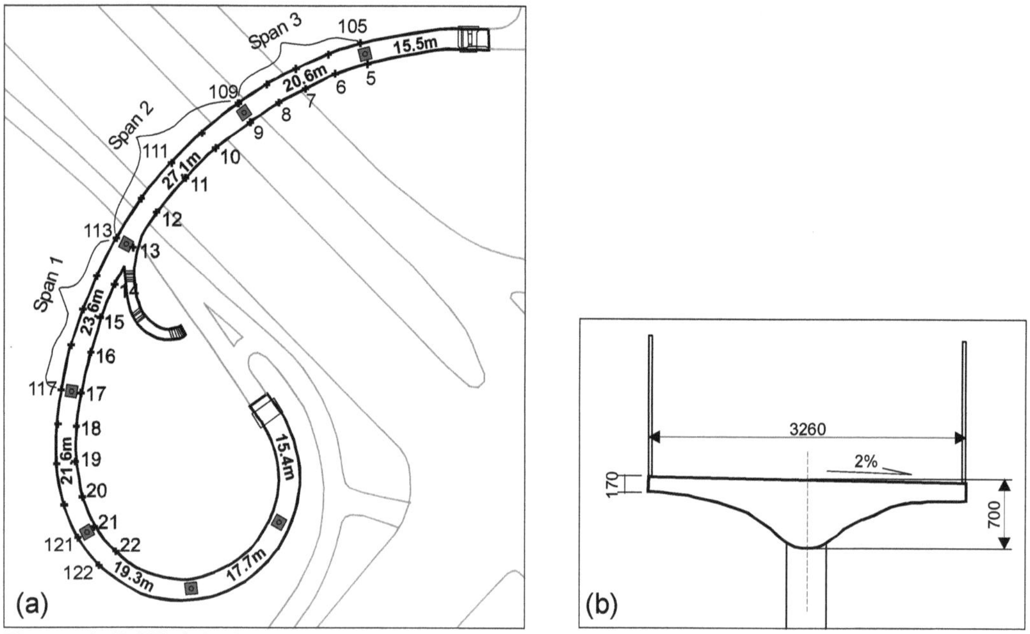

Figure 1: RCF: (a) plan view including test point numbers and (b) cross section (not to scale).

## 3 MODAL IDENTIFICATION

This section starts with the description of modal testing procedure. This is followed by the estimation of modal properties of the bridge in the frequency range up to 5 Hz and some comparison with an FE model.

### 3.1 FRF-based modal testing

To identify the modal properties of the RCF a set of frequency response functions (FRFs) was acquired experimentally. The modal testing was conducted using random excitation. The sampling frequency was 51.2 Hz. The data were acquired in 10 averages; each lasting 40 s. Hanning window with 50% overlapping was applied making overall duration of each measurement setup 220 s long.

The excitation was generated using an electrodynamic shaker, model APS 113. The force induced by the shaker was determined indirectly, by measuring acceleration of the reaction masses (using a piezoelectric ENDEVCO accelerometer) and multiplying this by the moving mass of the shaker (13.6 kg). Three force balanced QA accelerometers were available for measuring the responses in the vertical direction across measurement points shown in Figure 1a. The grid of 36 response points (TPs 5-22 and 105-122) was covered in 12 measurement setups. The shaker was placed at TP111 (centre of main span) to allow for identification of both vertical and torsional modes.

Figure 2 shows the shaker and an accelerometer in an actual setup during experiments. The bridge was closed for traffic during the measurements.

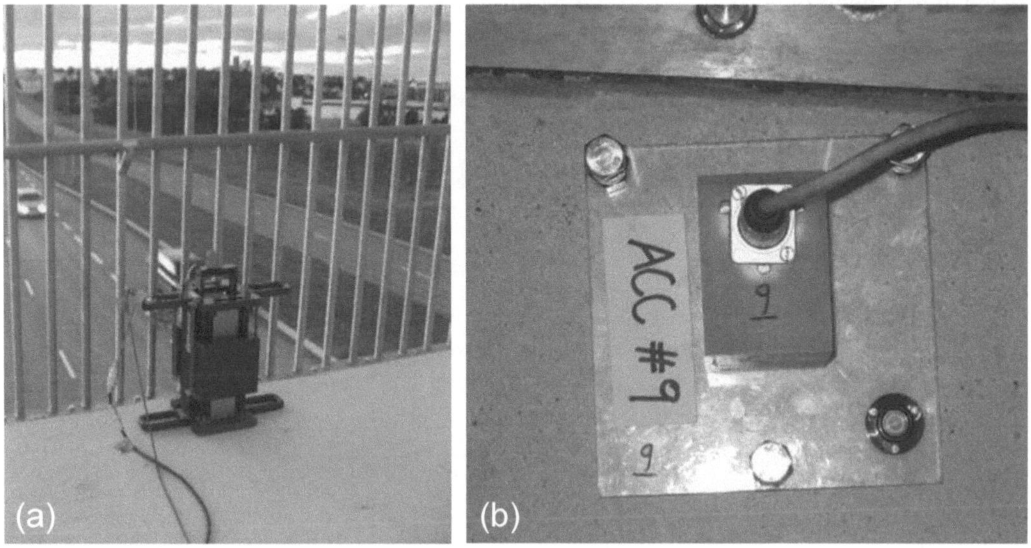


Figure 2: (a) Shaker at TP111 and (b) QA accelerometer at TP10.

### 3.2 Modal properties

The modal properties of the structure (mode shapes, modal damping ratios, natural frequencies and modal masses) were obtained using a MDOF curve fitting procedure available in MEscope software [12]. Figure 3 shows a very good matching between the measured and experimentally estimated FRF moduli for the first seven modes estimated between 0 and 5 Hz, the first mode natural frequency being at just over 2 Hz, as indicated by the first FRF peak observable.

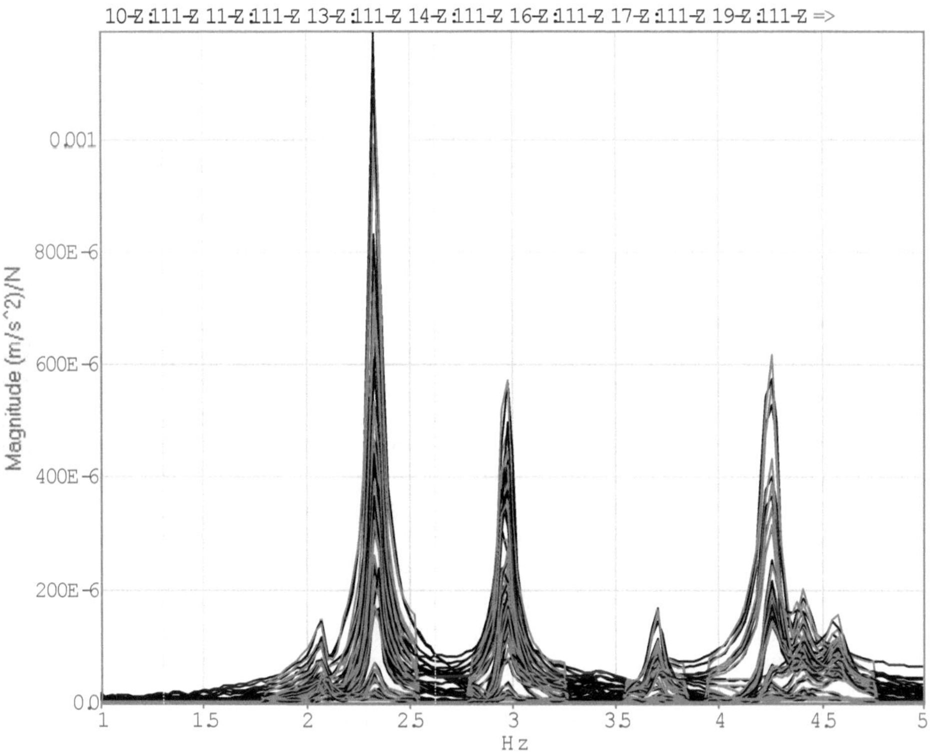


Figure 3: Experimentally estimated (black lines) and after curve-fitting regenerated (red lines) moduli of all measured FRFs.

Seven modes of vibration identified by this curve fitting are presented in Figure 4 for all natural frequencies between 0 and 5 Hz. The first two modes are most interesting from the vibration serviceability point of view due to their natural frequencies being in the range of normal pacing rates (1.5-2.5 Hz).

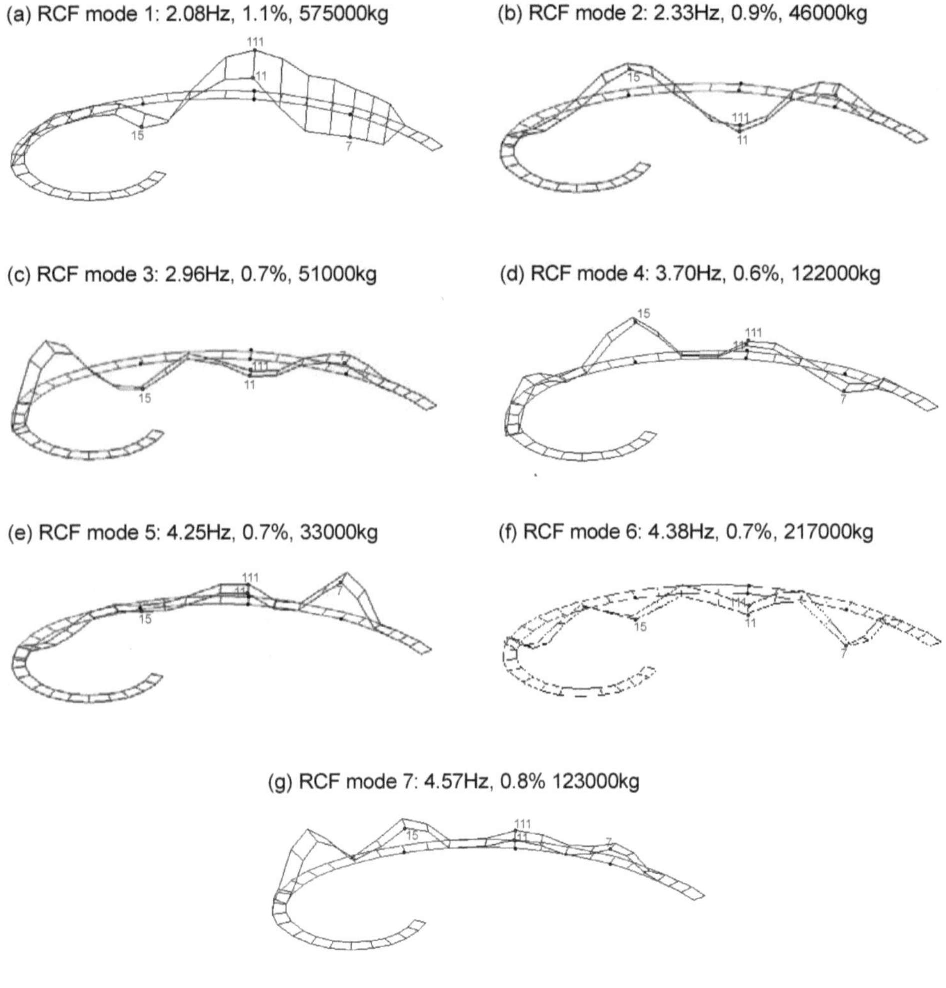


Figure 4: Experimentally identified vibration modes with natural frequency up to 5 Hz.

Looking at Figure 3, it is clear from the experimentally estimated FRF moduli that the first peak is more than five times lower than the second peak at approximately 2.3 Hz. Yet, Figure 4 indicates that the damping of the first and second mode are similar and within the range expected for a concrete footbridge. However, the estimated modal mass corresponding to the unity-scaled modeshape of the first mode of vibration is an order of magnitude greater than the modal mass of the second mode of vibration. Although this result could look suspicious, it actually makes sense considering the curved shape of the footbridge shown in Figure 1. Considering that this is a torsion-like mode and that the footbridge is curved in plan and supported by single columns, it appears that the first mode is actually not a torsional mode but a 'rocking' mode where the whole bridge deck moves both sideways and torsionally resulting in a motion which engages much more mass of not only the superstructure of the footbridge but also of the substructure and surrounding soil. Unfortunately, due to time limitations to carry out the modal testing in field conditions it was not possible to carry out additional FRF swipes with horizontal shaker excitation and accelerometer orientation. However, an FE model was developed prior to the modal testing using the commercial FE software SAP 2000 [13]. The bridge superstructure was modelled using shell elements to capture the three-dimensional behaviour of the cross section, whereas the columns where modelled with beams. The modal results from the FE model clearly indicated a trend for a potentially large difference between modal masses of the first and second mode of vibration: $M_1$=154400 kg versus $M_2$=42700 kg (Figure 5). The agreement of measured parameters for mode 2 with the measured data was quite good. However, without modelling the soil and flexibility of foundations, the FE model could not give more accurate estimate of the modal mass in the first mode.

Reasonable regenerated FRF moduli (Figure 3) and estimated modal damping values in all modes of vibration, as well as logical explanation for extremely large modal mass estimate in the fundamental mode of vibration indicate that good quality estimates of modal properties have been obtained via modal testing and subsequent FRF data analysis. In a rare exercise of this kind pertinent to civil engineering structures, these experimentally obtained modal properties could be used in some future analytical dynamic analysis of the structure under dynamic forces induced by human activities.

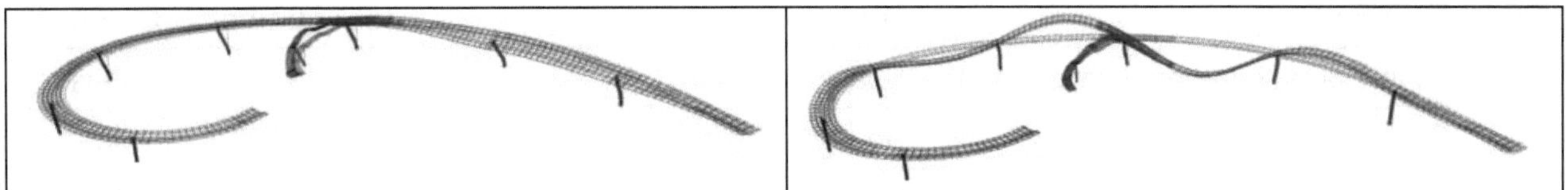

| Mode 1: Lateral and torsional vibration<br>$f_1$= 2.06 Hz, $M_1$ = 154400 kg | Mode 1: Vertical vibration<br>$f_2$= 2.33 Hz, $M_1$ = 42700 kg |
|---|---|

Figure 5: First two modes of vibration estimated prior to testing using FE analysis.

As already mentioned, the first two modes are expected to be most responsive to walking activities, i.e. walking excitation by the first forcing harmonic. The third mode has potential to be excited by the first harmonic of either jogging or jumping, as well as the second walking harmonic. The higher vibration modes could be excited mainly by second or higher harmonics of human activities, but their contribution to the responses is not significant, as will be shown in the next section.

## 4 VIBRATION PERFORMANCE UNDER HUMAN LOADING

This section presents the systematic tests including up to 38 test subjects, preformed to investigate vibration behaviour of the structure under a range of human activities. The results of these tests are commented with respect to vibration levels induced and their acceptability for footbridge users. The acceptable vibration limits are not well established in the current practice. A risk based approach to these is a way forward and is implemented in Setra guideline for footbridges [14]. This was the main reason to evaluate the footbridge vibration performance against the vibration levels recommended by Setra. This guideline defines frequency independent vibration limits in the form of peak acceleration value. The vibration in the range of up to 0.5 $m/s^2$ is considered to provide maximum comfort for footbridge users, the response level between 0.5 and 1.0 $m/s^2$ provides mean comfort, while vibration between 1.0 and 2.5 $m/s^2$ is defined as providing minimum comfort. Vibrations above 2.5 $m/s^2$ are considered unacceptable.

### 4.1 Description of tests involving humans

A comprehensive series of tests was conducted to test structural vibration response under wide range of traffic scenarios. Tests included three types of activities: walking, jogging and jumping. The testing area was limited to the main span of 27.1 m (Span 2 in Figure 1a), and its neighbouring spans (Span 1 and Span 3 in Figure 1a) due to multi-span nature of the footbridge and limited number of test subjects available. All tests were video recorded using two video cameras synchronised with the response data acquisition system.

In total 30 tests were executed, results of 28 being presented in this paper. The overview of the tests is presented in Table 1. In most tests the test subjects were asked to perform activities in their own free style (frequency), except for single person walking and jumping tests in which the test subject was asked to follow the metronome beat to match a natural frequency of the structure.

The bridge is not frequently used for the time being. However, based on the location of the bridge and the fact that the loading scenario involving traffic comprised of stream of pedestrians is potentially the most relevant traffic scenario, it would be expected that under this scenario the maximum comfort for footbridge users is provided (white rows in Table 1). Under single person walking at a resonance frequency, the mean comfort could be required since this loading is not expected to occur very often due to relatively fast pacing rates required to match the natural frequency of one of the first two vibration modes. Similarly, for people walking in groups, the mean comfort could be required due to the fact that people are generally less worried about vibrations when accompanied by friends. The tests with the mean comfort expected are marked as light grey rows in Table 1. Finally, it was decided that for group of people either jogging or jumping the minimum comfort should be required (dark grey rows in Table 1). This decision was made because people who are jogging are not likely to perceive strong vibrations due to their jogging action. Even if some other pedestrians are crossing the bridge at the same time and perceive strong vibrations, this would be for quite a short amount of time needed that the joggers cross the bridge and the pedestrians would have visual clue about the source of those strong vibrations. Regarding the vandal action by jumping, this is an extremely rare event and therefore anything apart from requiring minimum comfort could be too demanding and expensive. However, this load case should be treated in the ultimate limit state analysis as the load amplitudes from synchronised jumping can produce large stresses not considered in the static design of the bridge. Figure 6 shows two photographs related to some of the tests conducted.

Table 1: Description of tests with human-induced loading.

| Activity | Description | Direction | Test number | Comfort |
|---|---|---|---|---|
| WALK | *Single person (resonance loading) | Span 1 -> Span 3 | 29 | Mean |
| | Group of 5 | Span 1 -> Span 3 | 3, 4, 5, 6, 7 | Mean |
| | Group of 8 | Span 1 -> Span 3 | 16 | Mean |
| | Group of 10 | Span 1 -> Span 3 | 13, 14, 15 | Mean |
| | Stream (Poisson distribution) | Span 1 -> Span 3 (0.2ped/s) & Span 3 -> Span 1 (0.2ped/s) | 1 | Max |
| | Stream (Uniform distribution) | Span 1 -> Span 3 (0.33ped/s) & Span 3 -> Span 1 (0.33ped/s) | 2 | Max |
| | Stream (gradual increase of number of participants from 2 to 38) | Circular walking over Span 2 only | 25 | Max |
| JOG | Group of 5 | Span 3 -> Span 1 | 8, 9, 10, 11, 12 | Min |
| | Group of 8 | Span 3 -> Span 1 | 20 | Min |
| | Group of 10 | Span 3 -> Span 1 | 17, 18, 19 | Min |
| | Group of 38 | Span 3 -> Span 1 | 22, 24 | Min |
| JUMP | *Single person (resonance loading) | At the midspan of Span 2 | 30 | Min |
| | Group of 38 | At the midspan of Span 2 | 26 | Min |

**Well trained test subject*

Figure 6: (a) Test 25: all participants walking over Span 2 in circular manner. (b) Test 16: eight people walking across the bridge.

### 4.2 Results from tests involving humans

Test points 11 and 111, being in the middle of the longest span, had strongest response in all tests. In this section all time histories presented are those measured at TP111.

### 4.2.1 *Response to walking*

In Test 29 controlled walking of a single pedestrian over three spans was conducted with help of a metronome. First it was walking to match the natural frequency of the first, and then of the second mode (Figure 7a). It could be seen that the peak response reached 0.4 m/s$^2$ for resonance in mode 2. This is quite acceptable acceleration level, especially having in mind that the normal walking at a pace of 2.33 Hz could be rarely expected in real life, and that the walking test was performed by a test subject trained in matching the metronome beat. Therefore, under a single random pedestrian on this bridge, lower acceleration levels than 0.4 m/s$^2$ could be expected.

On the other hand, maximum response to five people crossing the bridge (without pacing rate control) was 0.2 m/s$^2$ (Figure 7b). This level of the response was never exceeded for larger groups of eight and ten people. Therefore, groups of 5-10 people could not induce more than half the response to single person walking to match the resonance. According the Setra guideline, these vibration levels provide maximum comfort to bridge users, which is better than required by specification in Table 1.

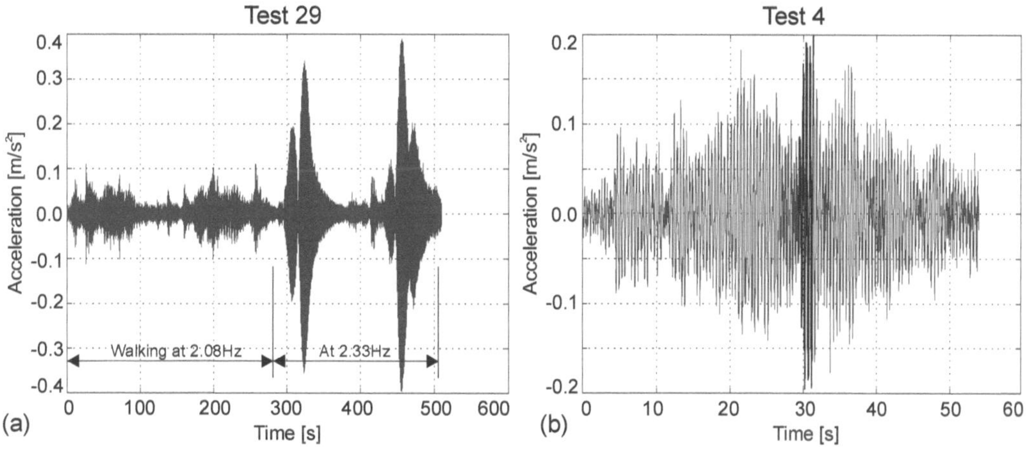


Figure 7: (a) Acceleration record at TP111 due to walking of: (a) a single person and (b) group of five people.

The first test with a stream of pedestrians was arranged in the way that people were entering the testing area (i.e. three spans of the bridge) following Poisson distribution of arrival times with mean arrival rate of 0.2 people/s from each end. Therefore total arrival rate was 0.4 people/s. Since typical time to cross the bridge is 50 s, then it follows that the total number of people on the three spans at any time during this 16 minute long test was around 20, i.e. 0.09 ped/m$^2$. The time history recorded at TP111 is shown in Figure 8a, with the corresponding spectrum in Figure 8b. Similar vibration levels were achieved under stream of pedestrians arriving on the bridge at a rate of 1 person each 3 s from each end of the testing area (Figure 8c,d). This scenario involved about 33 people on the bridge at any time (equivalent to 0.14 ped/m$^2$) during this 11 minute long test. The response in these two tests, being lower than 0.5 m/s$^2$, could be considered to provide maximum comfort to bridge users, as required.

Last walking test was conducted over Span 2 only, to increase the density of pedestrians on the bridge. The gradual increase in number of people on the bridge from two to 38 did not trigger vibration levels above 0.4 m/s$^2$ (Figure 8e,f). However, 38 people circulating on 27.1 m long span (corresponding to density of 0.43 ped/m$^2$) for about five minutes induced higher vibration levels that easily exceeded Setra limit for maximum comfort of 0.5 m/s$^2$, with maximum value measured being 1.1 m/s$^2$.

From the spectra, it is clear that first two modes are most responsive in these walking tests, which is expected having in mind their natural frequencies being excitable by the first harmonic of the walking force.

### 4.2.2 *Response to jogging*

In tests with 5, 8, 10 and 38 people jogging, the strongest response was measured in either mode 2 (2.33 Hz) or mode 3 (2.96 Hz). An example of the latter is given in Figure 9, for the test with 38 participants. This test induced the strongest vibration response of 1.16 m/s$^2$. The peak response to jogging in all tests ranged from 0.4 to 1.16 m/s$^2$ (Figure 10). The response generally increased with increasing number of participants. However, the vibration levels measured belong either to maximum or mean comfort level according to the Setra guideline, except for the test shown in Figure 9 corresponding to the minimum comfort. This is satisfactory having in mind that the minimum comfort was required in Table 1 for this loading scenario.

### 4.2.3 *Response to jumping*

Jumping activity can be linked with vandal loading, and therefore belongs to loading scenarios that occur rarely. A test subject well experienced in matching metronome beat tried to excite strongly the first, second and third mode while jumping (Figure 11a). He managed to generate response up to 1.44 m/s$^2$, when matching the frequency of the second mode, which is mode that was most responsive. On the other hand, large group of 38 people jumping (and not managing to jump in a well synchronised manner) induced vibrations up to 3.38 m/s$^2$ (Figure 11b). Therefore, with large group of vandals the response of the structure becomes unacceptable (according to Setra guideline), while in case of one person the response might be in the minimum comfort zone, as would normally be required for vandal action.

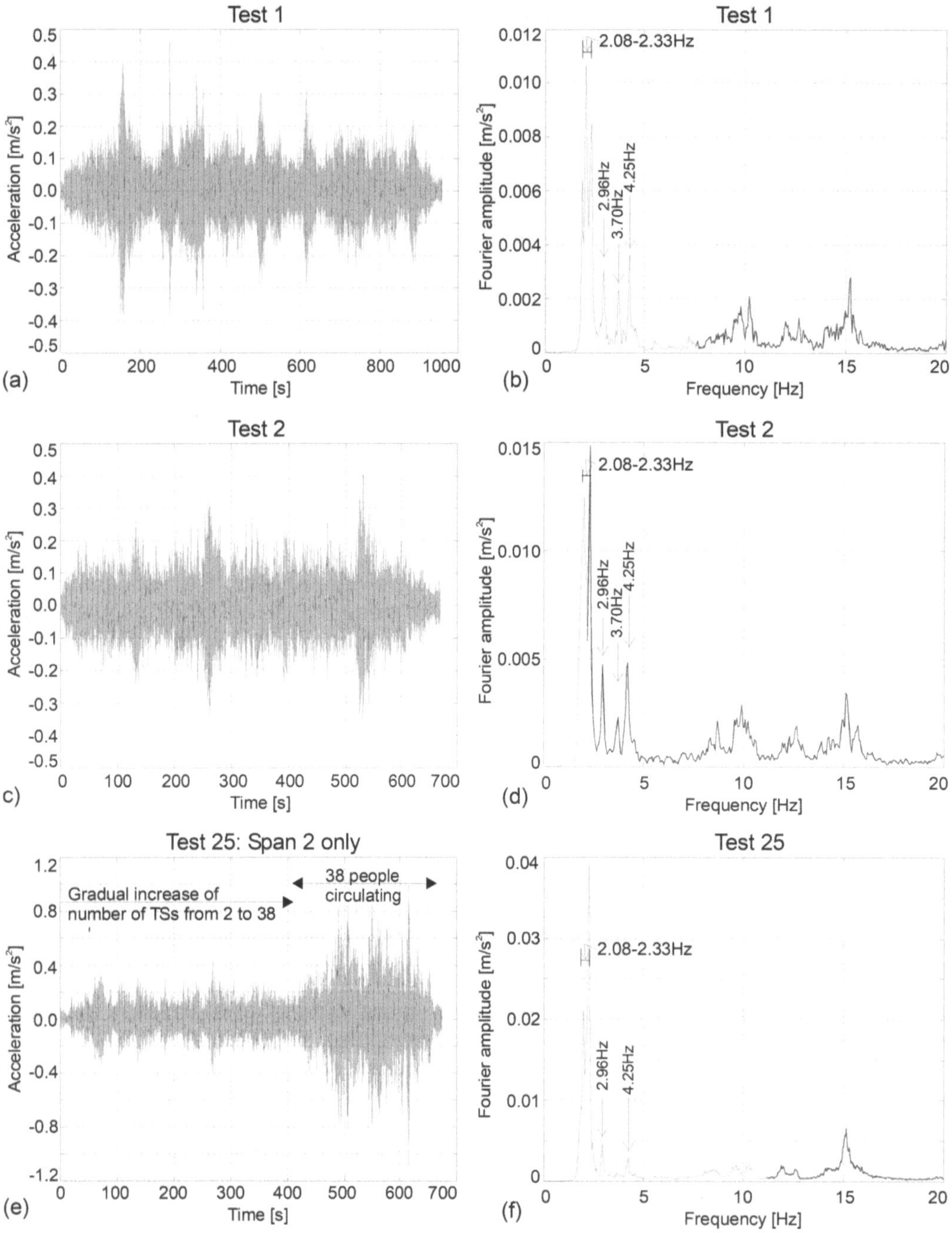


Figure 8: Acceleration responses and their spectra recorded in tests with stream of pedestrians.

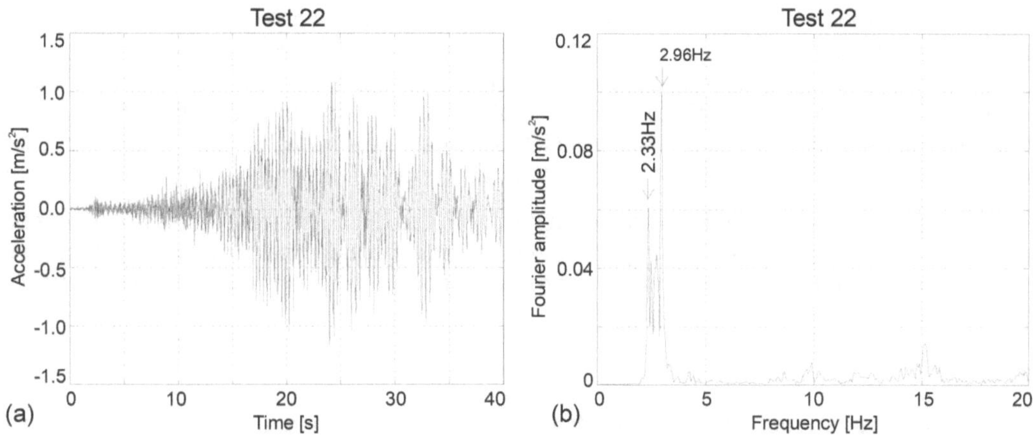


Figure 9: Response to 38 people jogging in (a) time domain and (b) frequency domain.

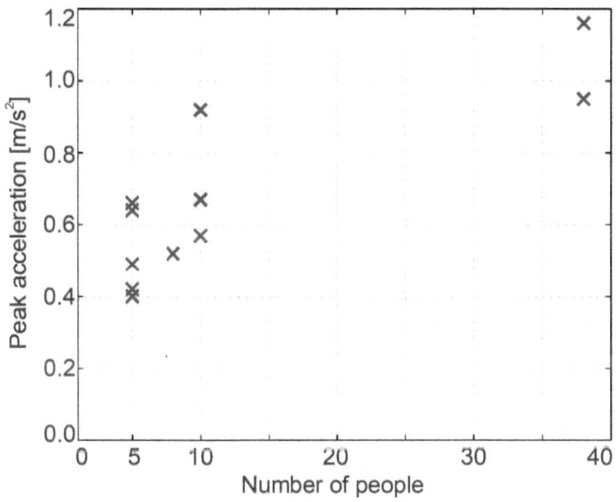


Figure 10: Peak acceleration level measured in jogging tests as a function of number of participants.

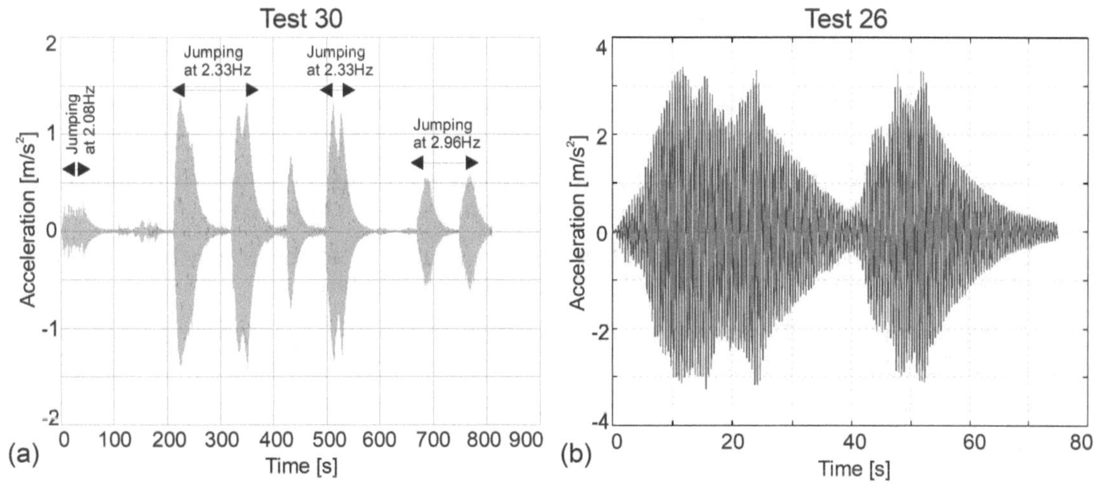


Figure 11: Acceleration induced by: (a) a single test subject jumping and (b) group of 38 people jumping.

## 5 CONCLUSION

This paper presents an extensive experimental study of the Reykjavik City Footbridge in Iceland. Modal testing of the bridge revealed several vibration modes that could potentially be strongly responsive under human-induced loading. To check this a series of tests with human test subjects performing walking, jogging and jumping actions on the bridge was conducted. These tests revealed that the bridge has quite a satisfactory vibration behaviour for most loading scenarios. The only exception was a large group of jumping people, which induced quite large acceleration levels. However, this scenario is not likely to occur in reality. Also it was found that the bridge could be quite lively if exposed to continuous traffic flow of density around 0.43 $m/s^2$. This traffic density is not expected to happen often however, especially not under a condition of a continuous supply of arriving pedestrians.

To conclude, the tests involving human-induced loading conducted on the RCF indicate that the vibration performance of the bridge is quite satisfactory for wide range of realistic loading conditions. However, it should be kept in mind that human loading is of random nature and that some variation in the results would be obtained if another group of test subjects took part in the testing programme.

## ACKNOWLEDGEMENTS

The authors acknowledge the financial support which came from the UK Engineering and Physical Sciences Research Council (EPSRC) for grant reference GR/T03000/01 (Stochastic Approach to Human-Structure Dynamic Interaction). The authors also thank the Icelandic Public Road Administration, Linuhonnun (now Efla) consulting engineers in Reykjavik for their financial and technical contribution in relation to the full scale testing of the bridge.

## REFERENCES

[1] McRobie, A. and Morgenthal, G. (2002) Risk management for pedestrian-induced dynamics of footbridges. In Proceeding of Footbridge 2002, Paris, France.

[2] Živanović, S., Pavic, A. and Reynolds, P. (2005) Vibration serviceability of footbridges under human-induced excitation: a literature review. Journal of Sound and Vibration, 279 (1-2), 1-74.

[3] Racic, V., Pavic, A. and Brownjohn, J. M. W. (2009) Experimental identification and analytical modelling of human walking forces: literature review. Journal of Sound and Vibration, 326 (1-2), 1-49.

[4] Venuti, F. and Bruno, L. (2009) Crowd-structure interaction in lively footbridges under synchronous lateral excitation: a literature review. Physics of Life Reviews, 6 (3), 176-206.

[5] Parkhouse, J.G. and Ewins, D. J. (2006) Crowd-induced rhythmic loading, Structures & Buildings, 159 (5), 247–259.

[6] Sim, J., Blakeborough, A., Williams, M. S. and Parkhouse, G. (2008) Statistical model of crowd jumping loads. Journal of Structural Engineering, 134 (12), 1852-1861.

[7] Racic, V. and Pavic, A. (2009) Mathematical model to generate asymmetric pulses due to human jumping. Journal of Engineering Mechanics, 135 (10), 1206-1211.

[8] Bachmann, H. And Ammann, W. (1987) Vibrations in structures induced by man and machines, IABSE-AIPC-IVBH, Zurich, Switzerland.

[9] Occhiuzzi, A., Spizzuoco, M. and Ricciardelli, F. (2008) Loading models and response control of footbridges excited by running pedestrians. Structural Control and Health Monitoring, 15 (3), 349-368.

[10] Kasperski, M. and Sahnaci, C. (2008) Serviceability of pedestrian structures exposed to vibrations during marathon events. In Proceedings of IMAC XXVI, Orlando, Florida, USA, 4-7 February.

[11] Gudmundsson, G. V., Ingolfsson, E. T., Einarsson, B. and Bessason, B. (2008) Serviceability assessment of three lively footbridges in Reykjavik. In Proceedings of the Third International Conference Footbridge 2008, Porto, 2-4 July.

[12] ME'scope VES 3.0. Vibrant Technology Inc.

[13] SAP 2000. Computers and Structures Inc.

[14] Setra (2006) Footbridges, assessment of vibrational behaviour of footbridges under pedestrian loading, Technical guide. The Technical Department for Transport, Roads and Bridges Engineering and Road Safety, Paris.

Proceedings of the IMAC-XXVIII
February 1–4, 2010, Jacksonville, Florida USA


# Baseline Model Establishment for a Bridge: I. Dynamic Testing

Kaoshan Dai, Shen-En Chen, David Boyajian, Jeremy Scott, and Yonghong Tong
Department of Civil & Environmental Engineering, University of North Carolina at Charlotte
9201 University City Blvd, Charlotte N.C. 28223

## Nomenclature

| | |
|---|---|
| $f'_c$ | compressive strength of concrete |
| $E_e$ | equivalent stiffness of concrete deck |
| $E_s$ | Young's Modulus for steel |
| $V_s$ | area ratio of reinforcing steel at typical cross section |
| $E_c$ | modulus of elasticity for concrete |
| $V_c$ | area ratio for concrete at typical cross section |
| $E_m$ | modulus of elasticity for elastomer in elastomeric bearing |
| $G$ | shear modulus of elastomer in elastomeric bearing |
| $k$ | elastomer hardness constant |
| $S$ | shape factor for one layer of elastomeric bearing |

## Abstract

A newly constructed bridge across interstate highway I-77 in Iredell County NC, is a hybrid steel girder bridge using both High Performance Steel (HPS) 70W and HPS 100W steel. The negative moment sections of the girders have HPS 100W flanges, whereas other parts of the girder are composed of HPS 70W steel. This bridge represents an increased effort in using high performance steel in civil construction. In order to ensure construction quality, a baseline finite element model was established. This model will be used as future reference for long-term structural performance monitoring. To validate the model, modal tests using impact excitation were conducted on the bridge prior to its opening to traffic. The baseline finite element model generated in this paper was further verified and used for the static load testing comparison, which is reported in a companion paper.

## 1. Introduction

In 1992 the Federal Highway Administration (FHWA) along with the American Iron and Steel Institute (AISI) and the United States Navy (USN) began a research and development program to create a new enhanced high strength steel called High Performance Steel (HPS) [1]. The goal of this research and development program was to develop a high strength steel with improved weldability, higher toughness, and enhanced atmospheric corrosion resistance or weathering properties [1]. There are several grades of HPS available, HPS 50W, HPS 70W, and HPS 100W with yield strengths of 50 ksi, 70 ksi, and 100 ksi, respectively [2]. HPS 70W is currently the mostly commonly used grade in bridges and was the first grade of HPS produced and commercialized through the research and development program [3]. An ideal design should select corresponding steel grades and member sizes based on strength and serviceability requirements determined from structural analysis and design calculations. However this is usually not realistic considering construction and economic factors. Performance based design balanced with other factors is more common. Hybrid HPS girders can be used to create more economical designs by using the higher grades of steel in the areas of the girder with the highest stresses and lower grades of steel in other areas. Hybrid designs are most economical when used in continuous girders in the negative moment regions [2].

The bridge of interest for this research is a two span continuous hybrid HPS 100W steel girder bridge. This bridge has several significant features. The steel plate girders have HPS 100W steel in the flanges of the negative moment region and HPS 70W steel in the flanges of the positive moment regions and the web. While HPS 70W is commonly used in bridges, HPS 100W is more recently developed and is not widely used [1]. The bridge is also oriented at an extreme skew angle of 132°22'30". The bridge was constructed in two stages due to the presence of an existing bridge that needed to remain in service until the new bridge was completed. The stage I portion of the bridge was constructed next to the existing bridge and once completed opened to traffic while the existing bridge was removed. The stage II portion of the bridge was then constructed. The intermediate bent, end bents, and concrete deck were spliced together longitudinally between the two stages. However, there is no connection directly between the girders of the two newly constructed structures.

In order to ensure construction quality, a series of tests were performed before the bridge was fully opened to traffic. Modal impact testing was performed first to determine the dynamic properties of the bridge superstructure. A static load test, as described in the companion paper, was performed using two 55,000lb tandem dump trucks as loading and a laser scanner to measure deflections. The data from the field tests was used to compare with a numerical model developed using ANSYS [4]. The purpose of this research is to develop a baseline finite element (FE) model, which will be used as a future reference for long-term structural performance monitoring. This model can also be used to aid periodical bridge inspection and can be used to improve the HPS bridge design.

In this paper, construction of the preliminary FE model for the bridge superstructure, as well as updating the model based on laser scanning field measurement, is mainly discussed. Impact modal testing conducted on the bridge is also briefly described and more results from the testing will be presented later. The FE model developed in this paper will be used to perform static analysis and compare with the truck loading tests in the second of this paper series.

## 2. Development of a bridge superstructure FE model

Baseline finite element models have been constructed for many bridge structures, for example, a long span bridge [5] and a cable-stayed bridge [6]. FE modeling is also an effective approach to study steel girder bridges [2, 7]. However, for large civil infrastructure, FE models constructed based on the design drawings may not exactly reflect the actual physical structure. There are several uncertainties for parameters used in building up the models. Updating the finite element model based on field measurements is widely used by researchers [8, 9]. This updating process is also a tool for parameter identification and damage detection. The goal of updating an FE model to better reflect the physical condition of a structure is to produce verified FE models, which then provide the benchmarks for long term monitoring. Accurate FE models are also significant for performing structural analysis to yield valid results used for design improvement.

### 2.1 Preliminary bridge FE model

The finite element model for this hybrid high performance steel bridge was first established based on the design drawings using commercially available software, ANSYS [4]. The bridge is composed of two equal length 146'-11.375" continuous spans and is 90'-7" wide. The layout of the bridge can be seen in Fig.1. The bridge has nine hybrid HPS steel plate girders, the flanges in the negative moment region between the two field splices are made up of HPS 100W steel whereas the web and flanges in the positive moment region are composed of HPS 70W steel. A typical girder and its cross section are shown in Fig.2. To account for the hybrid characteristic of steel girders, two different elements were used to model the girders: the flange is modeled by beam elements (BEAM188) and the web is modeled by shell elements (SHELL63). The section offset technique was applied to make sure there are no overlaps between the flange and the web.

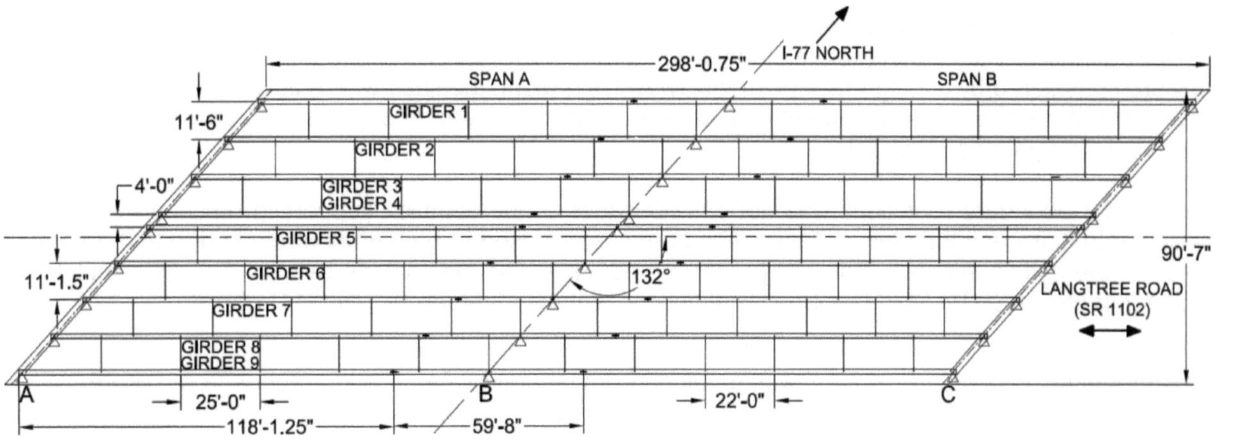


**Fig.1 Girder layout of the bridge**

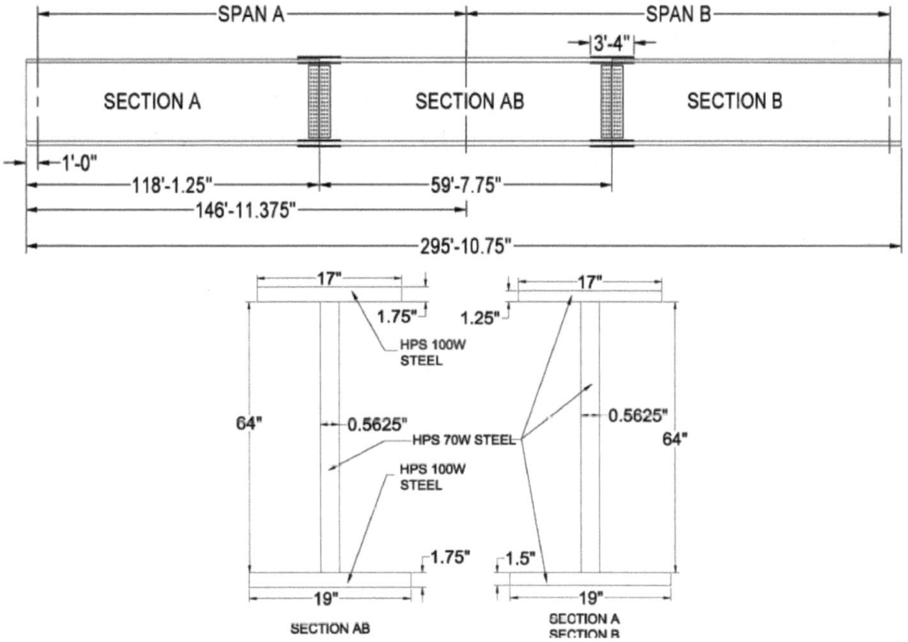


**Fig.2 Typical girder and its cross section**

All intermediate diaphragms are made up of AASHTO M270 grade 50W steel, which were modeled as link elements (LINK8) considering the bolted connections between these braces and the girders. Beam elements (BEAM188) were used to model the channel braces at both ends of the girders. To reduce the possibility of generating too many local modes in the dynamic analysis, the brace members were all modeled as one element.

The deck for this bridge is composed of NC DOT class AA 4500 psi concrete ($f'_c = 4500\,psi$) and is 9.5" thick between all girders and 1' thick on the parapets and longitudinal construction joints. Shell elements (SHELL63) were used to model the bridge deck and reinforcement was taken into account through calculating equivalent stiffness following the Rule of Mixtures for composite materials [10]:

$$E_e = E_s V_s + E_c V_c \tag{1}$$

where $E_e$ is equivalent stiffness; $E_s = 29000\,ksi$ is the Young's modulus of steel; $E_c = 3824\,ksi$ is the modulus of elasticity of concrete obtained using Eq. (2) from section 8.7.1 of the 16th edition of the AASHTO Standard Specifications for Highway Bridges [11]; $V_s = 0.011$ is the area ratio of reinforcement at a typical cross section; $V_c = 0.989$ is the area ratio of concrete at a typical cross section. Both $V_s$ and $V_c$ were calculated from the design drawings.

$$E_c = 57000\sqrt{f'_c} \tag{2}$$

The bridge deck and girders act compositely with shear studs along the top flange of each girder. The composite effects were simulated using rigid link elements (MPC184). The concrete parapets and aluminum rails on the bridge were considered by adding mass elements (MASS21) along the lines where they are located. The precast concrete traffic barrier, which weighed 400lb/ft, on the bridge deck used to separate two construction stages was also modeled through mass elements.

Modeling of the boundary conditions was not easy because of the complex supporting mechanism between the bridge superstructure and the piers. Elastomeric bearings are used at both ends of the bridge and a pot bearing was used at the middle support. The compressive modulus of the elastomeric bearing $E_m$ was calculated using section 14.2 of the AASHTO Standard Specifications for Highway Bridges 16th Ed. [11] where:

$$E_m = 3G(1 + 2\bar{k}S^2) \quad (3)$$

The shear modulus $G = 108\,psi$ at 73° F was obtained from laboratory certification reports from DS BROWN. The elastomer hardness constant $\bar{k} = 0.75$ was obtained from table 14.3.1 of the AASHTO standard specifications [11]. The shape factor for one layer of bearing $S = 8.876$ was calculated by dividing the plan area of the bearing by the area of the perimeter free to bulge according to section 14.2 of the standard specifications [11]. The compressive modulus of the elastomeric bearings was calculated to be $E_m = 3.862 \times 10^4 psi$. The elastomeric bearings were modeled using link elements (LINK10) with compression-only characteristics; while the anchor bolts at the supports were modeled with tension-only link elements (LINK10).

The bridge superstructure model is shown in Fig.3. The camber and slope of the HPS steel girder was considered in this preliminary FE model. The model has fine meshing in order to reduce incompatibilities between different element types [12]. There are a total of 4078 elements. Material property inputs are summarized in Table 1.

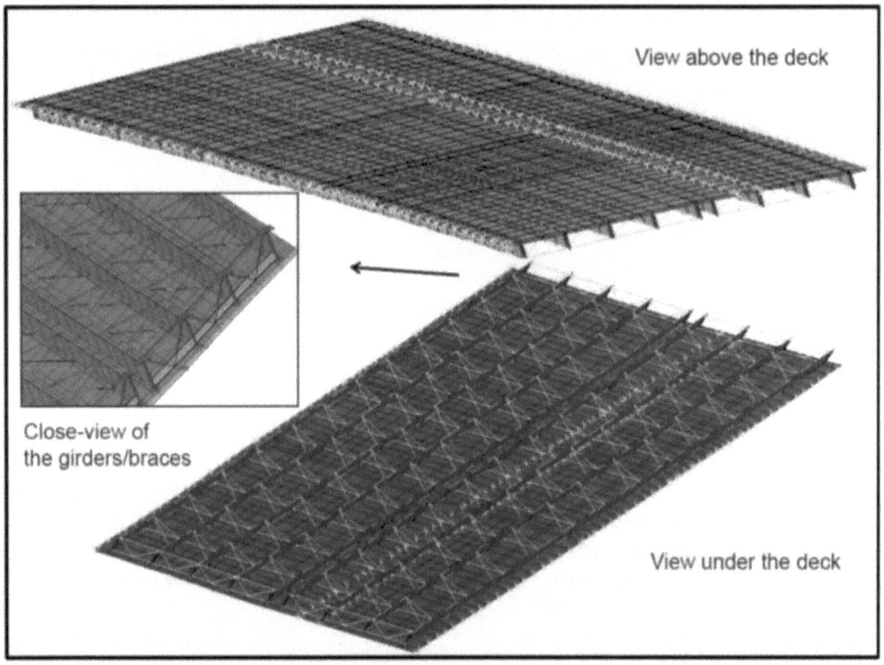


**Fig.3 FE model of the bridge superstructure**

**Table 1 Input material information for the preliminary FE model**

| Steel density (lbm/in$^3$) | Steel elastic modulus (ksi) | Deck density (lbm/in$^3$) | Deck elastic modulus (ksi) | Compressive Elastic modulus of end supports (ksi) | Compressive Elastic modulus of middle support (ksi) |
|---|---|---|---|---|---|
| 0.284 | 29000 | 0.087 | 4132 | 38.620 | 6.932 |

## 2.2 Updating the FE model

The FE model constructed based on the design drawings has some uncertainties, especially the parameters for boundary conditions. Model verification was performed through comparison with field measurements. A terrestrial 3D laser scan system called LiDAR (Light Detection and Ranging) was used to perform field measurements for this study. It is an optical remote sensing technology developed for range measurement. The uniqueness of this technique is that it is a noncontact sensing method and the equipment can be placed on the ground at the road side so that there is no interruption of traffic. The scanner measured the relative position of the object surface points when scanning along its line-of-sight. To obtain all of the surrounding surface information, a reflection mirror that rotates 360 degrees vertically is placed opposite of the scanner transmitter; the LiDAR head also rotates 360 degrees horizontally. A full scan is completed after the scanner head rotates 360 degrees horizontally. A typical scan in the current study uses a grid of approximately 9,000×4,000 points as measured with 360 degree horizontal and a 320 degree vertical travel (the latter being due to the blocking of the scanner underspan). Each point obtained registers a 3D coordinate value with the origin located at the position of the scanner head. The point cloud of the surrounding object can be measured and recorded in a single scan. During the test, the scanner was placed under the bridge (Fig.4). The dimensions of the bridge components was calculated and checked using the point coordinates offered by the scanner. Fig.4 is also given an example scan image for the test bridge.

**Fig.4 LiDAR scan testing and image of the test bridge**

With the in-situ LiDAR scan data, the FE updating process was primarily performed by manually tuning the boundaries of the model to match geometrical parameters of the girder obtained from analysis of the LiDAR scan data. Static analyses were performed with only self-weight of bridge superstructure considered. A typical deformed shape is shown in Fig.5. Theoretically, the deformation of the bridge girders under self-weight static loading should have curves the same as those measured on site with no traffic present. Based on this concept, the deformed shape of the girders measured using 3D LiDAR scanning were compared with the deformed shapes of the girders obtained from self-weight static analysis of the FE model. After several trials of adjusting the boundary conditions of the FE model, a reasonable matching deformed shape of the bridge under self-weight was obtained. Girder 8 curves are shown in Fig.6, in which the elevations were values relative to the middle support (support B). In the updated FE model, the compressive stiffness of the support A is 38.620 ksi, while the support C is 68.620 ksi; the stiffness of the middle support B is 5.932 ksi. This verified FE model will be used for dynamic analysis and truck loading analysis to compare with corresponding field tests.

**Fig.5 Deformed shape of the bridge superstructure under self-weight**

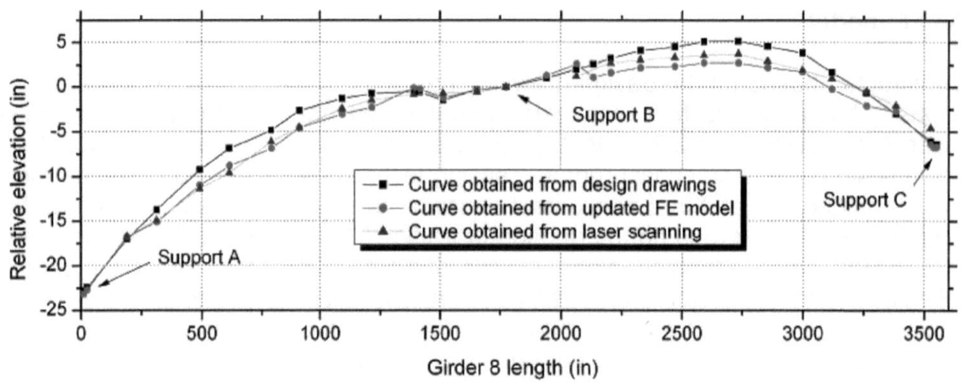


**Fig.6 Girder 8 relative elevation curve comparison**

## 3. Modal study of the hybrid HPS bridge

### 3.1 Modal analysis

Modal analysis is the fundamental study for determining the dynamic characteristics of a structure. It yields essential information in the frequency domain, such as natural frequencies and mode shapes. This is very important information to describe the bridge in the real world since modal characteristics are believed to represent physical behaviors of the real structure. The eigensolutions obtained from modal analysis are also used to design a structure to resist dynamic loads. Many studies have been performed to obtain modal behavior of a variety of bridges [13, 14, 15]. Modal analysis was conducted under ANSYS environment with the verified FE model in this research. Block Lanczos solver was used in modal analysis and the selected results are shown in Fig.7. It can be seen that global modes dominate at low natural frequency vibration modes. However, due to the complex structural system, there are many local modes for free vibration at high frequencies. For example, local girder vibration is the primary contribution to mode 44 shown in Fig.8.

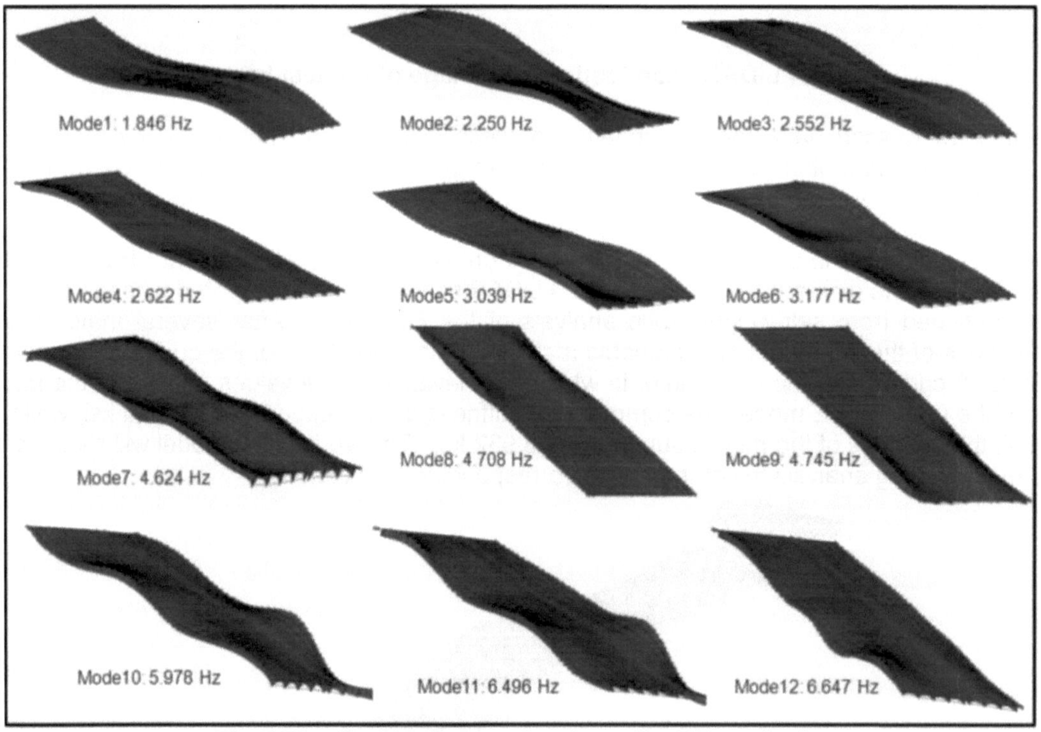


**Fig.7 The first twelve modes from the FE analysis**

### 3.2 Impact modal testing

Impact modal testing was used as a field testing method to determine the dynamic properties of the focus bridge. Impact points were laid out in a skewed grid pattern across the bridge deck (Fig.9). The impact points were spaced at 10' intervals along the bridge deck in the longitudinal direction and at 11' intervals along a diagonal parallel to the centerline of the bridge supports. Five impact measurements were performed at each test location for a total of 1450 data points. The testing took two days to complete.

The vibrations in the bridge superstructure were measured using an accelerometer placed under the bridge on girder 5 at 10'-10" from the support at end bent one. The hammer was instrumented to measure the impact at each point. The data from the accelerometer and the hammer were transmitted through cables to the WaveBook/516E 16-bit 1MHz data acquisition system. A PCB sensor signal conditioner was also used to amplify the signals from the accelerometer and hammer before they reached the data acquisition system. During the testing on the stage I portion of the bridge, traffic control was performed in order to minimize vibrations due to passing vehicles over the bridge. For the impact testing on the stage II portion of the bridge traffic control was not present but the impacts were timed so that there were no vehicles on the bridge when data was being recorded. The frequency transfer functions are processed to extract the modal parameters (natural frequencies and corresponding mode shapes). Analysis work is expected to be completed soon to yield more information about modal behaviors of this bridge. This information will be used to compare with the eigensolutions obtained from the FE analysis and further FE model updating will be performed based on the comparison.

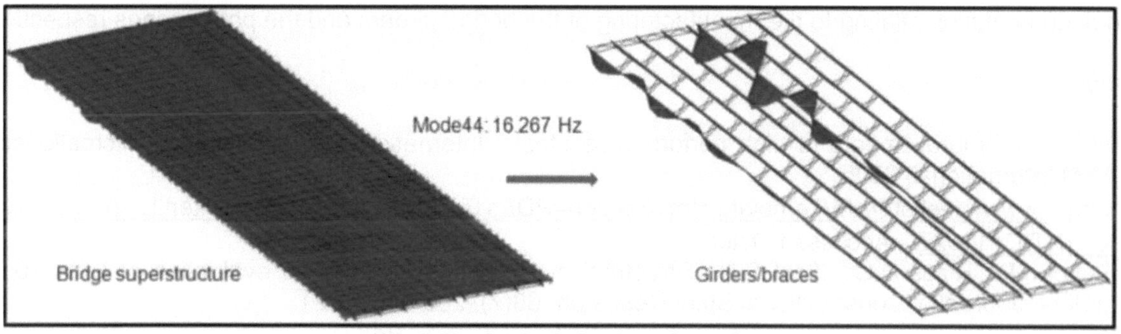


**Fig.8 Typical girder local mode**

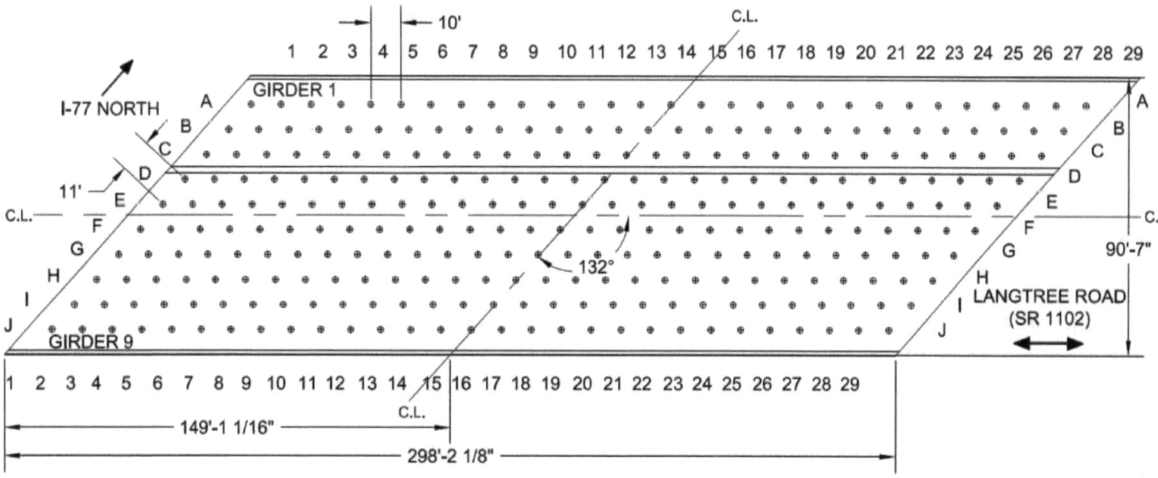


**Fig.9 Impact modal testing point locations**

### 4. Conclusions and future works

Through a series of field tests and finite element analysis, a hybrid high performance steel bridge was studied. The bridge has nine hybrid HPS steel plate girders: the flanges in the negative moment region between the two field splices are made up of HPS 100W steel whereas the web and flanges in the positive moment region are composed of HPS 70W steel. Dynamic behaviors of this bridge superstructure, including natural frequencies and mode shapes were the main topics discussed in this paper. The finite element model was built first based on the design drawings. The camber and slope of the bridge girders were considered in the model. To simulate the different steels used in the girders, both beam elements and shell elements were used in the model. Static analysis was performed by applying self-weight of the superstructure on this preliminary model. Bridge geometry was measured on site by a LiDAR scanner. By comparing the girder curves from LiDAR scanning with those obtained from self-weight static analysis of the bridge superstructure model, the FE model was updated by tuning the boundaries. Modal analysis was conducted using this updated FE model. Natural frequencies/mode shapes of this hybrid HPS bridge were obtained from the modal analysis. Impact modal testing was also performed on this bridge and data analysis is still in progress. The modal test results are expected to be used to verify the FE model and further model updating will be conducted by the authors to improve the model.

## Acknowledgment

The researchers from UNC Charlotte gratefully acknowledge the support of the North Carolina Department of Transportation for funding this research and providing the tandem dump trucks and traffic control for the static load testing of the I-77 Hybrid HPS bridge. The authors would also like to acknowledge the guidance and contributions of Dr. Mrinmay Biswas and Mr. Neal Galehouse of North Carolina DOT. Special thanks to Rea Contracting for allowing the research team to conduct field tests on the bridge before construction was completed. Lastly, the researchers would like to acknowledge the help of Structural Steel Products Inc and DS Brown for providing documentation relating to the manufacturing of the bridge girders and the pot bearings respectively.

## References

[1] Wilson A. Improvements to high performance steels. International Symposium on Microalloyed Steels, ASM International, 2005. <http://www.steel.org/AM/Template.cfm?Section=PDFs1&TEMPLATE=/CM/ContentDisplay.cfm&CONTENTID=12027.> Accessed October 1, 2009

[2] Felkel J.P., Rizos D.C., Ziehl P.H., Structural performance and design evaluation of HPS 70W bridge girders. Journal of Constructional Steel Research, 63(7), 2007:909–921.

[3] Wilson A. Production of high performance steels for U.S.A. Bridges. Conference of High Performance Steel Bridge, Nov 30-Dec 1, 2000, Baltimore Maryland. <http://www.nabro.unl.edu/articles/20002012/download/Wilson.pdf.> Accessed October 1, 2009.

[4] ANSYS 2009. ANSYS academic teaching advanced. Release 11.0SP1.

[5] Wang M. L., Heo G., & Satpathi D., Dynamic characterization of a long span bridge: a finite element based approach. Soil Dynamics and Earthquake Engineering, 16(7-8), 1997:503-512.

[6] Ren W., and Peng X., Baseline finite element modeling of a large span cable-stayed bridge through field ambient vibration tests. Computers & Structures, 83 (8-9), 2005:536-550.

[7] Bartha K.E., Haiyong W., Efficient nonlinear finite element modeling of slab on steel stringer bridges. Finite Elements in Analysis and Design, 42 (14), 2006:1304-1313.

[8] Brownjohn J.M.W., Xia P., Hao H., Xia Y., Civil structure condition assessment by FE model updating: methodology and case studies. Finite Elements in Analysis and Design, 37(10), 2001:761-77.

[9] Daniell W. E., Macdonald J.H.G., Improved finite element modeling of a cable-stayed bridge through systematic manual tuning. Engineering Structures 29 (3), 2007:358–371.

[10] Barbero E.J., Introduction to composite materials design. CRC press, Taylor & Francis Group, Boca Raton, FL, 1999.

[11] AASHTO Standard Specifications for Highway Bridges 16th ed. Washington DC. American Association of State and Highway Transportation Officials, 1996.

[12] Chung W., Sotelino E.D., Three-dimensional finite element modeling of composite girder bridges. Engineering Structures 28(1), 2006:63–71.

[13] Clemente P., Marulo F., Lecceb L. & Bifulco A., Experimental modal analysis of the Garigliano cable-stayed bridge. Soil Dynamics and Earthquake Engineering 17(7-8), 1998:485–493.

[14] Zong Z., Jaishi B., Ge J., Ren W., Dynamic analysis of a half-through concrete-filled steel tubular arch bridge. Engineering Structures 27(1), 2005:3–15.

[15] Živanović S., Pavic A., Reynolds P., Modal testing and FE model tuning of a lively footbridge structure. Engineering Structures 28(6), 2006:857–868.

Proceedings of the IMAC-XXVIII
February 1–4, 2010, Jacksonville, Florida USA


# Baseline Model Establishment for a Bridge: II. Static Load Testing

Jeremy Scott, Wanqiu Liu, David Boyajian, Kaoshan Dai, and Shen-En Chen
Department of Civil & Environmental Engineering, University of North Carolina at Charlotte
9201 University City Blvd, Charlotte N.C. 28223


## Abstract

Truck load testing is a field experimentation technique used widely for bridge load rating. For new bridge infrastructure, a baseline model developed through dynamic or static load testing can be used as a reference for long-term structural performance monitoring. This paper presents a study of truck load testing on a hybrid HPS steel girder bridge. The negative moment region of the girder is composed of HPS 100W steel in the flanges and HPS 70W steel in the web. A non-contact laser sensing technique was used to measure bridge deformation under static loads. The experimental results were compared with a numerical model for validation purposes. Static load testing, along with dynamic analysis reported in a companion paper, was used to create benchmark information for this structure that will be useful for future bridge inspection.


## 1. Introduction

The deflection of a girder in a bridge reflects the superstructure performance, including structural behavior of the deck and girders as well as composite action performance. Static load testing is a common technique used to evaluate bridge performance of newly constructed structures or aging ones. The National Cooperative Highway Research Program [1] developed the 'Manual for Bridge Rating through Load Testing' to assess bridge load carrying capacity [1]. Bridges under given static loads are often measured for displacement, stress, or strain [3, 4] at selected points. Several different instruments and techniques have been used in static load tests on bridges. Some of the most common instruments used to measure deflections and strains are linear variable differential transformers (LVDT), cable extension transducers [2], and strain gauges [3, 4]. While these instruments are effective, they require access to the bridge girders and possibly interruption of traffic. In certain situations such as bridges over water or over major highways with heavy traffic, installation of strain gauges and LVDT's are not feasible options. Therefore, a technique involving the use of a laser to measure bridge deflection was developed [5]. A study using a laser system to measure bridge deflections was performed on a bridge in Baltimore Maryland in April of 2000 at the I-695 and I-95 interchange [2]. Use of a laser to measure bridge deflections enabled traffic to continue underneath the bridge with no interruption; only traffic in the lanes where the loading was applied had to be closed off [2]. As part of the study, several other bridges in Maryland, Alaska, Virginia, Oregon, Michigan, and Ohio were tested using a laser [2].

A skewed hybrid high performance steel (HPS) bridge, located on SR1102 over I-77 in Iredell County, North Carolina (Fig.1), was just completed. The bridge consists of two 146'-11" spans with cast-in-place concrete decking. The negative moment region of the girder is composed of HPS 100W steel in the flanges and HPS 70W in the web. To assess the structural performance of this unique bridge, a series of studies were conducted, including impact modal testing, truck load testing and numerical analysis. In this paper, the truck load testing is described.

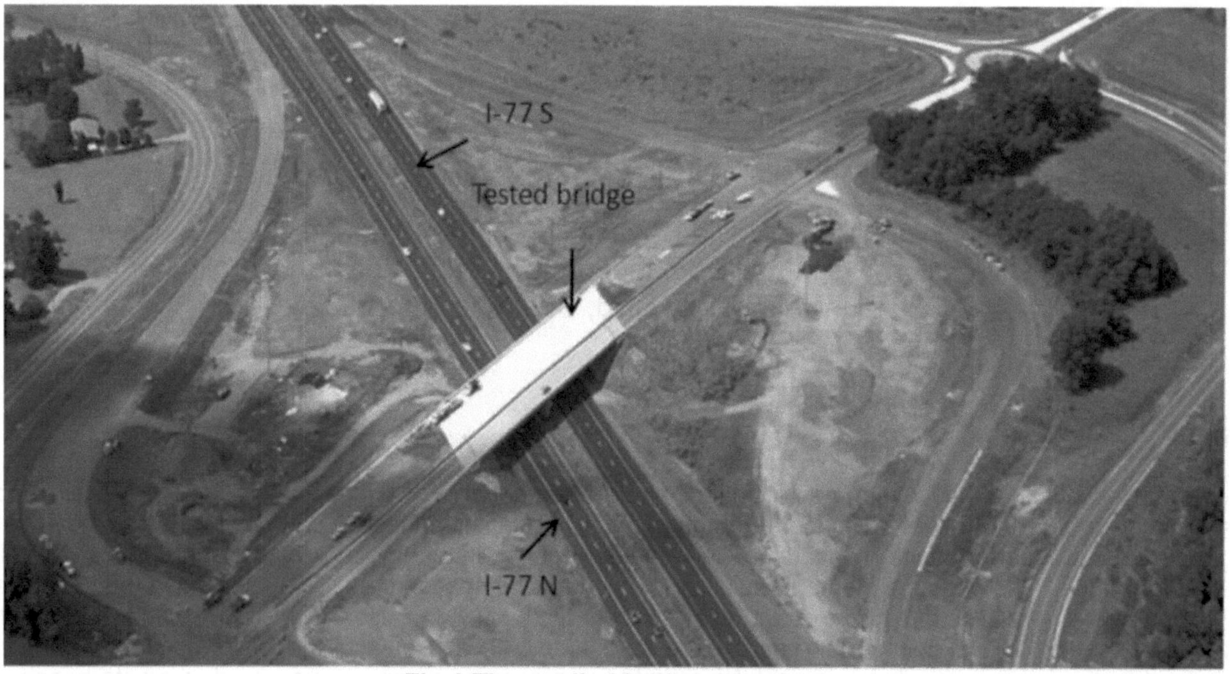


**Fig.1 The studied bridge over I-77**

The laser system used in this research to measure geometric properties of this bridge was also used to measure deflection under static truck loads. This commercially available Light Detection and Ranging (LiDAR) system is an optical remote sensing technology developed for range measurement. Most of the 3D laser scanners have the advantage of high speed data collection and large coverage area. However, due to the high cost and lack of investigation, currently, only a few research reports have used a 3D laser scanner for bridge inspection. Fuchs et al. [2]) introduced a laser system for bridge testing. This system can measure point displacement without the installation of light reflective targets. Liu et al. [6] have studied the applications of terrestrial LiDAR technology for bridge evaluation.

In this paper, girder deflections under three different loading scenarios, designed based on bridge influence lines, were obtained using a LiDAR scanner. Static load analysis was performed using the FE model developed in the companion paper. Comparison studies were carried out between these two results. Outcomes of the static loading comparison will be used to direct further model updating and verification in order to build up a reliable baseline FE model for this hybrid HPS 100W bridge.

## 2. Static truck load testing

The focus bridge was load tested using two tandem dump trucks, each weighing approximately 55,000 lbs. The trucks were provided by the North Carolina Department of Transportation. Three load configurations involving the pair of trucks are described herein. Truck configurations 1 and 3 were positioned to create a maximum negative moment in the girders at the interior supports. A preliminary analysis using influence lines was conducted to determine these load locations. The trucks for these positions were placed as close to the parapets as possible so that the exterior girders would carry the most load resulting in increased deflections. Truck configuration 2 was positioned so that the interior girder 7 would carry most of the load. For this configuration, the trucks were placed side by side in order to determine how the load is distributed between the interior girders. Fig.2 shows the two trucks in position 3 and Fig.3. shows the load configurations on the bridge deck.

**Fig.2 Trucks in load position 3**

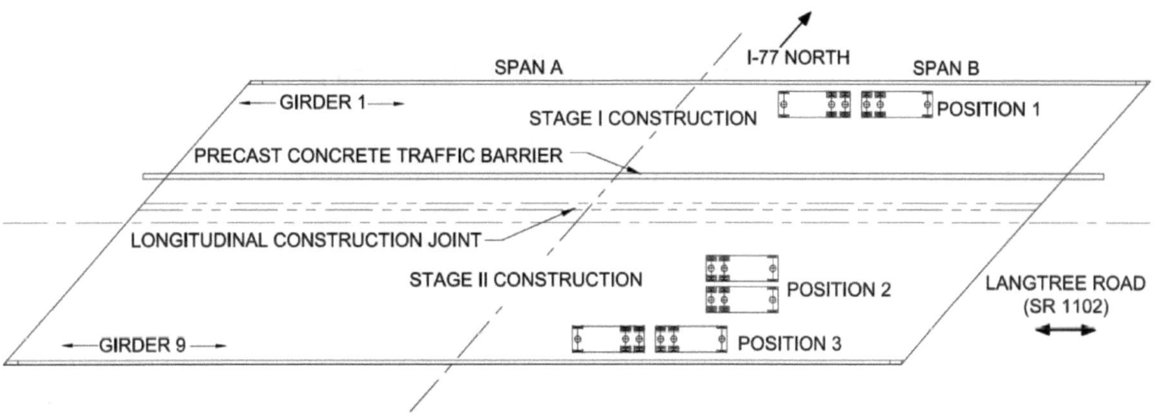


**Fig.3 Load testing truck configurations**

During the load tests, laser scanning was used to measure deflections of the bridges. The LiDAR scanner was placed under span B at the edge of the highway. It was located between girders six and seven. First, a test scan with the laser was performed with no truck loading on the bridge in order to ensure that all equipment was functioning properly. This measurement also yielded geometry information of the bridge under self-weight. The trucks were then placed in position 1 and traffic was stopped while the laser scanned the underside of the bridge. Once the laser scan was complete, the trucks were moved to the next position and the process repeated.

## 3. Load testing data analysis

An automatic bridge evaluation algorithm LiBE (LiDAR based bridge evaluation) based on LiDAR data has been developed and one of its functions is to determine the displacements of the bridge. With proper calibration, the Z coordinate value of each point collected through the LiDAR scan can represent the relative height of the point to the scanner. The displacement of this point on the bridge can be calculated by comparing the change of z values of the point. However, after bridge deformation, the scanner will no longer record the same point in each scan direction. Therefore, a matching technique was developed in the LiBE system to find the points on bridge surface at the same horizontal position from each pair of scans. The displacement can then be calculated by comparing the z values of the matching points.

The displacements of all the girders in each loading case were calculated by LiBE system and the results are given in Figs.4-6. From these figures, we can see that the maximum displacement of the bridge happened near the truck locations. The error of the measurements will increase when approaching the object boundary.

Therefore at the edge of the girders, discontinuous displacement values will be found. Also for the relatively thin elements such as the intermediate diaphragms, the measured errors may be higher. In most cases, the edge effects will result in much larger displacement values on the boundary than that of the inner element surfaces. To make the display of the displacement on the girders more obvious, 1.181 inch and -0.0394 inch were used for the maximum and minimum deflection values on in all figure legends. Warm colors in the figure indicate downward deformation (positive values) of the girders. It is also noted that in Figs.4-6, only girders of span B, where the trucks were positioned, were calculated and shown.

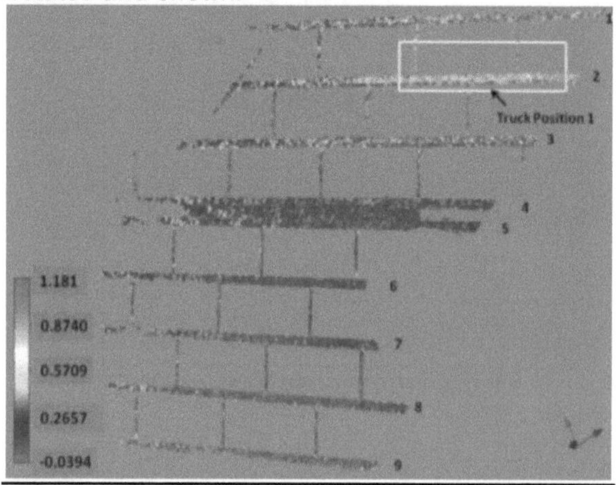


**Fig.4 Bridge girder deflection (in) under truck load position 1**

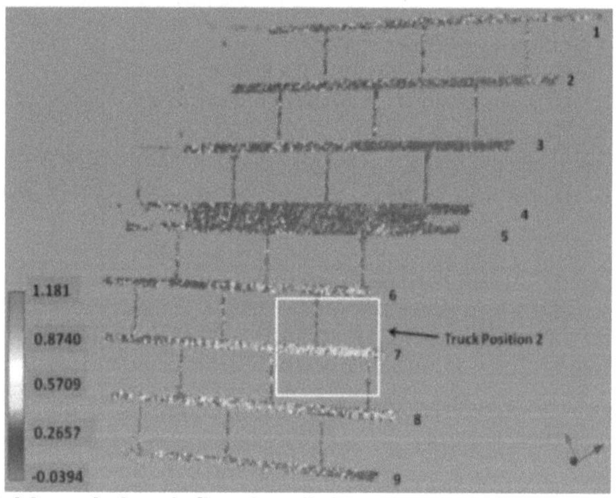


**Fig.5 Bridge girder deflection (in) under truck load position 2**

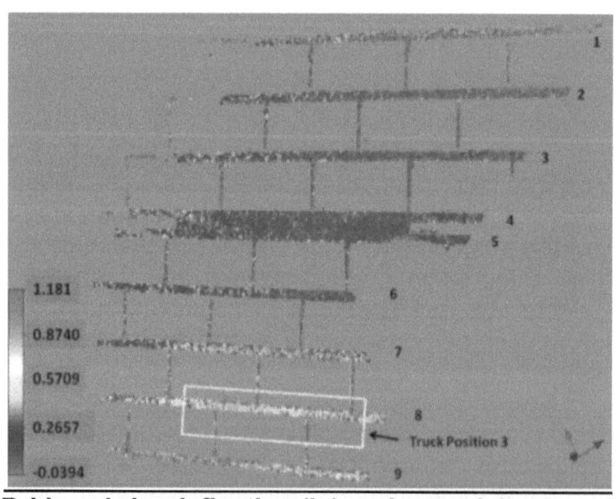


**Fig.6 Bridge girder deflection (in) under truck load position 3**

## 4. Truck load static analysis using the FE model

The verified bridge FE model constructed in the companion paper was used for linear elastic static analysis under truck loads. Truck loads were generated in the FE model as forces [7] distributed on the nodes generated exactly where the truck wheels were placed during the field tests. Nodal deflections of the entire bridge superstructure and girder deflections in the direction normal to the bridge deck under different loading cases are shown in Fig.5. Negative values (cold colors in the figures from the FE model) are deflections along the gravity direction while positive values (warm color in the figures) are deformations opposite to gravity. Please note that this is different from the display of LiDAR results (Figs.4-6). It is obvious from the FE result (Fig.7) that the maximum deflections take place where the trucks are positioned. This has also been indentified during the LiDAR scanning. Due to the two-span structural characteristics, some parts of the bridge superstructure in the span A deforms in the opposite direction (upwards) when the span B deflects downwards, where the trucks were placed. However, upward deformations are all small with a maximum value of 0.185 inches.

First principal stress distributions of the bridge superstructure under three different load cases were obtained and are shown in Fig.8. The maximum 1st principal stress for all cases is 3876 psi when the trucks were positioned at the bridge edge close to girder 9. The maximum stress along x direction for girder 9 is also 3786 psi, which is lower than the steel yield strength (100 ksi and 70 ksi for girders and 50 ksi for the intermediate diaphragms).

Comparison of the deflections between the laser scan results and the FE analysis results is listed in Table 1. It can be seen that these deformations are close for loading case 1, when the trucks were positioned near girder 1. It is also noted that there is a nearly 45% difference for truck position 2 although both maximum values are found on girder 7. In this research, the resolution of laser scan results is around 0.2 in. This may be one of the reasons that account for the difference. The reasons for this difference may also lie in FE model parameters. Further FE model checking and updating work will be performed based on this truck loading tests.

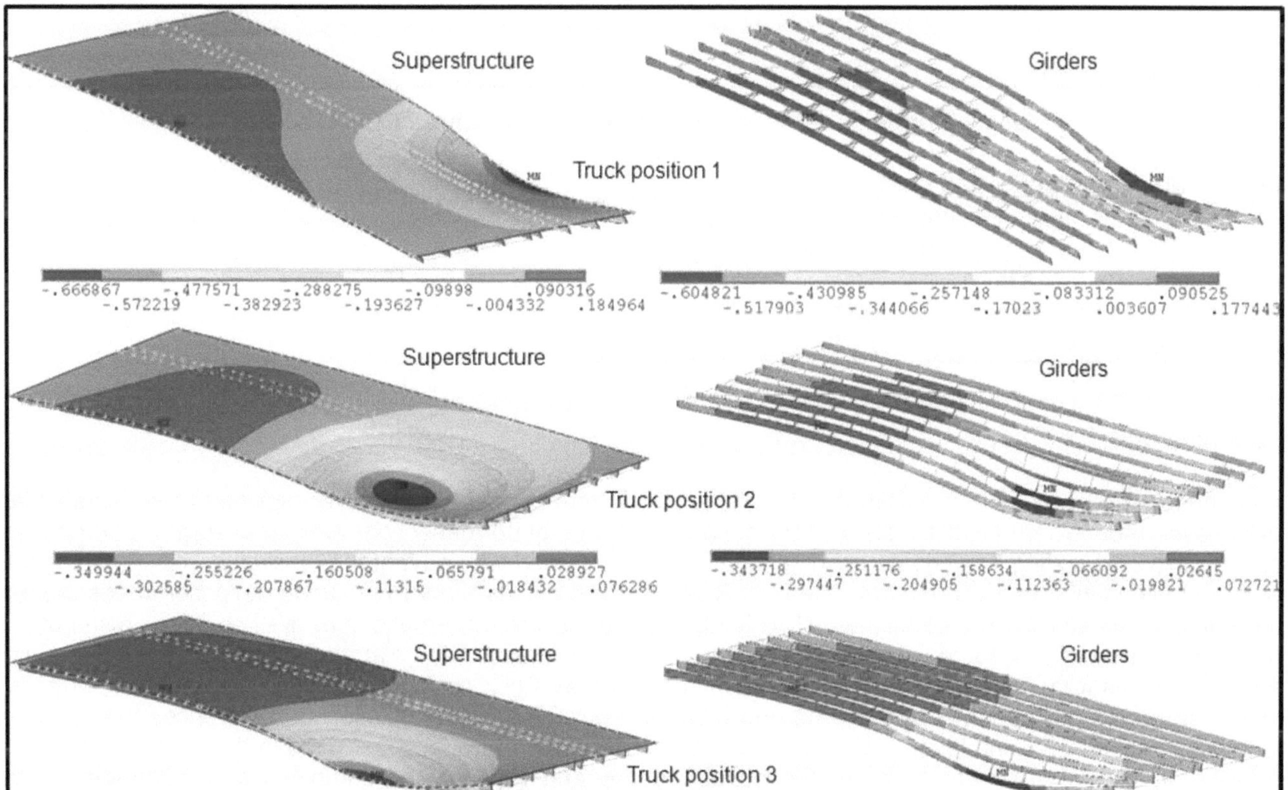


**Fig.7 Bridge superstructure deformations under three load cases**

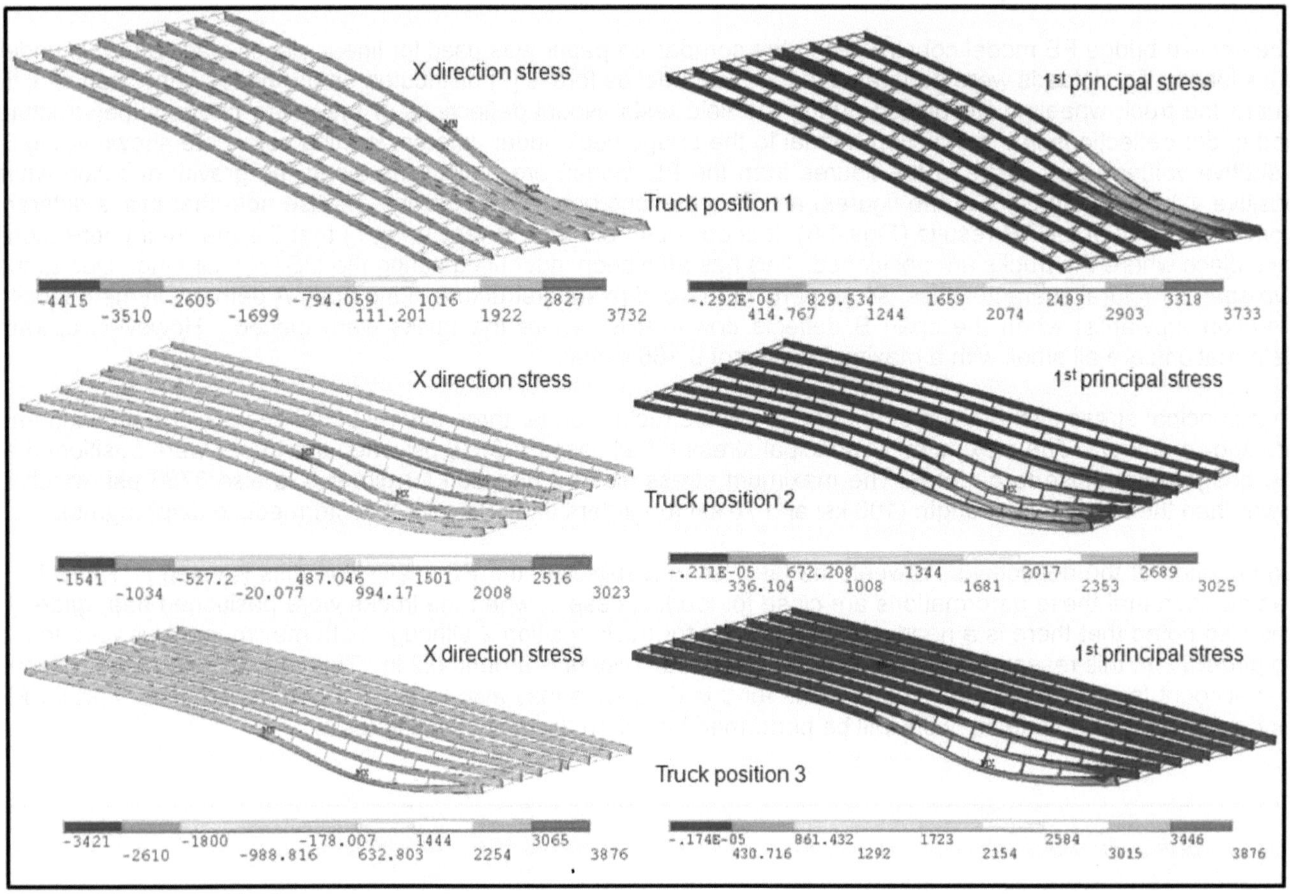


**Fig.8 Bridge girder stress distribution under three load cases**

**Table 1. Maximum deflection comparison**

| Loading cases | Maximum deflection | | Girder No. |
|---|---|---|---|
| | Laser scanning | FE model | |
| Truck position 1 | 0.663 in | 0.605 in | 1 |
| Truck position 2 | 0.624 in | 0.344 in | 7 |
| Truck position 3 | 0.780 in | 0.588 in | 9 |

## 5. Conclusions

Truck load testing was carried out on the hybrid high performance steel bridge over highway I-77. Three truck positions were considered in the tests and corresponding bridge girder deflections were measured. Using a laser scanner, the load testing procedures become significantly simplified with advantages including minimal impact to traffic and elimination of traffic control. The LiDAR scan also provides higher resolution deflection measurement than photogrammetric based techniques. The bridge superstructure deformations as well as stress distributions under these three truck load cases were obtained through linear elastic finite element analysis in this paper. The deflection results from both these studies were compared. Under truck position 1, girder 1 maximum deformation is 0.663 in (downward) from laser scanning and 0.605 in from FE analysis. Deflection differences among other load cases between FE and laser scan, if not due to the resolution of the scanner used in this paper (0.2 in), will be further studied and used to direct further model updating and verification. Combined with the studies in the first paper and the proposed updating work based on the modal testing data, it is expected that a reliable baseline FE model for this hybrid HPS bridge canl be developed.

**Acknowledgment**

The researchers from UNC Charlotte gratefully acknowledge the support of the North Carolina Department of Transportation for funding this research and providing the tandem dump trucks and traffic control for the static load testing of the I-77 Hybrid HPS bridge. The authors would also like to acknowledge the guidance and contributions of Dr. Mrinmay Biswas and Mr. Neal Galehouse of North Carolina DOT. Special thanks to Rea Contracting for allowing the research team to conduct field tests on the bridge before construction was completed. Lastly, the researchers would like to acknowledge the help of Structural Steel Products Inc and DS Brown for providing documentation relating to the manufacturing of the bridge girders and the pot bearings respectively.

**Reference**

[1] National Cooperative Highway Research Program (NCHRP). Manual for Bridge Rating Through Load Testing, Research Result Digest 234, Transportation Research Board, Washington, D.C., 1998.
[2] Fuchs P.A., Washer G.A., Chase S.B. and Moore, M. Laser-based Instrumentation for Bridge Load testing. Journal of Performance of constructed facilities, 18(4), 2004:213-219.
[3] Chan T. H. T., Li Z. X., and Ko J.M. Fatigue analysis and life prediction of bridges with structural health monitoring data — Part II: application. International Journal of Fatigue, 23(1), 2001:55-64.
[4] Klowak C.S. and Mufti A. A. Behaviour of bridge deck cantilever overhangs subjected to a static and fatigue concentrated load. Construction and Building Materials, 23(4), 2009:1653-1664.
[5] Nassif H., Gindy M., Davis J. Comparison of laser Doppler vibrometer with contact sensors for monitoring bridge deflection and vibration. NDT&E International 38(3), 2005:213-218.
[6] Liu W., Chen S., and Hauser, E. Remote sensing for bridge health monitoring, SPIE Optics + Photonics, San Diego, California, No.7456-13, 2009.
[7] ANSYS academic teaching advanced. Release 11.0SP1, 2009.

Proceedings of the IMAC-XXVIII
February 1–4, 2010, Jacksonville, Florida USA


# Sensitivity Study of Stochastic Walking Load Models

Lars Pedersen, Christian Frier

Aalborg University
Department of Civil Engineering
Sohngaardsholmsvej 57
DK-9000 Aalborg

## ABSTRACT

On flexible structures such as footbridges and long-span floors, walking loads may generate excessive structural vibrations and serviceability problems. The problem is increasing because of the growing tendency to employ long spans in structural design. In many design codes, the vibration serviceability limit state is assessed using a walking load model in which the walking parameters are modelled deterministically. However, the walking parameters are stochastic (for instance the weight of the pedestrian is not likely to be the same for every footbridge crossing), and a natural way forward is to employ a stochastic load model accounting for mean values and standard deviations for the walking load parameters, and to use this as a basis for estimation of structural response. This, however, requires decisions to be made in terms of statistical distributions and their parameters, and the paper investigates whether statistical distributions of bridge response are sensitive to some of the decisions made by the engineer doing the analyses. For the paper a selected part of potential influences are examined and footbridge responses are extracted using Monte-Carlo simulations and focus is on estimating vertical structural response to single person loading.

## NOMENCLATURE

| | | | | | |
|---|---|---|---|---|---|
| $f_0$ | Bridge frequency | $f_s$ | Step frequency | $f$ | Walking load |
| $l_s$ | Stride length | $m$ | Weight of pedestrian | $p$ | Prob. density function |
| $q$ | Modal load | $L$ | Bridge length | $M$ | Bridge modal mass |
| $P$ | Prob. distribution function | $a$ | Bridge acceleration | $\alpha$ | Dynamic load factor |
| $\zeta$ | Bridge damping ratio | $\mu$ | Mean value | $\sigma$ | Standard deviation |

## 1. INTRODUCTION

For the studies of this paper, walking-induced vibrations of footbridges are considered (but problems with walking-induced vibrations may also be encountered on flooring-systems). That vibration problems can occur in footbridges due to the action of walking is known to many after the closure of the London Millennium Bridge [1]. In codes of practise for footbridges it is often stated that the vibration serviceability limit state related to walking loads is to be checked. Generally, footbridge vibration problems may occur as a result of vertical or horizontal excitation. The present paper considers the vertical action.

As for the vertical action, most codes suggest to model it as a moving harmonic force. This approach is also taken in this paper, but whereas most current codes model the parameters of the harmonic force (amplitude and frequency) deterministically, the present paper models these parameters as random variables. This would seem sensible as research has documented that the excitation frequency (step frequency) is a random variable ([2,3]), in that the step frequency varies from one pedestrian to the next. Nevertheless, a number of current codes [4,5] suggests a load model in which the excitation frequency is to be set equal to the bridge frequency for computing bridge response. This approach definitely does not consider the probability of the modelled resonant excitation and the procedure does not provide information on the probability of reaching the calculated bridge vibration level; although this would seem to be valuable information for the bridge operator.

As for the amplitude of the harmonic force it is determined from pedestrian weight ($m$), and the dynamic load factor ($\alpha$). These are parameters which fundamentally are also stochastic, but in a number of codes these characteristics are modelled deterministically as well. For instance, some codes assume the weight of the pedestrian to be 75 kg.

An approach to a probability-based estimation of bridge vibration levels was introduced in [3], and this paper adapts the general idea behind it. It allows statistical distributions of bridge response (to actions of walking of a single pedestrian) to be determined, recognising that walking parameters are stochastic variables. However, in [3], the pedestrian weight ($m$) was modelled deterministically and a stochastic model for $m$ was not considered. For the studies of this paper an addition is made in which the pedestrian weight is modelled as a random variable (along with modelling the other parameters of the load model as random variables in the way suggested in [3]).

It is not quite obvious how to model the statistical distribution of pedestrian weight, but nevertheless it is one of the inputs needed for computing statistical distributions of bridge response. Generally not much information is available about weights of pedestrians, but a distribution type need be selected and then characteristics of the distribution are to be decided upon (such as mean value and standard deviation of the random variable).

In light of the fact that information available for the decisions is sparse, the paper approaches the problem the other way around. It seeks to examine how sensitive statistical distributions of bridge response are to choices made about distribution type, mean value and standard deviation of pedestrian weight; this in the hope that the findings in terms of sensitivity will be useful for the engineer making such decisions.

To facilitate the investigations a footbridge model is required, and to this end a pin-supported footbridge (idealised as a single-degree-of-freedom system) is employed. The response characteristic given focus is the midspan peak accelerations. For the sensitivity study, three different distribution types, mean values, and standard deviations for pedestrian weight are considered and used for calculating statistical distributions of footbridge response. For reference purposes a deterministic model for pedestrian weight is also employed.

The bridge excited by pedestrians in computations is introduced in section 2 along with the walking load model. Section 3 outlines study assumptions in terms of walking parameters, and section 4 describes how statistical distributions of bridge response are obtained. Section 5 presents results. The results are discussed, and a conclusion is provided.

## 2. MODEL OF BRIDGE AND BRIDGE EXCITATION

The modal characteristics of the bridge considered for the studies of this paper are shown in Table 1.

| $f_0$ | $M$ | $\zeta$ |
|---|---|---|
| 2.00 Hz | 39.500 kg | 0.3% |

**Table 1.** Dynamic characteristics.

The frequency of the bridge ($f_0$) is chosen such that it represents a bridge prone to react lively to actions of walking. As can be seen, the bridge damping ratio ($\zeta$) is quite low. The modal mass ($M$) is believed to be quite realistic considering the frequency of the bridge, as is the length of the bridge, $L$, which is assumed to be 43 m (between the two pin supports).

For the paper (and as often done for modelling the vertical excitation generated by a pedestrian [6,7]), the dynamic load acting on the bridge, $f(t)$, is modelled as shown in eq. 1 in which $t$ is time.

$$f(t) = mg\,\alpha\,\cos(2\pi f_s t) \qquad (1)$$

It is a harmonic load with a frequency, $f_s$, representing the step frequency of walking. The step frequency is assumed constant during the locomotion of the pedestrian whilst crossing the bridge. This is an idealisation, but the paper considers that the value of $f_s$ will change from one pedestrian to the next in order to model randomness in the action of walking. This approach is also taken for the amplitude of the harmonic excitation, in that the dynamic load factor, $\alpha$, will be modelled as a random variable, as will the pedestrian weight, $m$ (in kg). The parameter $g$ represents acceleration of gravity. Generally, there would also be super-harmonics of the action of walking worth considering, but for the bridge considered in this paper it would be the first harmonic of the action (eq. 1) that is of interest (the bridge is modelled as a SDOF system and it is the first harmonic that can cause resonance as this is the load harmonic in close vicinity of the bridge frequency).

It can be shown that the modal load on the bridge (first vertical bending mode) may be computed using eq. 2:

$$q(t) = mg\,\alpha\,\cos(2\pi f_s t)\,\sin(\pi f_s l_s / L) \qquad (2)$$

In brief, this equation assumes that the mode space function of the first vertical bending mode of the bridge corresponds to a half-sine. Furthermore it assumes, that the pedestrian traverses the bridge with a locomotion style in which the stride length, $l_s$, (or step length) is constant. For the studies of this paper, randomness in stride length is considered (from one pedestrian to the next).

## 3. MODELS FOR WALKING PARAMETERS

For the studies of this paper primary focus is on various ways of modelling pedestrian weight, and implications hereof. The various assumptions made are outlined in section 3.1, and section 3.2 outlines study assumptions for other walking parameters.

### 3.1 Pedestrian weight

For pedestrian weight ($m$) three different stochastic models are assumed; a normal distribution, a log-normal distribution, and a uniform distribution. Mean values and standard deviations assumed for each of the distributions are listed in table 2.

| Variable | Unit | $\mu_m$ | $\sigma_m$ |
|---|---|---|---|
| $m$ | kg | 50 | $0.1n\mu$ $(n = 0,1,2,3)$ |
| | | 75 | $0.1n\mu$ $(n = 0,1,2,3)$ |
| | | 85 | $0.1n\mu$ $(n = 0,1,2,3)$ |

**Table 2.** Mean values $\mu$ and standard deviations $\sigma$

As it would appear, three different study assumptions are made for the mean value, and for each assumption in terms of mean value, four different assumptions are made for the standard deviation ($n$ = 0, 1, 2, 3). The value of $n$ equal to zero is the deterministic approach. In this model, the weight of all pedestrians crossing the bridge is assumed to equal the mean value defined for the population (e.g. 50 kg). For values of $n$ larger than zero, a stochastic model is assumed for pedestrian weight. As the value of $n$ increases (from 1 over 2 to 3), so does the standard deviation of pedestrian weight.

As an example, Figure 1 shows the normal, log-normal and uniform distribution functions of pedestrian weight when assuming $\mu_m$ = 75 kg and $\sigma_m = 0.1\cdot 2\cdot 75$ kg = 15 kg (i.e. assuming $n$ = 2).

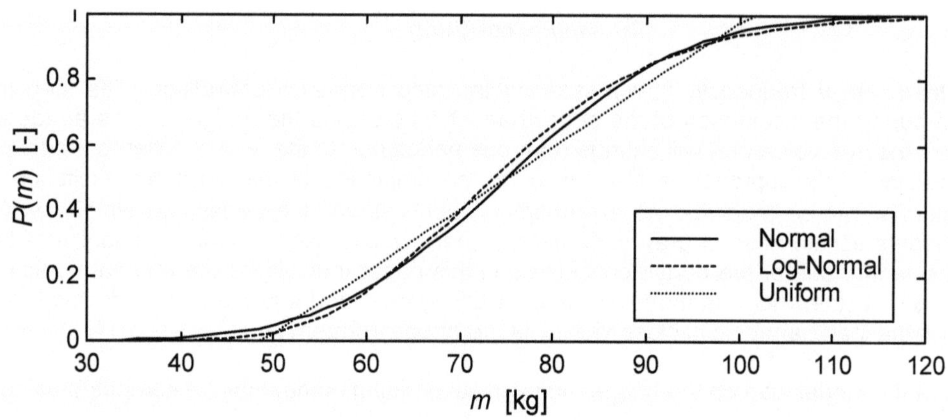


**Figure 1.** Probability distribution functions for $m$. Assumed is: $\mu_m$ = 75 kg and $\sigma_m$ = 15 kg.

As can be seen, the uniform distribution does not accept that pedestrian weight can be lower than, say 50 kg, and larger than, say 100 kg. Such restriction is not enforced for the normal and log-normal distributions.

### 3.2 Other walking parameters

For the dynamic load factor ($\alpha$), step frequency ($f_s$), and stride length ($l_s$) randomness is modelled. The study assumptions, covering mean values ($\mu$) and standard deviations ($\sigma$) for the individual random variables and associated distributions, are outlined in table 3. Table 3 represents stochastic models suggested in literature (references are stated) and they all rely on normal distributions.

| Variable | Unit | $\mu$ | $\sigma$ | Reference |
|---|---|---|---|---|
| $f_s$ | Hz | 1.99 | 0.173 | Matsumoto [2] |
| $\alpha$ | - | eq. 3 | $0.16\mu_\alpha$ | Kerr [9] |
| $l_s$ | m | 0.71 | 0.071 | Živanovic [3] |

**Table 3.** Mean value and standard deviation for $\alpha$, $f_s$, and $l_s$

In table 3 a reference is made to eq. 3, which describes the modelled relationship between the dynamic load factor (its mean value, $\mu_\alpha$) and the step frequency, $f_s$ (to be inserted in Hz):

$$\mu_\alpha = a\, f_s^3 + b\, f_s^2 + c\, f_s + d \tag{3}$$

where

$$a = -0.2649 \qquad b = 1.306 \qquad c = -1.7597 \qquad d = 0.7613 \tag{4}$$

Generally eq. 3 indicates that $\mu_\alpha$ is conditioned on $f_s$. The relationship is calibrated to measurement results in the frequency range 1 Hz < $f_s$ < 2.7 Hz.

Figure 2 illustrates some of the study assumptions defined in this section.

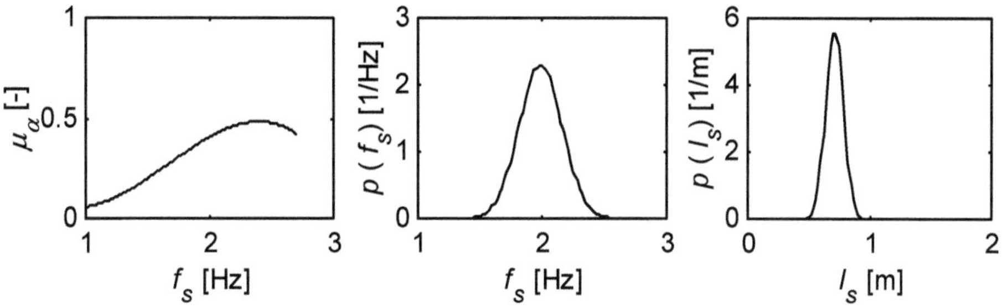


**Figure 2.** Relationship defined by eq. 3 (left), probability density function for step frequency, $p(f_s)$ (middle) and density function for stride length, $p(l_s)$ (right).

## 4. CHARACTERISTICS OF BRIDGE RESPONSE

For the calculations of bridge response, the bridge is assumed to be at rest when a pedestrian enters the bridge. The response considered is the vertical peak acceleration response encountered at midspan of the bridge, as this response characteristic is often used to evaluate the vibration serviceability limit state. Considering the various possible ways of modelling pedestrian weight interest is on results of characteristics of the statistical distribution of bridge peak response calculated on different assumptions. Here focus is on quantiles of bridge acceleration response, $a$. High quantiles, such as $a_{95}$ are believed to be of primary interest for the bridge designer and operator and the notation indicates that in 1 out 20 bridge crossings, the acceleration level $a_{95}$ is expected to be exceeded. For completeness, some other quantiles are also monitored.

The quantiles are extracted from statistical distributions of bridge response computed using MonteCarlo Simulation methods considering the modelled randomness in walking parameters. As many as 500,000 simulations runs (each emulating a pedestrian crossing the bridge) were made to provide confidence in the computed statistical distributions. For computing bridge response a Newmark time integration scheme was employed.

## 5. RESULTS

In terms of different quantiles of bridge acceleration response, the calculations gave the results shown in figure 3.

Focusing on the quantile $a_{75}$ (upper 3 plots) it appears that the mean value of pedestrian weight has a significant bearing on bridge response. An increase in mean value increases bridge loading and therefore its response. This is not surprising considering eq. 2. It can be shown that the calculated values of $a_{75}$ are close to linearly linked with the mean value of pedestrian weight ($\mu_m$) assumed for the calculations.

The upper 3 plots also reveal that for a given value of mean weight of the pedestrians (whether $\mu_m$ = 50, 75 or 85 kg), almost identical results in terms of $a_{75}$ are obtained whether a normal, a log-normal or a uniform distribution is assumed for pedestrian weight. The results also suggest that it is not important whether one or the other standard deviation is assumed ($n$ = 0, 1, 2, or 3). An almost identical result is obtained anyway. To give some perspective in terms of a realistic value of $n$ it might be 1.8. At least this was the value obtained by weighting a large number of students at Aalborg University. But in terms of an estimate of $a_{75}$ for the bridge considered in this paper, the value of $n$ appears to be of marginal importance. Basically, the deterministic model ($n$ = 0) would provide a sufficiently accurate estimate of $a_{75}$.

Turning to the quantiles $a_{95}$ and $a_{97.5}$ ($a_{97.5}$ especially), a tendency is seen in which an increase in standard deviation of pedestrian weight (increase in the value of $n$) results in gradually, but slightly, increasing values of the quantiles. This observation suggests that the random nature of pedestrian weight (specifically the standard deviation assumed) has some bearing on the uppermost quantiles of bridge acceleration response. This is observed regardless of whether a normal, log-normal or uniform distribution is assumed for pedestrian weight. In fact almost identical results in terms of $a_{95}$ and $a_{97.5}$ are obtained for the three different distributions. It is also seen that the value of $n$ (and thus the size of the standard deviation) only slightly influences, for instance, $a_{95}$.

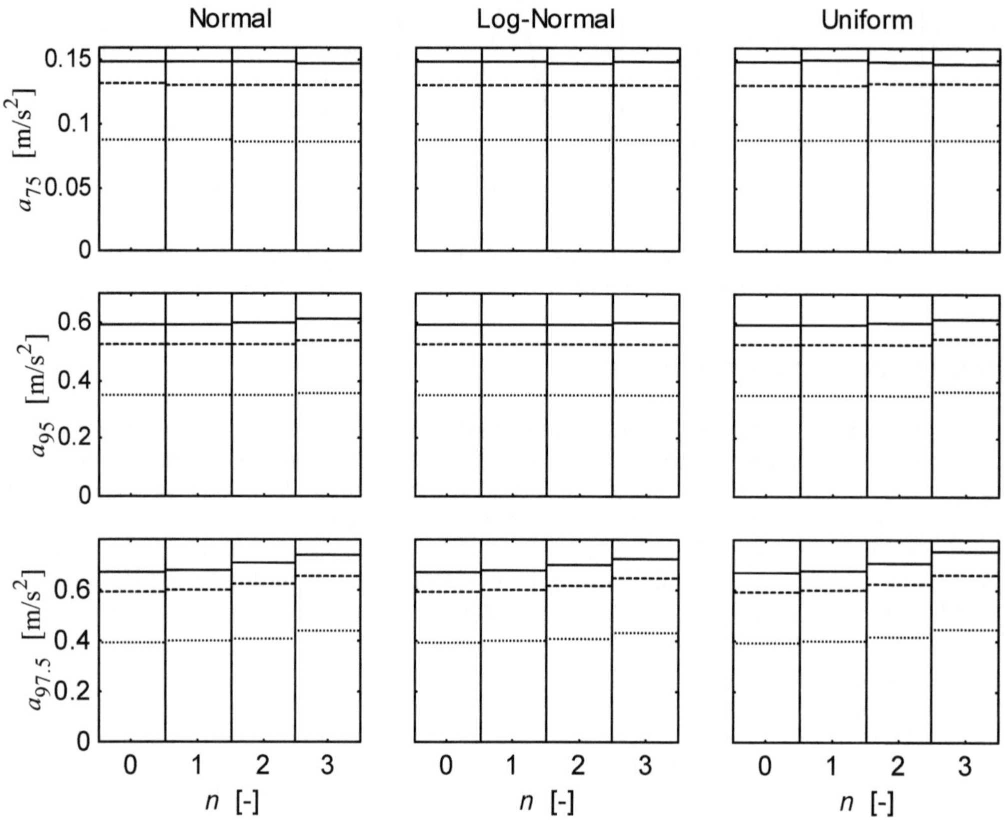


**Figure 3.** Three quantiles of bridge response and how they depend on $n$, $\mu_m$, and distribution type for $m$. Solid line: $\mu_m$ = 85 kg, Dashed line: $\mu_m$ = 75 kg, Dotted line: $\mu_m$ = 50 kg.

A different way of illustrating that the statistical distribution of bridge response only to some extent is influenced by the assumption of the size of the standard deviation is also illustrated in figure 4. The plot in figure 4 shows the probability distribution functions for bridge accelerations calculated assuming $n$ = 0 and $n$ = 3, respectively.

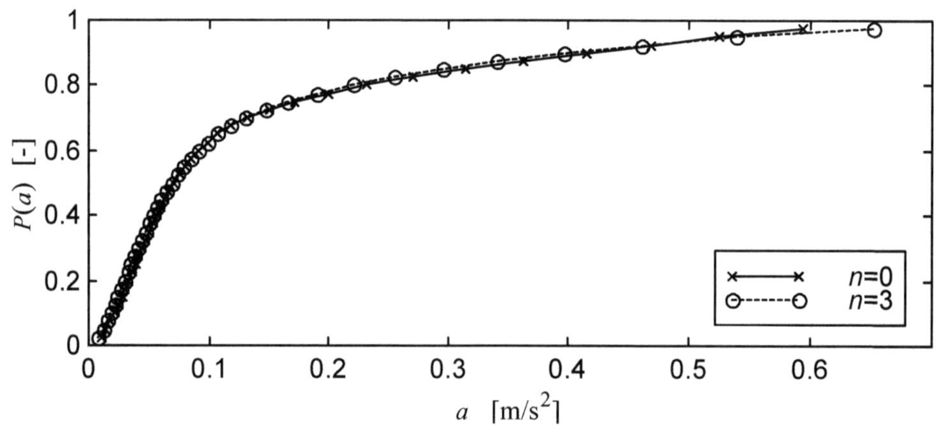


**Figure 4.** Probability distribution functions of bridge acceleration. $\mu_m$ = 75 kg and a normal distribution for $m$ is assumed (it is thus an example).

In figure 4 it can be seen that the highest standard deviation considered ($n$ = 3) result in a statistical distribution of bridge response which is fairly identical to the distribution calculated assuming $n$ = 0, which is the deterministic model for pedestrian weight. Minor differences can be noticed.

Figure 1 illustrated the three different distributions assumed for pedestrian weight (normal, log-normal, and uniform), and figur 5 illustrates the statistical distributions of bridge response calculated assuming the three different distributions for pedestrian weight.

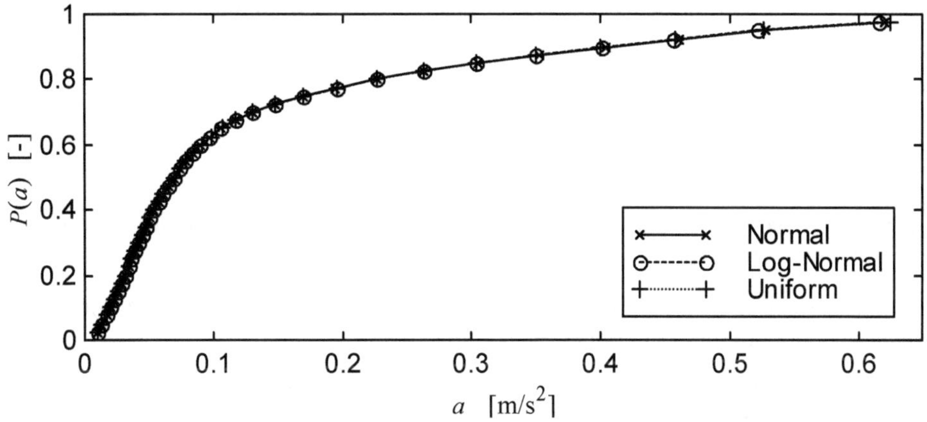


**Figure 5.** Probability distribution functions of bridge acceleration. Assumed is: $\mu_m$ = 75 kg and $\sigma_m$ = 15 kg (it is thus an example, as more combinations of $\mu_m$ and $\sigma_m$ have been investigated).

As can be seen, the output of calculations (the statistical distribution of bridge response calculated on three different assumptions about the distribution type for pedestrian weight) becomes almost identical even though the input for the calculations is in fact different.

## 6. CONCLUSION AND DISCUSSION

It was investigated how some of the choices that need be made, related to modelling pedestrian weight, influenced the statistical distribution of vertical bridge response for a particular footbridge. Three different statistical distributions were considered for pedestrian weight and for each distribution, different mean values and standard deviations were considered. It appeared that the statistical distribution of bridge response was not sensitive to whether a normal, log-normal, or uniform distribution was assumed for pedestrian weight. The results also showed that the statistical distribution of bridge response was very sensitive to the mean value of pedestrian weight but only slightly sensitive to the standard deviation of pedestrian weight applied for the calculations. A fully deterministic model of pedestrian weight (standard deviation set to zero) showed to provide a statistical distribution which only differed slightly from those calculated when pedestrian weight was modelled as a random variable.

This would immediately suggest that it might be unnecessary to model pedestrian weight as a random variable, but that it is quite important to employ a value of pedestrian weight that well represents the mean value of pedestrian weight found in the actual population of pedestrians expected to traverse the footbridge.

It should be recalled that these conclusions are reached studying only a single footbridge, and thus that they may not be valid for any footbridge. It is also important to recognise that the quite simplistic distributions of pedestrian weight employed for the studies of this paper might not be representative for actual populations of pedestrians. Not much data are available on pedestrian weight, but the paper may be considered a baseline study which can be extended considering other bridges and more complex distributions of pedestrian weight.

## REFERENCES

[1] Dallard, P., Fitzpatrick, A.J., Flint, A., Le Bourva, S., Low, A., Ridsdill-Smith, R.M., and Wilford, M., *The London Millennium Bridge*, The Structural Engineer Vol. 79, pp. 17-33, 2001.

[2] Matsumoto, Y., Nishioka, T., Shiojiri, H., and Matsuzaki, K., *Dynamic design of footbridges, IABSE Proceedings*, No. P-17/78: pp. 1-15, 1978.

[3] Živanovic, S., *Probability-based estimation of vibration for pedestrian structures due to walking*, PhD Thesis, Department of Civil and Structural Engineering, University of Sheffield, UK, 2006.

[4] Ontario Highway Bridge Design Code, Highway Engineering Division; Ministry of Transportation and Communication, Ontario, Canada, 1983.

[5] British Standard Institution, *Steel, concrete and composite bridges. Specification for loads*, BS 5400: Part 2, 1978.

[6] Ellis, B.R., *On the response of long-span floors to walking loads generated by individuals and crowds*, The Structural Engineer, Vol. 78, pp. 1-25, 2000.

[7] Bachmann, H., and Ammann, W., *Vibrations in Structures – Induced by Man and Machines*, IABSE Structural Engineering Documents 3e, Zürich, Switzerland, 1987.

[8] Rainer, J.H., Pernica, G., and Allen, D.E., *Dynamic loading and response of footbridges*, Canadian Journal of Civil Engineering, Vol. 15., pp. 66-78, 1998.

[9] Kerr, S.C., and Bishop N.W.M., *Human induced loading on flexible staircases*, Engineering Structures, Vol. 23, pp. 37-45, 2001.

Proceedings of the IMAC-XXVIII
February 1–4, 2010, Jacksonville, Florida USA


# IDENTIFICATION OF HIGH-ORDER LOCAL VIBRATION PROPERTIES OF RC VIADUCT

Kodai MATSUOKA, Student, Department of Civil Engineering, Osaka University,
k-matsuoka@civil.eng.osaka-u.ac.jp
Kiyoyuki KAITO, Associate Professor, Frontier Research Center, Osaka University,
kaito@ga.eng.osaka-u.ac.jp
Tsutomu WATANABE, Researcher, Railway Technical Research Institute
masamichis@rtri.or.jp
Masamichi SOGABE, Senior Researcher, Railway Technical Research Institute
tsutomuw@rtri.or.jp

## ABSTRACT

In order to sustain the development of high-speed trains, it is necessary for railway bridges to grasp the dynamic behavior of bridge members precisely, by identifying not only whole bridge vibration properties but also local ones, and low-order to high-order mode vibration properties. Actually, there are common concerns over resonance and noises caused by the vibration of bridge members due to high-speed trains. However, there have been few vibration measurement experiments for carrying out identification of high-order member vibration properties. In this circumstance, the authors conducted the passing train experiment, measuring vibration concurrently at several points on the intermediate and projecting slabs of a RC viaduct for high-speed trains. Through the vibration measurement experiment, the authors verified the possibility of detection of high-order vibration modes of members and identified the outstanding vibration mode when a high-speed train passes. This paper also includes a few discussions about the relation between vibration properties and the outstanding frequency induced by the multi-axle load of trains. By accumulating the data of actual measurements of local vibration, it would be possible to contribute to the sophistication of finite element analysis, etc.

## 1. INTRODUCTION

Recently, the speed of trains that run on arterial railroads across Japan has significantly increased, and the technology for higher speed is being developed[1)]. The major problems with the structures caused by high-speed trains include resonance from the viewpoint of safety, and ground vibration and noise from the viewpoint of usability.

Among these vibration problems, the research into noise has been advanced especially in the field of acoustics. As the technology for identifying the sound source has been developed, it is now possible to identify the members that cause noise when a train passes to some degree. As for structure vibration, most of previous studies are focused on resonance, because it is essential to secure safety. Resonance control has been discussed based on the measurements of actual bridges and numerical calculation, and most studies are focused on the first-order mode of the entire bridge. From now on, it will be necessary to tackle usability problems also in the structure field, and so it is required to grasp the vibration properties of each member when a train passes. By grasping the characteristics of member vibration, it will become possible to conduct local structural health monitoring, apply data to partial dynamic numerical calculation, and study member resonance when a train passes at a higher speed. These are difficult with the data on the entire system only. However, a bridge is constituted by several kinds of members, and all members interact with one another. At a low order, the vibration mode of the entire system is excited, and it is necessary to discuss the high-order vibration components to study the vibration mode of a member. However, there have been few researches into high-order member vibrations.

In this circumstance, the authors attempt to identify local vibration characteristics of members from the acceleration response to the loading of a passing train, and precisely grasp the actual behavior of members in response to a passing train, while focusing on a RC bridge, which is the most common type in Japan. Chapter 2 describes the basic position of this study. Chapter 3 outlines the experiment on a RC viaduct under a passing train, and identifies and discusses vibration characteristics.

## 2. BASIC POSITION OF THE STUDY

### (1) Local vibration problems

The structural vibration of a railway bridge is closely related to the "speed effect of multi-axle loading," and dates back to the research by Fryba et al.[2] in the 1970s. In Japan, theoretical studies, model experiments, and bridge measurements, etc. have been conducted continuously since the age of national railways. To secure safety, resonance has been taken into account in the designing process since 1983, and the methods of using the nomogram was proposed while considering high-speed running in 2004[3]. In representative researches, Sogabe et al. [4] pointed out the resonance of the first-order mode of the entire bridge, which is a simple girder structure, after numerical analyses and actual bridge experiments, and proposed the above dynamic designing method. As for the vibration problems regarding usability, Yoshioka and Hara et al. analyzed the relation between ground vibration and the vibration of an RC viaduct, based on accumulated vibration measurements[5]. As for noise, it has been studied mainly in the field of acoustics, as mentioned above. Recently, Nagakura briefly summarized the characteristics of railroad noise and countermeasures against it[6]. The level of structural noise is lower than that of a rail, etc. but its frequency component is the lowest among noises, and distance decay is low. Therefore, it spreads in the broad area along railways as low-frequency noise. In addition, it was pointed out that the power of structural noise is over the cube of train speed, and it is anticipated that future high-speed trains will cause significant noise.

In this situation, it is meaningful to analyze the vibration of a railway bridge from the structural viewpoint, to improve usability. In addition, it is considered important to grasp high-order member vibrations due to a passing train. Then, this study is aimed at providing the findings regarding member vibrations due to a passing train, which are indispensable for coping with vibration problems, and the authors measured vibrations due to a passing train and studied the characteristics of member vibrations, focusing on two types of railway bridges. In addition, the major vibration modes are elucidated, by comparing them with the predominant frequency when a train passes. The studied bridge type is a RC viaduct. This bridge was built after the construction of Sanin Shinkansen, and a slab track was installed, and it is inferred that this bridge has different vibration properties from those of previously-built railway bridges.

### (2) Identification method and prerequisites

This study is focused on the vibration properties of a viaduct due to a passing train. As a method for identifying vibration characteristics, the cross-spectral method, in which a measurement point is used as a reference point, was adopted, and the natural frequency and vibration modes were identified. With this method, the damping ratio cannot be identified, but by using the cross-spectral method, the effects of noise components can be removed, and so it is possible to identify vibration modes precisely. However, it is necessary to consider the two conditions: (1) the damping ratio of the bridge is small, and modes are not close to one another, and (2) the input is white noise, which does not have frequency components. This study is conducted under the assumption that the above conditions are satisfied. For the details of the identification method, refer to the reference [7]. When vibration characteristics are identified using the acceleration response due to a passing train, the results do not represent the vibration characteristics of a bridge only (it actually represents the complex system of a bridge and a train), and the vibration state is like forced vibration. However, since the objective of this study is to grasp the actual behavior of bridge members when a train passes, the above problems are disregarded. In order to clarify this, the vibration characteristics identified from the response to a passing train are hereinafter represented by equivalent vibration characteristics, equivalent natural frequencies, and equivalent vibration modes, to differentiate them from those of a single bridge. At each measurement point, a piezoelectric accelerometer was adopted, and only low-frequency components were measured, using a 2-kHz anti-aliasing filter.

## 3. Vibration properties of a RC viaduct

### (1) Outline of the target bridge and vibration measurement

This chapter is focused on a RC viaduct, whose aspect ratio per span is low and in which member vibrations would occur ata lower order. Fig. 1 shows the schematic diagram of the bridge. This bridge is a 3-span RC bridge constructed in 1979, and is constituted by two 7.88-m spans at both ends and a 8.1-m span at the center. The central span is longer than the other spans. This study is focused on the central span. This bridge is a symmetrical double track, except the shorter overhanging slab width at the downbound side.

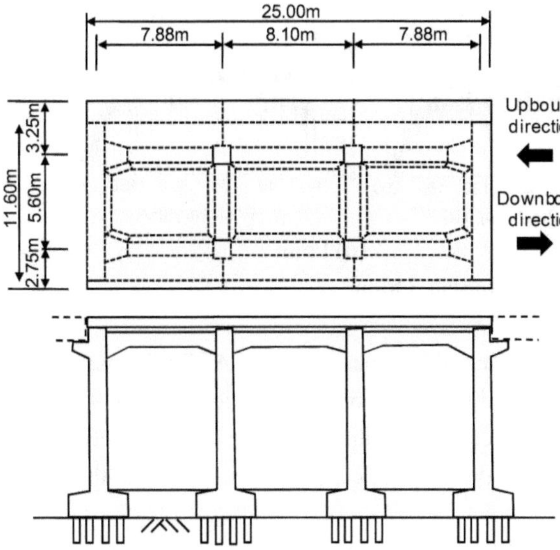


Fig. 1 Schematic diagram of the target bridge

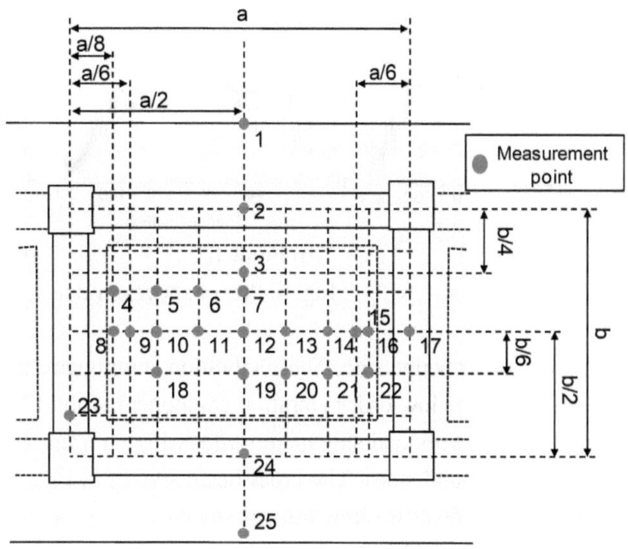


Fig. 2 Layout of accelerometers

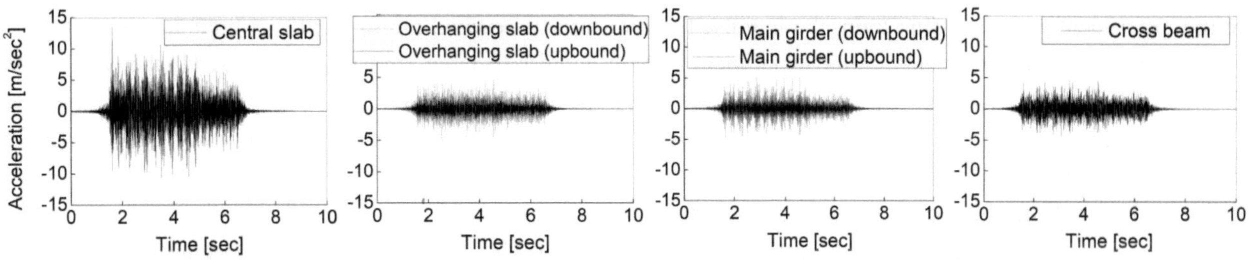


(a) Point 12 (b) Point 1, 25 (c) Point 2, 24 (d) Point 17

Fig. 3 Acceleration response due to a passing downbound train

Mainly the members of slabs (intermediate slabs and overhanging slabs) were studied. About 20 measurement points are set at each target member, and measurement points are also set at other members, to discuss the differences in vibration amplitude and frequency characteristics. Fig. 2 shows the layout of the measurements points on the central slab. The layout of measurement points varied among target members, and the acceleration response of each member to a passing train was measured. The next section compares members in acceleration amplitude and frequency characteristics due to a passing train.

### (2) Comparison of members in vibration properties

As for the vibration of each member of the target span, Fig. 3 shows the acceleration response to a passing train at each measurement point on the central and overhanging slabs, the main girder, and the cross beam shown in Fig. 2. As for the central slab, the data of the measurement point 12 is used. The target 16-car train passed in the downbound direction at a speed of 267 km/h. As shown in Fig. 3-(a), the acceleration amplitude due to a passing train reached 14 m/sec$^2$ in the central slab. As shown in Fig. 3-(b), (c), and (d), the acceleration amplitude in the train-passed side of the overhanging slab, the main girder, and the cross beam reached about 5 m/sec$^2$, and that in the no-train side of the overhanging slab and the main girder reached about 2 m/sec$^2$. The acceleration amplitude of the central slab is remarkably high, but the amplitude of the overhanging slab is near to that of the main girder. Since a train passes on one of the two railways on a bridge, the amplitudes of the overhanging slab and the main girder vary considerably between the upbound and downbound railways.

Next, Fig. 4 shows the power spectrum of measured acceleration response. Each member has the line spectrum-like peak due to multi-axle loading. Considering the span length, the vibration component of less than 5Hz is the entire vibration component of the three spans. For every member, the vibration components of less than 10Hz were weak. As shown in Fig. 4-(a), the central slab showed several peaks in the broad range from 50Hz to 150Hz. As for the overhanging slab shown in Fig. 4-(b), the spectra of the upbound and downbound railway sides have similar shapes, but the overhanging slab at the downbound side where a train passed (represented by the red line) showed larger vibration components around 40Hz, 60Hz,

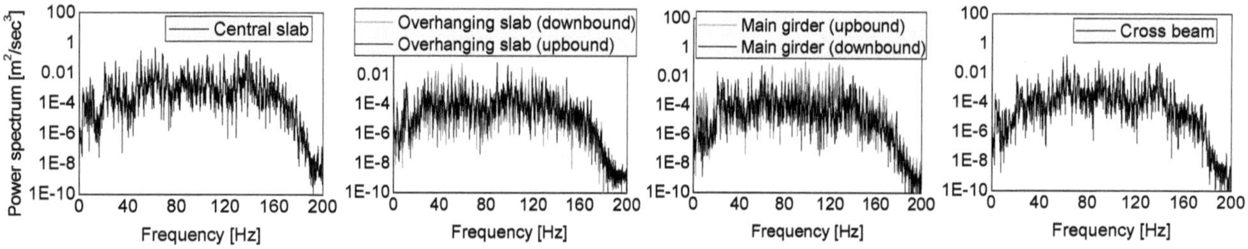


(a) Point 12 (b) Point 1, 25 (c) Point 2, 24 (d) Point 17

Fig. 4 Power spectrum due to a passing downbound train

and 120Hz than the upbound side. These vibration components are probably caused by a passing train. As for the main girder shown in Fig. 4-(c), the low-order vibration component around 10Hz and the high-order vibration component from 90-120Hz were strong in the downbound side main girder where a train passed, and spectral configurations were different between the upbound and downbound sides. The cross beam shown in Fig. 4-(d) showed the spectral configuration similar to that of the central slab, and the power spectral value is small.

The following sections identify the equivalent vibration characteristics of the central slab and the overhanging slab, and discuss the relation between the entire and partial vibrations of the RC bridge, and the outstanding mode when a train passes.

Before the experiment with a passing train, the impulse hammer test was carried out to identify vibration characteristics. However, only the identification results for the experiment with an impulse hammer are written, because (1) the error in the equivalent natural vibration characteristics identified in the results of the identification and the experiment with a passing train is several %; (2) there is the mode that cannot be excited in the impulse hammer; and (3) the purpose of this study is to grasp the vibration characteristics when a train passes.

### (3) Equivalent vibration characteristics of the central slab

The equivalent vibration characteristics of the central slab are identified. The layout of measurement points is as shown in Fig. 2. Here, the mode of the flat plate whose 4 sides are simply supported was defined as the theoretical mode.

#### a) Identification of equivalent vibration characteristics with the experiment with a passing train

Like Figs. 3 and 4, the equivalent natural frequencies and equivalent vibration modes of the central slab were identified, using the response to a passing train obtained at each measurement point. The results are summarized in Table 1. Table 1 includes the diagram representing the equivalent vibration modes of the central slab. For this diagram, the central slab was divided with 8 meshes in the bridge axis direction and 6 meshes in the vertical direction, and the value of each contact point was gridded with respect to measurement points. In addition, it was assumed that the mode displacement of the upper part of each column (the points a, b, c, and d in Fig. 13) is zero. The equivalent vibration mode in the vertical direction represents the cross section of the center of the span. The modes in the vertical direction include the overhanging slab. Table 1 indicates that the 3 low-order modes from 22.6Hz to 43.0Hz can be said to be the entire modes with the amplitude of the main girder being large. As the mode of the central slab whose amplitude is outstanding, the first-order bowl-shaped mode was identified at 52Hz. The following modes can be considered to be the vibration modes of the members of the central slab, because the amplitudes of the main girder and the overhanging slab are smaller than the mode amplitude of the central slab. At 60.6Hz, the second-order mode whose maximums appear on the diagonal line of the central slab was identified. At 63.6Hz, the second-order mode in the vertical direction of the central slab was identified. Around 60Hz, the second-order mode in the bridge axis direction was also identified. Therefore, there is a possibility that the second-order modes of the central slab interact with each other. As for higher-order modes, there were the second-order modes in the bridge axis and vertical directions at 72.1Hz, the third-order mode of the central slab at 93.7Hz, the fourth-order mode at 120.8Hz, the third-order mode in the bridge axis direction and the second-order mode in the vertical direction at 161.1Hz. From a total of 9 experiments, it was found that the variance of the mode was about 5% regardless of train direction and speed.

#### b) Outstanding mode when a train passes

The speed of the trains passing on the bridge can be classified into the two ranges: around 260 km/h and around 140 km/h. The configuration of a power spectrum varied with train speed, but not with train direction. Fig. 5 shows the representative power spectrum due to a passing train for each train speed. The values in the figure are the averages of values obtained at

Table 1 Results of identification by the experiment with a passing train (central slab)

| | |
|---|---|
| 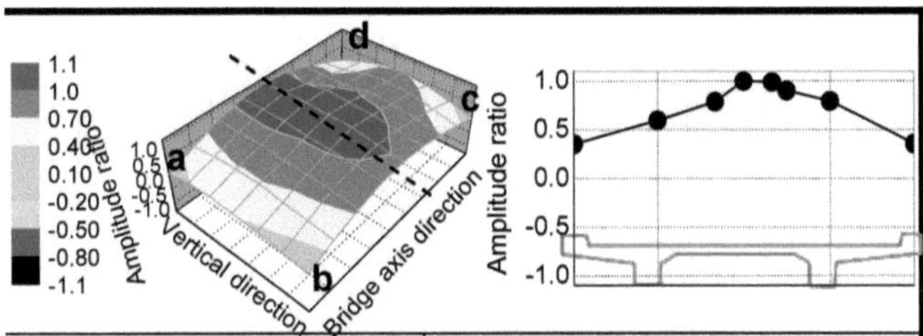 | 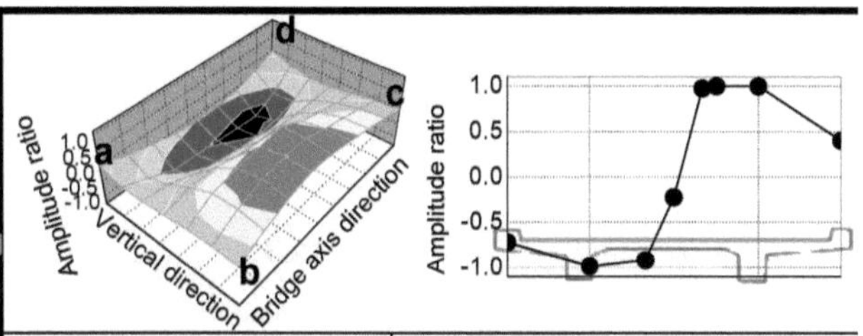 |
| Contour map / Vertical direction [mm] | Contour map / Vertical direction [mm] |
| First-order deflection mode of the entire bridg | First-order torsional anti-symmetric mode of the entire bridge |
| Equivalent natural frequency : 22.6Hz | Equivalent natural frequency : 30.3Hz |
| 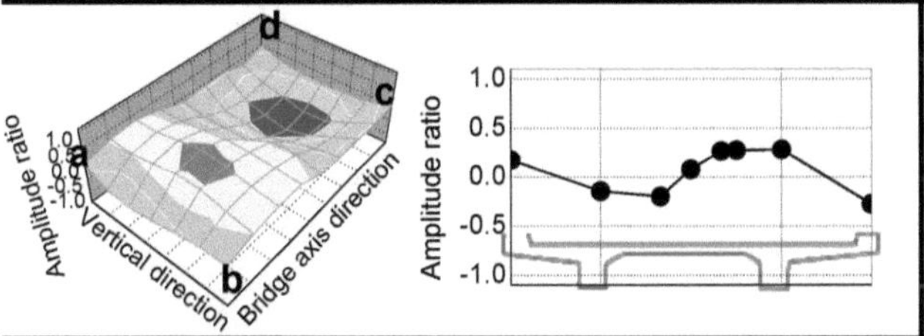 | 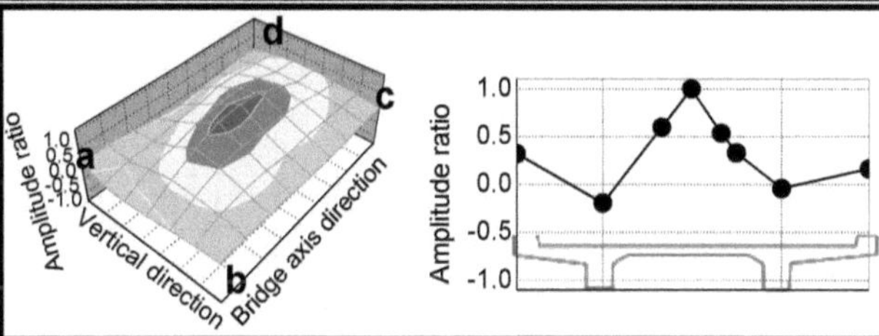 |
| Contour map / Vertical direction [mm] | Contour map / Vertical direction[mm] |
| Second-order deflection mode of the entire bridge | First-order mode of the central slab |
| Equivalent natural frequency : 43.0Hz | Equivalent natural frequency : 52.7Hz |
| 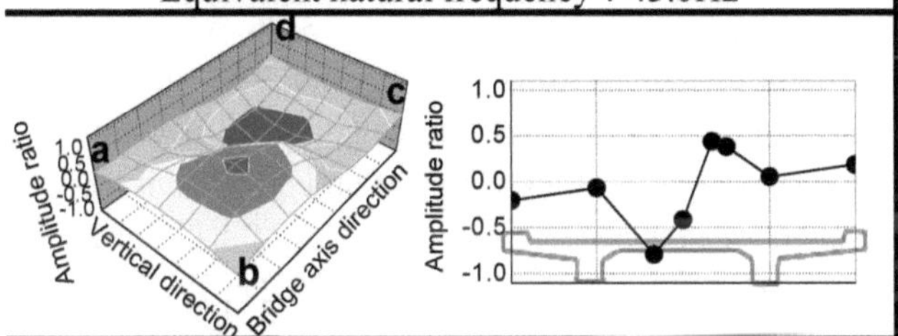 | 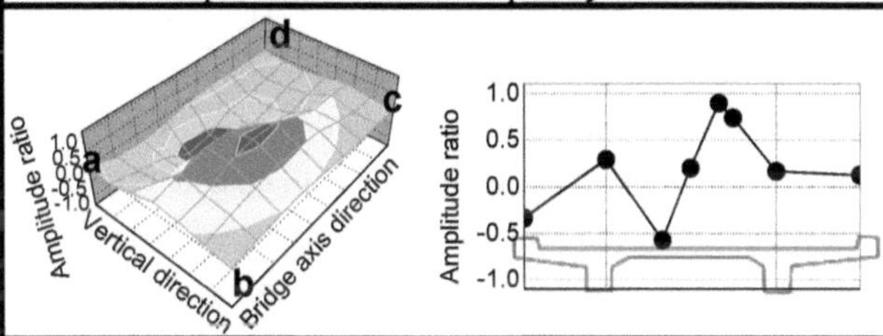 |
| Contour map / Vertical direction [mm] | Contour map / Vertical direction[mm] |
| Second-order diagonal mode of the central slab | Second-order mode of the central slab in the vertical direction |
| Equivalent natural frequency : 60.6Hz | Equivalent natural frequency : 63.6Hz |
| 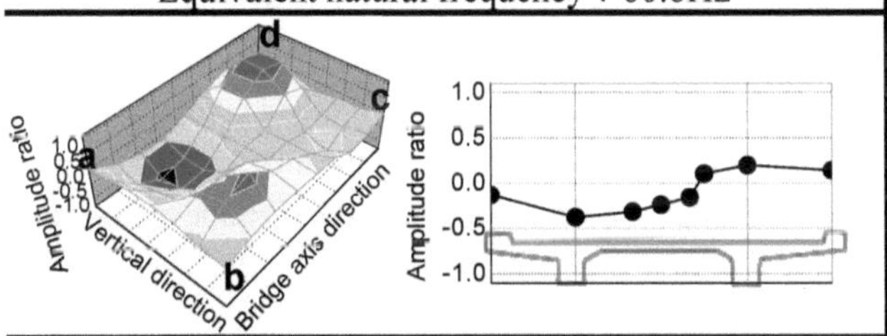 | 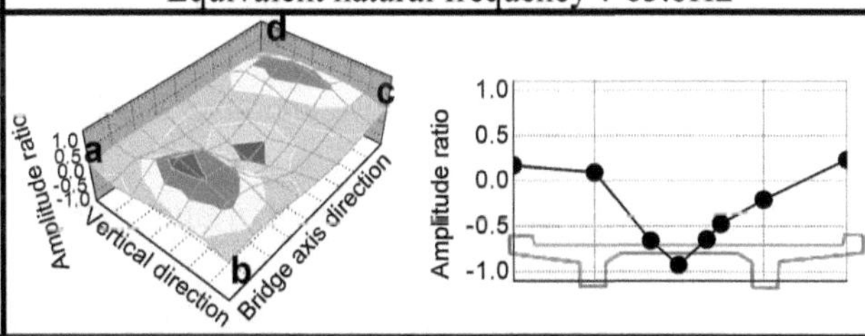 |
| Contour map / Vertical direction [mm] | Contour map / Vertical direction[mm] |
| Second-order modes of the central slab in the bridge axis and vertical directions | Third-order mode of the central slab |
| Equivalent natural frequency : 72.1Hz | Equivalent natural frequency : 93.7Hz |
| 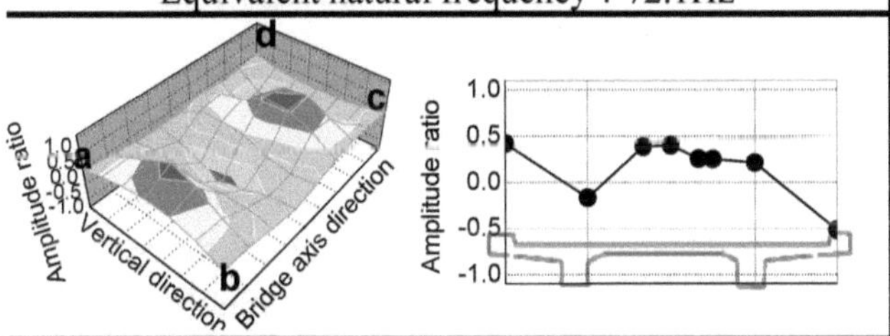 | 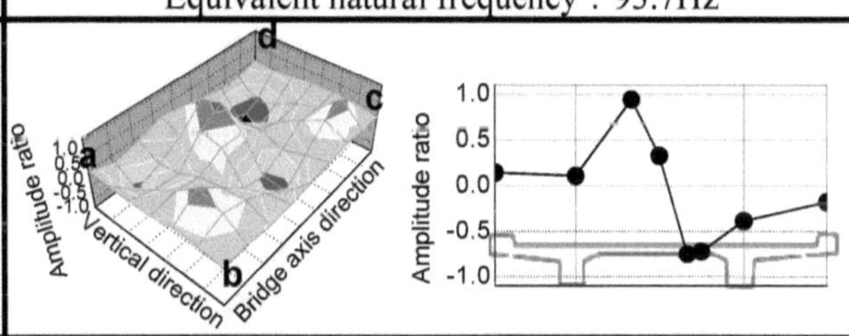 |
| Contour map / Vertical direction [mm] | Contour map / Vertical direction[mm] |
| Fourth-order mode of the central slab | Third-order mode in the bridge axis direction and second-order in the vertical direction of central slab |
| Equivalent natural frequency : 120.8Hz | Equivalent natural frequency : 161.1Hz |

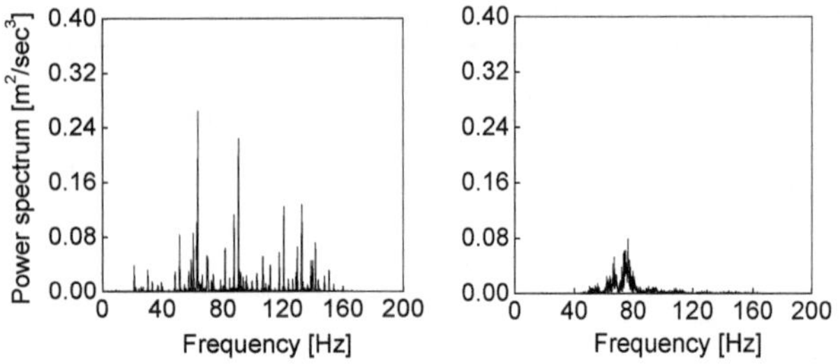


(a) Train Speed 266km/h (b) Train Speed 146km/h

Fig. 5 Power spectrum due to passing train

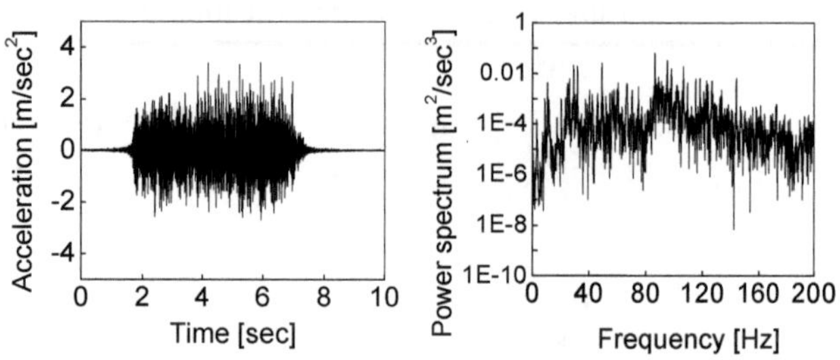


(a) Acceleration response (b) Power spectrum

Fig. 7 Measurements of vibrations due to a passing train

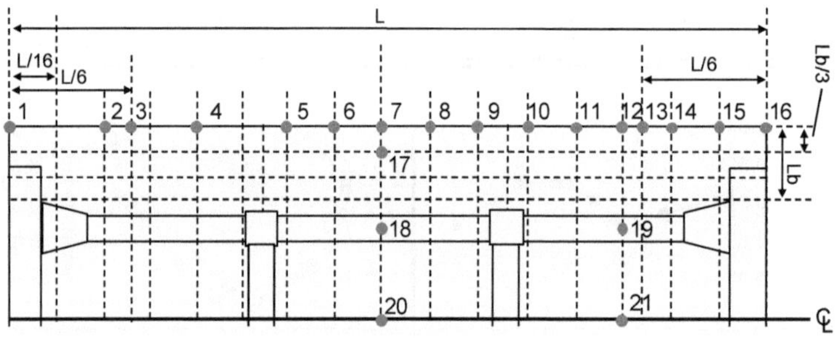


Fig. 6 Layout of accelerometers focused on overhanging slab

measurement points 5, 7, 10, 12, 14, 19, and 21, to discuss the outstanding components of the central slab. When train speed is around 140 km/h, the outstanding components can be found around 60Hz only. Meanwhile, when train speed exceeds 260km/h, the power spectral value around 60Hz is over three times. In addition, outstanding components of mainly high-order modes can be found around 90Hz, 120Hz, and 150Hz, etc. It is considered that the vibration around 60Hz is due to the second-order mode of the central slab regardless of train speed. As mentioned above, the second modes of the central slab in the diagonal, bridge axis, and vertical directions were all identified around 60Hz, and it is inferred that the outstanding components are produced when these transit and interact, but these outstanding components could not be separated in this study. It is considered that the outstanding component around 90Hz is caused by the third-order mode of the central slab and the outstanding component around 120Hz is caused by the fourth-order mode. On the other hand, there were no peaks in the low-order components of less than 50Hz, which is caused by the entire mode; except that there was a small peak around 20Hz where the first-order deflection mode was excited when train speed was 266 km/h.

In summary, when a train passes, the second-order mode of the central slab is outstanding around 60Hz compared with the entire and first-order modes. When train speed increases to 260 km/h, the vibration components of the members of the central slab become outstanding around 90Hz, 120Hz, and 150Hz. Around 150Hz, the equivalent vibration mode could not be identified.

### (4) Equivalent vibration characteristics of the overhanging slab

This section is focused on the overhanging slab whose structure is similar to the central slab. The layout of measurement points is as shown in Fig. 6. In the overhanging slab, edge cutting was made for each span, and the spans are not connected structurally. In addition, there are no cross beams, and so it is expected that the panel of each span vibrates in conjunction with one another. Then, a total of 16 measurement points were set at the three spans.

#### a) Identification of equivalent vibration characteristics with the experiment with a passing train

As the examples of the acceleration variation and power spectrum obtained at measurement points in the experiment, Fig. 7 shows the results of the measurement point 7 when a downbound train passed at 272 km/h. Using the response to a passing train obtained at measurement points, the equivalent natural frequencies and equivalent vibration modes of the overhanging slab were identified. Table 2 summarizes the results of the identification. Table 2 shows the mode amplitude of the main girder, under the assumption that the maximums for the overhanging slab and the main girder are 1 because the mode amplitude of the central slab is larger than that of other members. Table 2 shows the mode in which the entire overhanging slab vibrates with the node being at the main girder, at 11.5Hz. The mode of the entire overhanging slab appeared at 16.1Hz

Table 2 Results of the identification by the experiment with a passing train (overhanging slab)

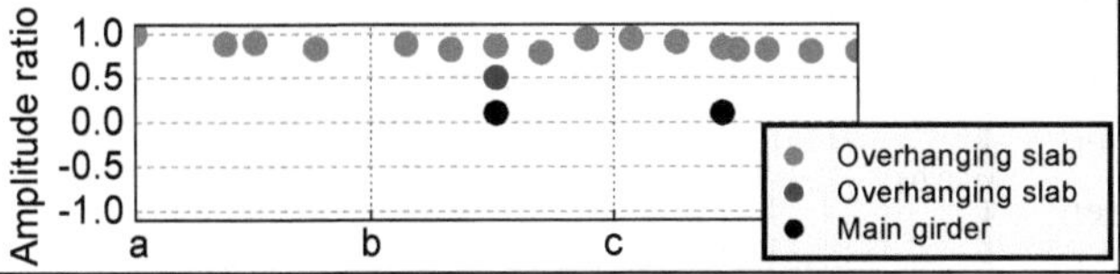


Vertical direction [mm]

First-order mode of the overhanging slab

Equivalent natural frequency : 11.5Hz

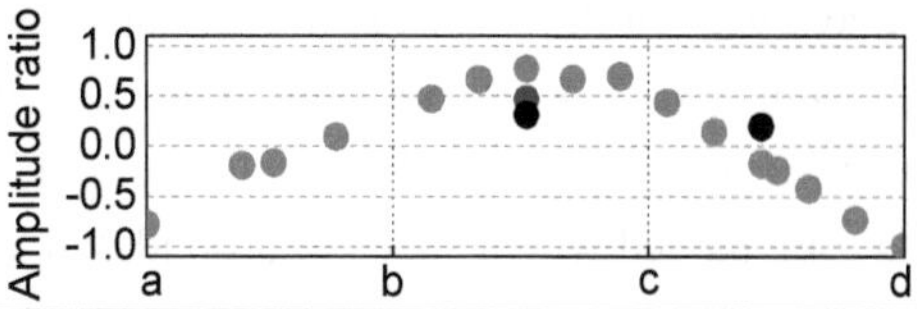


Vertical direction[mm]

Second-order mode of the overhanging slab

Equivalent natural frequency : 16.1Hz

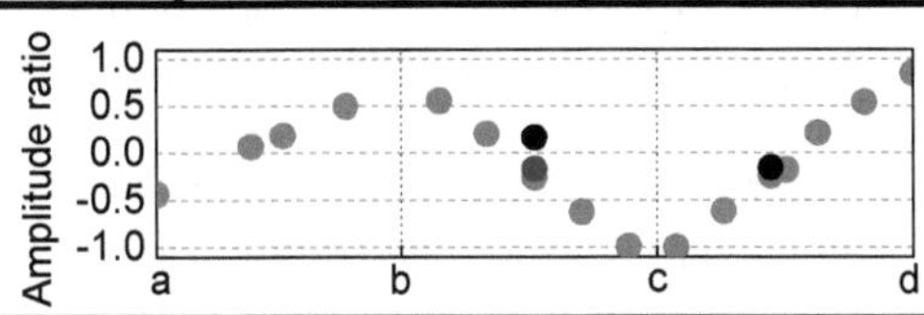


Vertical direction [mm]

Third-order mode of the overhanging slab

Equivalent natural frequency : 17.7Hz

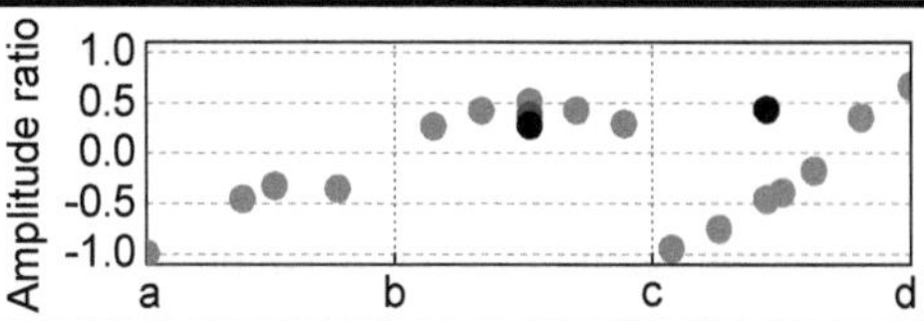


Vertical direction[mm]

First-order deflection mode of the central span

Equivalent natural frequency : 22.8Hz

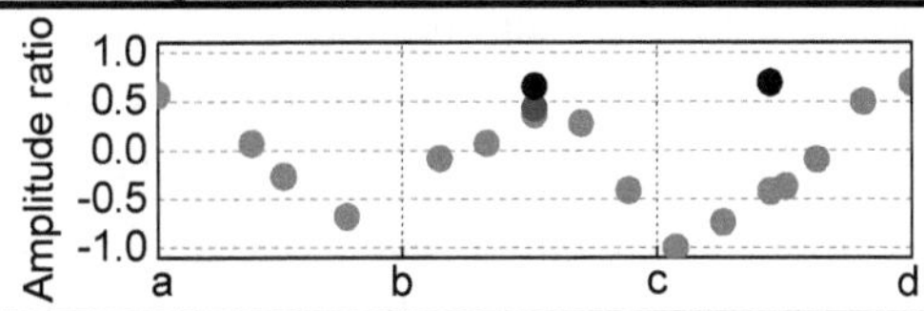


Vertical direction [mm]

Fifth-order mode of the overhanging slab

Equivalent natural frequency : 51.6Hz

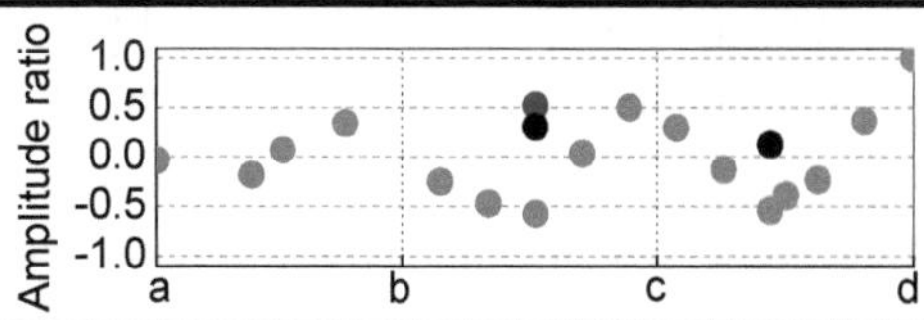


Vertical direction[mm]

Second-order mode of each panel

Equivalent natural frequency : 98.8Hz

and 17.7Hz. In these modes, this bridge vibrates although edge cutting was made, which is indicated by the large curvature of the mode. On the other hand, around 20Hz, where the entire deflection mode of the central span appeared, each panel of the overhanging slab vibrates. As shown in Table 2, the overhanging slab of the central span showed a deflection mode at 22.8Hz, and the overhanging slab between c and d showed a torsional mode in which the phases of both ends are reversed. This tendency was also observed around 45Hz, where the second-order deflection mode of the central span is excited. In the fifth-order mode of the overhanging slab identified at 51.6Hz, only the overhanging slab at the center showed a deflection mode, and the overhanging slabs at both ends showed a torsional mode. However, the mode displacements at the measurement points crossing the edge-cutting parts are close, and the entire vibration of the overhanging slab is significant. Around 50Hz, the first-order deflection mode of the central slab was identified. The vibration of members of the central slab does not vibrate the panels like the entire mode around 20Hz. The same tendency can be seen in the mode over 50Hz, and the third-mode of the central slab was identified at 93.7Hz, and the sixth-order mode of the entire overhanging slab was identified at 98.8Hz. The above mentioned characteristics were seen in all of the 7 measurements regardless of train direction and speed.

The member vibration of the overhanging slab is excited, including all of the 3 spans. When the mode of the central span is excited, the overhanging slabs at both ends vibrate differently. In each mode, the central slab has large mode amplitude, but the mode amplitude of the main girder is small, and so it can be considered that the member vibrations of the overhanging slab and the central slab are excited separately with the main girder being the node.

This study is focused on the central span, and the entire consideration was discussed with limited data. From now on, it will be necessary to improve the reliability of this research by discussing the outstanding modes in each span and the behavior of each panel of the overhanging slab.

**b)** Outstanding mode due to a passing train

As mentioned above, train speed can be classified into the two ranges: around 260 km/h and around 140 km/h. As for the overhanging slab, the configuration of a power spectrum varies according to train speed and direction. As the downbound side of the overhanging slab was observed, the maximum amplitude was observed when a train passed with a speed of around 260 km/h, and the minimum amplitude was observed when an upbound trai n passed with a speed of around 140 km/h. Fig. 8

shows the power spectrum due to the two trains. The values in the figure are the averages of the values at the measurement points 2, 4, 6, 8, 11, and 14. Fig. 8 indicates that the outstanding components appear around 30Hz, 60Hz, 90Hz, and 120Hz, when a downbound train passed with a speed of around 260 km/h. When an upbound train passed with a speed of 140 km/h, the peak was observed around 60Hz only. Such characteristics of the power spectrum can be seen also in the case of the central slab. In addition, since any modes of the overhanging slab were not identified at these outstanding frequencies, the outstanding vibrations are excited by the vibration of the central slab rather than the overhanging slab when a train passes on the overhanging slab. The value of each outstanding component of the overhanging slab is about one-tenth of the value of the central slab, which indicates that the vibration of the central slab is dominant.

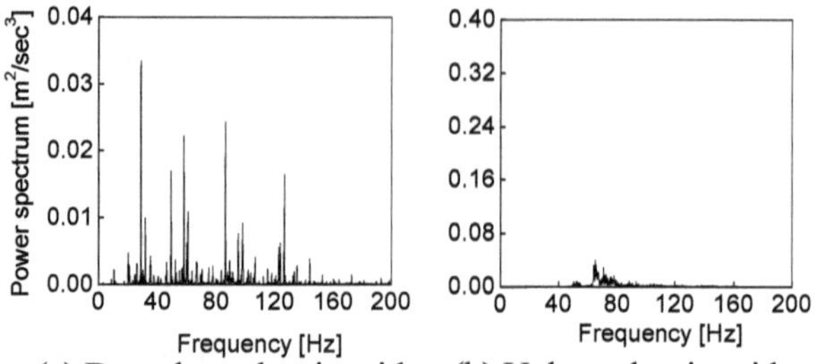

(a) Downbound train with a speed of 266 km/h (b) Upbound train with a speed of 146 km/h

Fig. 8 Power spectrum due to a passing train

## 4. Conclusion

In this study, the authors summarized the local vibrations of a bridge due to a passing train from the viewpoints of safety and usability, and conducted vibration measurement and identification analysis for each member of the RC viaduct. As a result, the authors could observe the modes in which the vibration of each member is outstanding, and found that the vibration of the RC bridge due to a passing train is mainly caused by the vibration of the central slab, although there are low-order vibrations of the overhanging slab. It was also found that the second-order mode contributes to outstanding components.

The future mission would be to reflect these results in actual bridges. It is considered necessary to identify the cause of such vibration problems and clarify influential modes and vibration modes, by discussing the relations with noise and ground vibration. In this study, the authors adopted the most fundamental method for identification. It would be necessary to adopt more advanced methods for obtaining more detailed information on the damping and phase of vibrations due to a passing train.

## Acknowledgment

Part of this study was carried out at Frontier Research Base for Global Young Researchers, Graduate School of Engineering, Osaka University based on the project of the Ministry of Education, Culture, Sports, Science and Technology titled "Promotion of Suitable Research Environment for Young Scientists' Independent Research."

## References

[1] GOTO Haruo, The pulling test run of a linear vehicle is begun in the Yamanashi linear experiment line, Journal of the Japan Society of Civil Engineers,Vol.82, No.2, pp.44,1997.2 (in Japanese)

[2] Fryba, L., Vibration of solids and structures under moving load, Noordhoff International, 1972.

[3] Railway Technical Research Institute, Design standard and explanation such as railway structures – Concrete Structures, Maruzen, 2004. (in Japanese)

[4] SOGABE Masamichi, Dynamic response and design of continuous concrete railway bridge in resonance area, Journal of Structure Mechanics and Earthquake Engineering, Vol.724, I-62, pp.83-102, 2003. (in Japanese)

[5] HARA Tsuneo, et al, Development of a New Method to Reduce SHINKANSEN-Induced Wayside Vibrations Applicable to RigidI Frame Bridges, Journal of Structure Mechanics and Earthquake Engineering, Vol.766, I-68, pp.325-338, 2004. (in Japanese)

[6] NAGAKURA Kiyoshi, Present situation and future prospects of railway noise problem, The Journal of the Acoustical Society of Japan, Vol.64(10), pp.624-628, 2008. 10 (in Japanese)

[7] MATSUOKA Kodai, KAITO Kiyoyuki, SUGISAKI, Koichi, WATANABE Tsutomu, SOGABE Masamichi, Idetification of Vibration Properties of Open Deck Viaducts under Passing Train Loads, Journal of Applied Mechanics JSCE, Vol.12, pp.983-994, 2009.8 (in Jpanese)

Proceedings of the IMAC-XXVIII
February 1–4, 2010, Jacksonville, Florida USA


# The Influence of Single Progressive Off-centered Crack on the Dynamics of Beam

by

Ahmad Azlan Mat Isa
Zamri Abdul Rahman
Faculty of Mechanical Engineering,
MARA University of Technology,
40450 Shah Alam, Selangor,
Malaysia

email: ahmadazlan@salam.uitm.edu.my

Abstract

*Structural defects alter the dynamic characteristics of mechanical structure. They reduce the service life of mechanical structures which may lead to catastrophic failure. These defects are characterized by changes in their eigen-parameters. Thus, detection of defect using reliable and effective method is an important consideration to guarantee the component safety and prolong the service life of the structure and hence save costs. This paper discusses the effect of variations in the dynamic properties with respect to the present of a single progressive crack in beam with special considerations on free-free end condition using operational modal analysis. The validity of the results is verified with the well-established experimental modal analysis*
*Keywords*

Progressive off-centered crack, natural frequency, damping ratio, experimental and operational modal analysis

## 1.0 INTRODUCTION

In the past two decades, significant amount of research has been carried out using non destructive detection method via changes in the dynamic responses of the structure. These studies are necessary to monitor constantly the health of its critical components to ensure continuous safety operations. The change of modal parameters often indicates the appearance of structure damage[1]. The dynamic characteristic of structural components varies whenever there are changes in its structural properties i.e. mass, stiffness, and damping values. This calls for continuous assessment of changes in their static and/or dynamic behavior to ascertain the operation of the component is not interrupted. The presence of a crack in a structural member reduces the stiffness and increases the damping of the structure. As a consequence, there is a decrease in natural frequencies and modification to the modes of vibration [2-4]. The development of a crack does not necessarily make a component instantly unusable, but it signals a message that its behavior needs constant monitoring before eventual replacement. This has made the vibration based monitoring of components with cracks or crack-like defects during service very important and therefore the study of vibration of components with crack becomes important [5, 6]. Attempt has been made in [7] to characterize the effect of crack present on dynamic behaviour. However, only single constant crack was investigated.

Cracks identification and detection using vibration signal was investigated as an alternative non-destructive evaluation (NDE) methods [8, 9] since hidden cracks inside the structure is rather difficult to detect using conventional methods [10]. Relatively larger cracks located away from sensor can be detected by means of change in the vibration response of the structure. However, interpretation of the vibration response is often difficult, and detection of a crack at its initial stage also difficult since it has a small effect on the vibration response [11]. Some approaches store pre-damage or healthy vibration responses, and detect cracks by changes in the structural modes of vibration. These techniques require storage of large data sets and the response can vary due to changes in the environment such as temperature, and also changes in the boundary condition of the structure. Another difficulty was to discriminate between changes due to the environment and the crack itself [12, 13].

Modal testing or experimental modal analysis (EMA) is an important engineering technique to experimentally determine the structural modal parameters. The EMA has some limitations, such as test must be carried out in static conditions, controllable input is required and curve fitting using numerical methods is involved. To overcome this limitations operational modal analysis (OMA) has become a popular alternative. The OMA was developed from the time-domain decomposition (TDD) [14] using Ibrahim time domain method (ITD) that performs singular value decomposition (SVD) on sdof transient response to determine structural modal model suited for random input excitation. Then, Brincker et. al.[15] adopted the frequency-domain decomposition (FDD) method to extract modal parameters by processing the output power spectrum. Another frequently adopted approach to perform OMA is the stochastic subspace identification (SSI) method where the empirical decomposition of the measured data in time domain is based[16].

This paper presents the effect of single progressive of off-centered crack on the dynamic behaviors of beam with free-free ends. In order to identify this influence, classical modal analysis and operational modal analysis techniques were applied.

## 2.0 OPERATIONAL MODAL ANALYSIS AS A DIANOSTIC TOOL

This section provides a brief theory of operational modal analysis techniques applied to vibration of a beam. Special consideration is given to the end condition with free-free ends. The classical experimental modal analysis techniques are well known and will not be discussed. The detailed discussion of these techniques can be found for instance in [17] and [18].

### 2.1 Frequency Domain Decomposition (FDD)

Output-only identification of structures is normally associated with the identification of modal parameters from the natural responses of civil engineering structures, space structures and large mechanical structures. The FDD method is an extension of the Basic Frequency Domain (BFD) method or commonly known Peak-Picking method. The technique is a non-parametric in nature that estimates the modal parameters directly from signal processing data calculations. This approach utilizes the properties that the mode shapes can be estimated from the calculated spectral density for the condition of random noise input or stochastic input applied to lightly damped structure where the modes are well separated [19].

In contrast to the classical techniques where the power spectral density (PSD) matrix is directly and easily estimated via Fast Fourier Transformation (FFT), in the frequency domain decomposition (FDD), it is not directly processed, but decomposed using the singular value decomposition (SVD) at each spectral line. The PSD matrix is decomposed into auto spectral density functions consist of single degree of freedom systems. The modes are simply picked by locating the peaks in SVD plots. The accuracy of the estimated natural frequency depends on the FFT resolution with no modal damping is calculated [20].

### 2.2 Enhanced Frequency Domain Decomposition (EFDD)

The EFDD technique is an extension to the FDD technique. The technique is closely related to the classical frequency domain techniques where the modes are identified by picking the peaks in the spectral diagrams. It gives an improve estimate of both the mode shapes and the natural frequencies and also provides modal damping. The SDOF auto spectral densities are identified using the modal assurance criterion (MAC),and are transformed back to time domain, and the damping and more accurate natural frequency estimates are identified from the corresponding free decays. In EFDD, the SDOF Power Spectral Density function which is identified around a resonance peak is transformed back to time domain using Inverse Discrete Fourier Transform (IDFT). The frequency is obtained through the number of zero crossing as function of time and the damping by logarithmic decrement of the normalized auto correlation function. The SDOF function is predicted using shape previously determined using FDD which is used as reference vector in the correlation analysis based on Modal Assurance Criteria (MAC). This value is calculated between the FDD vector and a single vector for each frequency line [21].

## 3.0 EXPERIMENTAL METHODS

The experimental procedure was carried out to obtain the mobility in the form of a frequency response functions (FRF) using both classical and operational modal tests. In the experiment, a progressive single off-centered crack was investigated with free-free beam end condition.

### 3.1 Experimental Modal Analysis

#### 3.1.1 Modal Test

In this test, the excitation is exerted to the test structure by applying an impulse force from a roving impact hammer in a single Z- direction, see Figure 1. Table 1 gives details of the geometric and material properties of the beam used. The force was applied at 11 different locations while the fixed accelerometer at a single location captured the response signal. The FRFs at different positions of the impact hammer are stored for further post processing. ME'Scope was then used as post processing software to extract dynamic properties such as natural frequencies and damping ratio, and to simulate the vibration modes of a beam. The measurement utilized Bruel&Kaer PULSE Frontend Type 3560D Analyzer data acquisition system.

Coherence is also taken into consideration when selecting the placement of the sensor so that the influence of noise to the signal is minimized. In general, coherence decreases as the accelerometer is closer to the boundary or edge of the structure. For all measurements, coherence over 0.98 is considered for a reasonable accuracy since the structure is very simple.

Modal testing or experimental modal analysis is an important engineering technique to experimentally determine the structural modal parameters. The EMA has some limitations, such as test must be carried out in static conditions, controllable input is required and curve fitting numerical methods is involved. To overcome this limitations operational modal analysis (OMA) has become a popular alternative. The time-domain decomposition (TDD) was developed by Kim et. al. [14] using Ibrahim time domain method (ITD) to perform singular value decomposition (SVD) on sdof transient response to determine structural modal model suited for random input excitation. Then Brincker et. al.[15] adopted the frequency-domain decomposition (FDD) method to extract modal parameters by processing the output power spectrum.

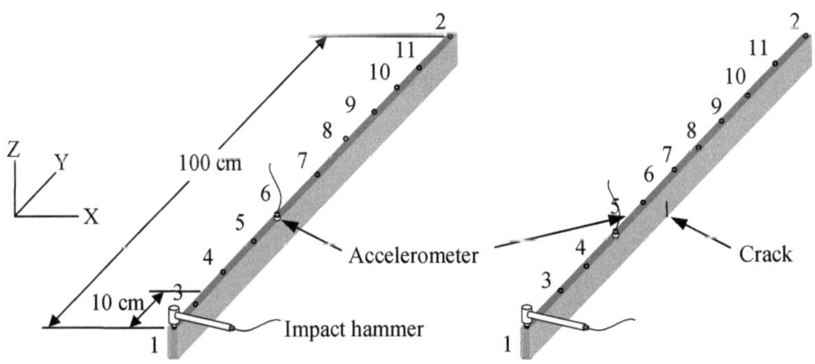


Figure 1 – A roving hammer and a measurement point

Table 1: The properties of beams used for the experimental study

| | |
|---|---|
| Material | Aluminium |
| Young's Modulus E | 69.79 GN/m$^2$ |
| Mass Density, $\rho$ | 2600 kg/m$^3$ |
| Poisson Ratio, $v$ | 0.33 |
| Beam length, $L$ | 1000 mm |
| Beam width, $w$ | 12.55 mm |
| Beam depth, $d$ | 50.6 mm |
| Boundary condition | Free-free |

### 3.1.2 Modal Identification

The modal identification was performed using the well established curve fitting and direct parameter estimation techniques in the frequency range of zero to 3200 Hz. Six distinct modes of vibration were extracted along with the corresponding natural frequencies and damping ratios.

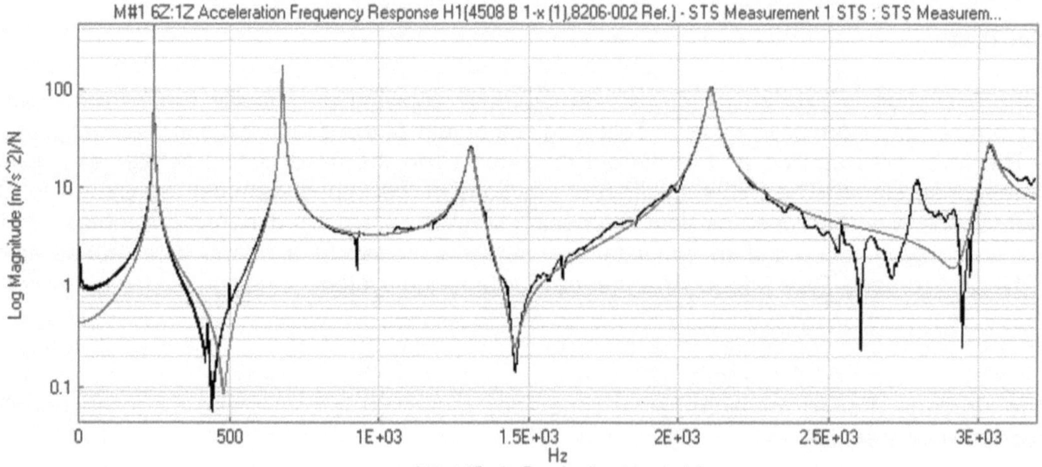


Figure 2 – FRF with curve fitting and Modal peaks function

## 3.2 Operational Modal Analysis

### 3.2.1 Operational Modal Test

To assess the relevant structural vibrations of both healthy and cracked beams in operation using the operational modal analysis technique, the beams were randomly excited at various locations in the z direction. Accelerations were measured with 11 uniaxial accelerometers simultaneously with the accelerometer at point or node 1 was selected as a reference since it exhibited most information about the vibration, see Figure 3. The measurement utilized the same Bruel&Kaer PULSE Frontend Type 3560D Analyzer data acquisition system. A total of 11 measurement degrees of freedom were taken in order to obtain the overall modes of vibration.

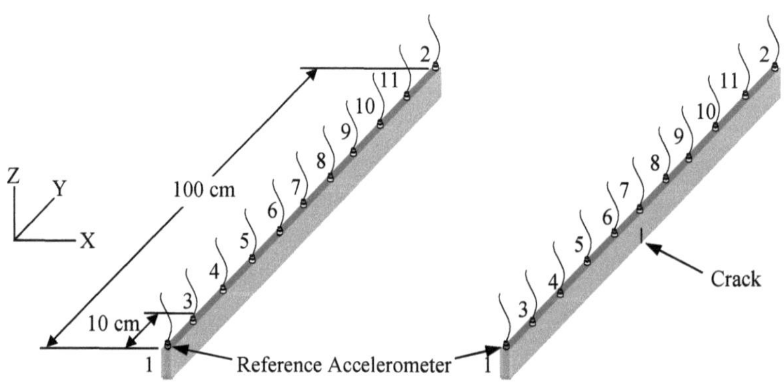


Figure 3 – Measurement points and a reference accelerometer

The measurements were repeated several times in order to obtain an accurate observation of the structural properties of the beam and also to exhibit the effectiveness of the OMA technique. Figure 4 shows a typical time record of the history of the random excitation signal.

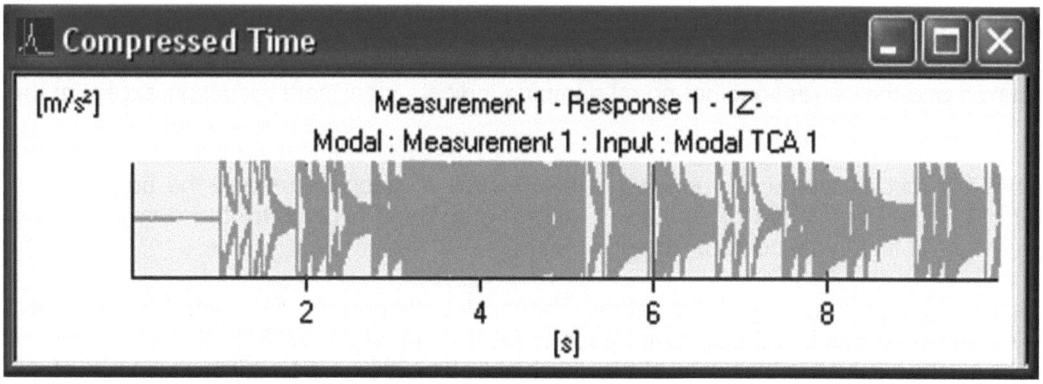


Figure 4 – Typical time record of random excitation

### 3.2.2 Operational Modal Identification

To identify the operational modes of vibration governing the response of the beam, three different operational modal analysis techniques were applied: 1) FDD, 2) EFDD and 3) CFDD. The time data were calculated in all techniques where the spectral density matrix was determined utilizing a 2048 point FFT. The overall operational natural frequencies of vibration were obtained from two different measurement sets. Both techniques also exhibited good modal coherence.

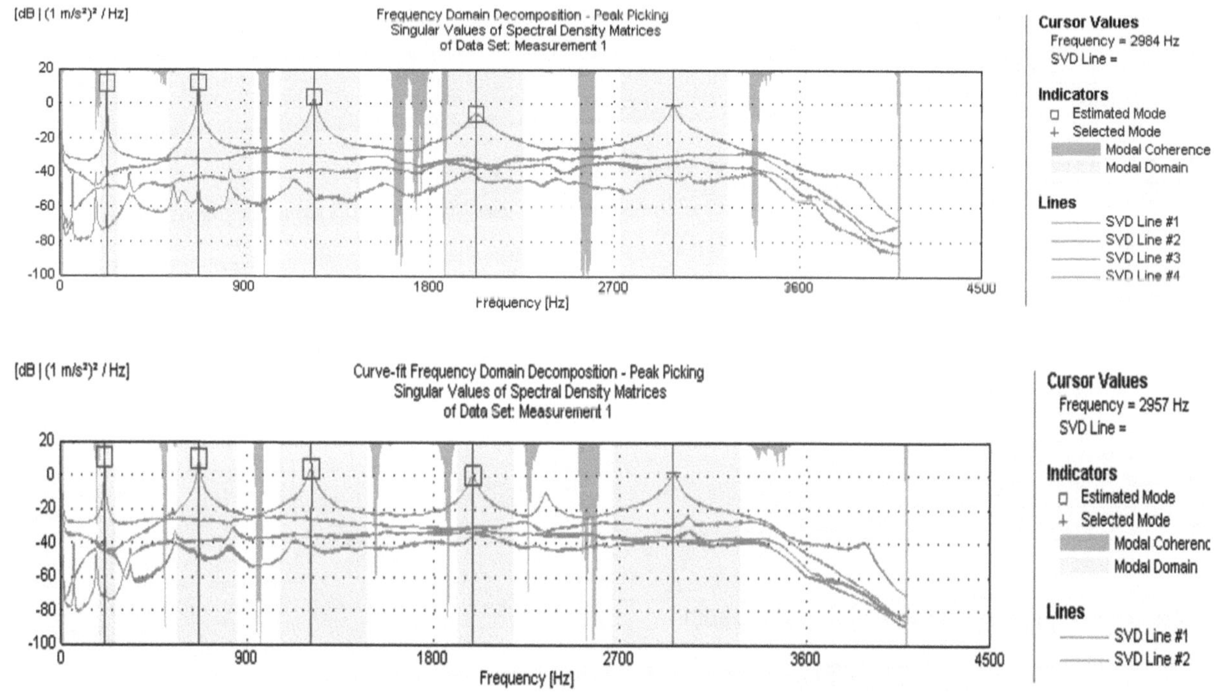


Figure 5 – FRFs of a good beam using FDD and CFDD

## 4.0 RESULTS AND DISCUSSION

All the results from the experimental studies were tabulated and comparisons were presented. The results for different cracked depth on a beam were compared between EMA and OMA techniques.

A comparison of results from the three operational modes of vibration using FDD , EFDD and CFDD and the modes identified by classical experimental modal analysis techniques on the healthy and cracked beams clearly indicate changes in dynamic properties. The three techniques yield consistent results. Both the OMA and EMA techniques show no change in the mode shapes for the two beams at all modes, however, the natural frequencies at which these modes occurred and the corresponding modal damping indicate significant reduction except at the fifth mode.

Table 2 to Table 6 compare the modal natural frequencies of the healthy or good beam and the beam with a single off-centered crack for the first five modes of vibration. Table 7 and Table 8 compare the damping ratios EMA and OMA, respectively for the first five modes of vibration.

Furthermore, the identification of mode shapes and the natural frequencies utilizing the FDD and EFDD yielded the same modes with very small differences in the natural frequencies. It is understandable that the frequencies for the uncracked or healthy beam are higher than that of the cracked beam. This is due to the reduction of the stiffness at the cracked location.

Table 2: Natural Frequencies for the Uncracked or healthy Beam

| MODE | EMA (Hz) | OMA (Hz) | | |
|---|---|---|---|---|
| | | FDD | EFDD | CFDD |
| 1 | 252.00 | 240 | 253.7 | 256 |
| 2 | 681.00 | 688 | 681.8 | 683.3 |
| 3 | 1310.00 | 1312 | 1308 | 1308 |
| 4 | 2110.00 | 2112 | 2114 | 2114 |
| 5 | 3040.00 | 3040 | 3042 | 3041 |

Table 3: Natural Frequencies for the 10 % crack depth

| MODE | EMA (Hz) | OMA (Hz) | | |
|---|---|---|---|---|
| | | FDD | EFDD | CFDD |
| 1 | 250.00 | 250 | 250.3 | 250.7 |
| 2 | 681.00 | 682 | 681 | 681 |
| 3 | 1300.00 | 1312 | 1306 | 1306 |
| 4 | 2100.00 | 2112 | 2105 | 2106 |
| 5 | 3020.00 | 3030 | 3032 | 3032 |

Table 4: Natural Frequencies for the 20 % crack depth

| MODE | EMA (Hz) | OMA (Hz) | | |
|---|---|---|---|---|
| | | FDD | EFDD | CFDD |
| 1 | 239.00 | 238 | 238.5 | 238.8 |
| 2 | 677.00 | 678 | 676.9 | 676.9 |
| 3 | 1270.00 | 1278 | 1275 | 1276 |
| 4 | 2070.00 | 2062 | 2059 | 2061 |
| 5 | 2990.00 | 3000 | 3000 | 2999 |

Table 5: Natural Frequencies for the 40% crack depth

| MODE | EMA (Hz) | OMA (Hz) | | |
|---|---|---|---|---|
| | | FDD | EFDD | CFDD |
| 1 | 227.00 | 228 | 227.7 | 228.1 |
| 2 | 674.00 | 674 | 674.3 | 674.2 |
| 3 | 1250.00 | 1240 | 1240 | 1241 |
| 4 | 2040.00 | 2028 | 2030 | 2033 |
| 5 | 2970.00 | 2984 | 2985 | 2985 |

Table 6: Natural Frequencies for the 50 % crack depth

| MODE | EMA (Hz) | OMA (Hz) | | |
|---|---|---|---|---|
| | | FDD | EFDD | CFDD |
| 1 | 211.00 | 212 | 211.4 | 211.5 |

| | | | | |
|---|---|---|---|---|
| 2 | 669.00 | 668 | 667.9 | 667.9 |
| 3 | 1210.00 | 1214 | 1212 | 1212 |
| 4 | 1990.00 | 1994 | 1991 | 1991 |
| 5 | 2950.00 | 2956 | 2958 | 2957 |

Table 7: Modal Damping using EMA

| MODE | Damping Ratio | | | | |
|---|---|---|---|---|---|
| | Uncracked | 10 % | 20% | 40% | 50% |
| 1 | 0.111 | 0.184 | 0.185 | 0.206 | 0.219 |
| 2 | 0.31 | 0.352 | 0.358 | 0.402 | 0.5 |
| 3 | 0.961 | 0.549 | 0.529 | 0.927 | 0.703 |
| 4 | 0.633 | 0.964 | 0.984 | 0.742 | 0.827 |
| 5 | 0.609 | 0.991 | 0.809 | 0.695 | 0.926 |

Table 8: Modal Damping the using OMA

| MODE | Damping Ratio | | | | |
|---|---|---|---|---|---|
| | Uncracked | 10 % | 20% | 40% | 50% |
| 1 | 3.721 | 0.4446 | 0.4754 | 0.4911 | 0.5533 |
| 2 | 1.4 | 0.3688 | 0.3653 | 0.3029 | 0.5138 |
| 3 | 1.234 | 0.7477 | 0.7756 | 0.6498 | 0.9899 |
| 4 | 0.9091 | 1.279 | 1.475 | 1.817 | 0.7928 |
| 5 | 0.5711 | 0.6179 | 0.575 | 0.6343 | 0.588 |

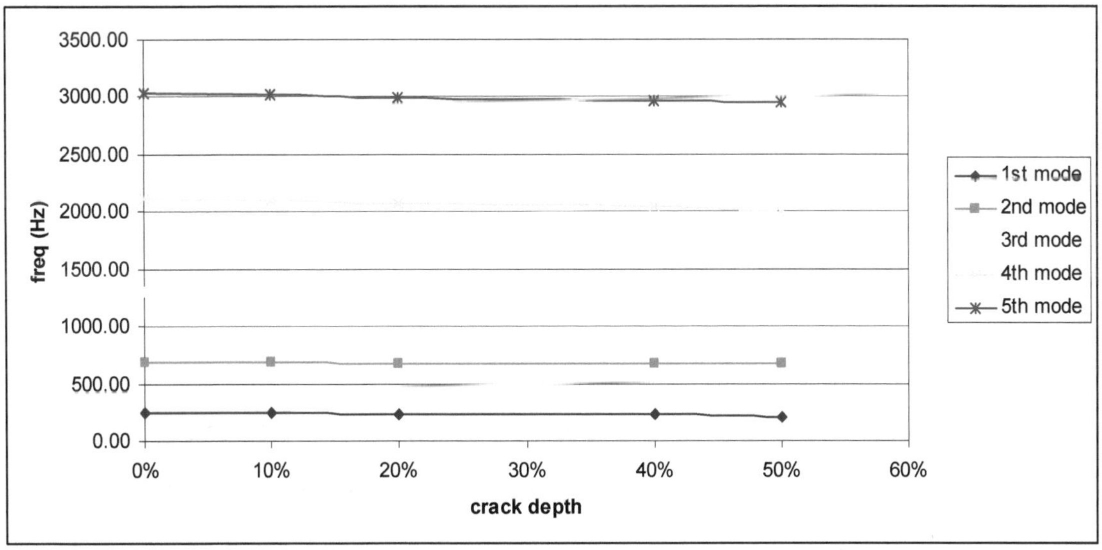


Figure 6: Frequency reduction using EMA technique

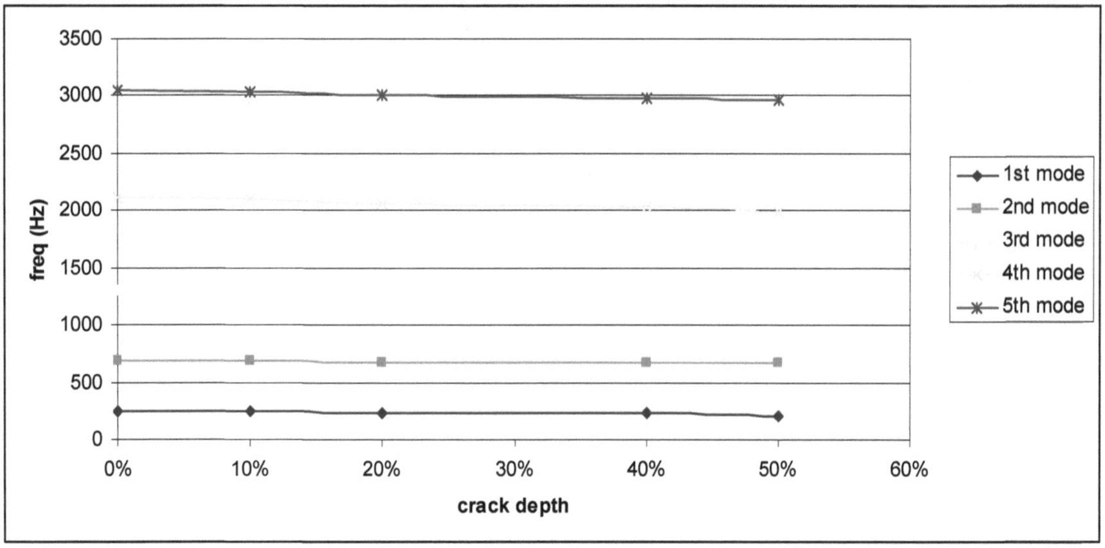


Figure 7: Frequency reduction using OMA technique

Though different excitations were made, the OMA method yielded consistent structural properties of the beam.

## 5.0 CONCLUSIONS

The measured modal natural frequencies and mode shapes of the rectangular beam using EMA and OMA for both the healthy beam and the beam with a single off-centered progressive crack are in good agreement. The size of the crack clearly influences the dynamic properties of the beam with free ends. The modal damping frequencies of the cracked beam are generally lower than the frequencies of the uncracked beam. The lower values are due to the reduction of the stiffness at the cracked area. The mode shapes of the beam, however, are not affected by the presence of the crack and that the two techniques, OMA and EMA, yield essentially the same modes of vibration.

## 6.0 ACKNOWLEDGEMENTS

The authors would like to thank to all individuals who are directly or indirectly involve in making this project a success, particularly Mr Mohd Fauzi, for the good cooperation with respect to all data acquisition issues.

## 7.0 REFERENCES

1. Qi, K., et al., *Vibration based operational modal analysis of rotor systems.* Measurement, 2008. **41**: p. 810-816.
2. Schulz, M.J., et al., *Structural damage detection using transmittance functions*, in *IMAC-XV Conference.* 1997: Orlando, Florida.
3. Martin, W.N., et al., *Detecting damage on symmetric structures using vibration measurement*, in *5th Intl. Conference on Composites Engineering.* 1998: Las Vegas, Nevada.
4. Friswell, M.J. and J.E. Penny. *The practical limits of damage detection and location using vibration data.* in *11th Symposium on Structural Dynamics and Control.* 1997. Blacksburg, VA.
5. Schulz, M.J., et al. *Structural health monitoring using frequency response function and sparse measurement.* in *IMAC-XVI.* 1998. Santa Barbara, CA.
6. Zheng, D.Y. and N.J. Kessissoglou, *Free vibration analysis of a cracked beam by finite element method.* Journal of Sound and Vibration, 2004. **273**(3): p. 457-475.
7. Isa, A.A.M. and Z.A. Rahman, *The Influence of Off-centered Crack on the Dynamics of Beam*, in *Proceeding of the 3rd International Operational Modal Analysis Conference (IOMAC)* 2009: Portonovo (Ancona), Italy.

8. Shifrin, E.I. and R. Ruotolo., *Natural frequencies of a beam with an arbitrary number of cracks.* Journal of Sound and Vibration, 1999. **222**(2): p. 409-423.
9. Zheng, D.Y. and S.C. Fan, *Natural frequency changes of a cracked Timoshenko beam by modified Fourier series.* Journal of Sound and Vibration, 2001. **246**(2): p. 297-317.
10. Bamnios, Y., E. Douka, and A. Trochidis, *Crack identification in beam structures using mechanical impedance.* Journal of Sound and Vibration, 2002. **256**(2): p. 287-297.
11. Fernandez-Saez, J., L. Rubio, and C. Navarro, *Approximate calculation of the fundamental frequency for bending vibrations of cracked beams.* Journal of Sound and Vibration, 1999. **225**(2): p. 345-352.
12. Sekhar, A.S., *Vibration characteristics of a cracked rotor with two open cracks.* Journal of Sound and Vibration, 1999. **223**(4): p. 497-512.
13. Rizos, R.F., N. Aspragathos, and A.D. Dimarogonas, *Identification of crack location and magnitude in a cantilever beam from the vibration modes.* Journal of Sound and Vibration, 1990. **138**(3): p. 381-388.
14. Kim, B.H., N. Stubbs, and T. Park, *A new method to extract modal parameters using output-only response.* Journal of Sound and Vibration, 2005. **282**: p. 215-230.
15. Brincker, R., L. Zhang, and P. Andersen. *Modal identification from ambient responses using frequency domain decomposition.* in *Proceedings of 18th International Modal Analysis Conference*. 2000. San Antonio, TX.
16. Yu, D.J. and W.X. Ren, *EMD-based stochastic subspace identification of structure from operational vibration measurement.* Engineering Structures, 2005. **27**: p. 1741-1751.
17. Ewin, D.J., *Modal Testing: Theory, Practice and Application.* 2nd Edition ed. 2000: Research Studies Press, England.
18. Allemang, R.D., *Vibrations: Experimental Modal Analysis.* 1999.
19. Møller, N.R.B., H. Herlufsen, P. Andersen. *Modal Testing Of Mechanical Structures Subject To Operational Excitation Forces.* in *Proceedings of The 19th International Modal Analysis Conference (IMAC)*. 2001. Kissimmee, Florida,
20. Brincker, R.L.Z., P. Andersen. *Output-Only Modal Anlysis By Frequency Domain Decomposition.* in *Proceedings of The ISMA25 Noise And Vibration Engineering*. 2000. Leuven, Belgium.
21. Jacobsen, N.J., P. Andersen, R. Brincker. *Eliminating the Influence of Harmonic Components in Operational Modal Analysis.* in *Proceedings of The 25th International Modal Analysis Conference (IMAC)* 2007. Orlando, Florida

MIX
Papier aus verantwortungsvollen Quellen
Paper from responsible sources
FSC® C105338

If you have any concerns about our products,
you can contact us on
**ProductSafety@springernature.com**

In case Publisher is established outside the EU,
the EU authorized representative is:
**Springer Nature Customer Service Center GmbH**
**Europaplatz 3, 69115 Heidelberg, Germany**

Printed by Libri Plureos GmbH
in Hamburg, Germany